CROP MANAGEMENT AND POSTHARVEST HANDLING OF HORTICULTURAL PRODUCTS

Volume I — Quality Management

CROP MANAGEMENT AND POSTHARVEST HANDLING OF HORTICULTURAL PRODUCTS

Volume I — Quality Management

Editors

Ramdane Dris
Department of Applied Biology
University of Helsinki
Helsinki, Finland

Raina Niskanen
Department of Applied Biology
University of Helsinki
Helsinki, Finland

Shri Mohan Jain
Plant Breeding and Genetics Section
Division of Nuclear Techniques in Food and Agriculture
International Atomic Energy Agency
Vienna, Austria

Science Publishers, Inc.
Enfield (NH), USA   Plymouth, UK

SCIENCE PUBLISHERS, INC.
Post Office Box 699
Enfield, New Hampshire 03748
United States of America

Internet site: *http://www.scipub.net*

sales@scipub.net (marketing department)
editor@scipub.net (editorial department)
info@scipub.net (for all other enquiries)

Library of Congress Cataloging-in-Publication Data

Crop management and postharvest handling of horticultural products/editors, Ramdane Dris, Raina Niskanen, and Mohan Jain.
    p. cm.
  Includes bibliographical references (p.)
  Contents: v. 1. Quality management.
  ISBN 1-57808-140-8
    1. Horticultural crops. 2. Horticultural crops—Postharvest technology. I. Dris, Ramdane, II. Niskanen, Raina, III. Jain, Mohan.
  SB318 .C76 2000
  635'.046–dc21

00-063494

ISBN 1-57808-140-8

© 2001, Copyright Reserved

All rights reserved. No part of this publication may be reproduced, stored in a retrieval system, or transmitted in any form or by any means, electronic, mechanical, photocopying or otherwise, without the prior permission from the publisher. The request to produce certain material should include a statement of the purpose and extent of the reproduction.

Published by Science Publishers Inc., Enfield, NH, USA
Printed in India.

Preface

An ever increasing world population and poor postharvest storage technologies are gradually leading to the shortage of plant products. Nevertheless, we should not view this state as an emergency, but rather as a challenge that can lead us to develop better technologies for improving crop yield and quality and minimize postharvest crop losses. The need of the hour is to produce high yielding crops and maintain higher postharvest quality including appearance, texture, flavor, nutritional quality and wholesomeness. Broadly speaking, the assimilation of nutrients and their functions related to metabolism, growth and yield are considered a part of the plant production. Apart from this, high productivity as well as quality can be achieved by proper fertilization, treatment and storage.

Over the last several years, there was a lack of interest in cultivation, nutrition, production, postharvest and marketing of horticultural products. Furthermore, deterioration is a cumulative process and occurs at all the stages during harvest and consumption. By the time the produce is harvested, cleaned, packed and shipped, its value will have increased several folds. This clearly shows that more serious losses occur during postharvest deterioration. Now there is a growing interest among scientists in this area of research. In the recent past, researchers have focused on the problems affecting yield, nutrition, and the postharvest life of commodities. A continuous supply of good quality food products to the consumers throughout the year, adequate post-harvest activities such as transport, handling and storage are of prime importance. Horticultural crops are mostly perishable and create problems in long-term storage and on long distance transportation from the producing countries to the major export markets. Often, the process of mineral nutrition becomes erratic, e.g. inadequate or excessive mineral nutrition can have serious implications on postharvest quality of horticultural crops.

Plant nutritionists and postharvest technologists will have to put in a great amount of work to minimize quality loss, e.g. this could be done by providing more information on recent progress on minimizing the quality losses, nutritional status, maturity or ripening stages, physiological changes, harvest, handling and storage operations as well as other related problems. The producers and researchers must be aware of the rapid and continuous technological changes to assure good crop quality and get a higher price premium in a wide range of crops.

Ramdane Dris
Raina Niskanen
Shri Mohan Jain

Contents

Preface	*v*
List of Contributors	*ix*
1. The Quality Cycle Daryl Joyce	1
2. Testing Soils for Determining Fertilizers Needs of Horticultural Crops C.D. Tsadilas and N. Barbayiannis	13
3. Nutritional Management of Vegetables and Ornamental Plants in Hydroponics Dimitrios Savvas	37
4. Managing Crop Load Sally A. Bound	89
5. Low-alcohol Grape and Fruit Wine Gary J. Pickering	111
6. Paprika Spice Production Andreas Klieber	133
7. Factors Associated with Peach Yield, Quality and Postharvest Behavior Salvador Pérez González,	157
8. Influence of Calcium Nutrition on the Quality and Postharvest Behaviour of Apples Ramdane Dris	175
9. Economic Analysis of Postharvest Systems in Root and Tuber Crops Gerd Fleischer and Stefan Agne	189
10. Postharvest Handling of Strawberries for Fresh Market José M. Olias[†], Carlos Sanz and Ana G. Pérez	209
11. Postharvest Biological Changes and Technology of Citrus Fruit Giovanni Arras	235

12. Ethylene Biosynthesis—Role in Ripening and Quality of 'Hayward' Kiwifruit after Harvest, During Storage and Shelf-Life 263
    E. Sfakiotakis, M.D. Antunes, G. Stavroulakis and N. Niklis

13. Postharvest Pests and Physiological Disorders of Cassava 289
    Weston Msikita, Henry T. Wilkinson and Robert M. Skirvin

14. Environmental Control for Storage of Rooted Propagation Material 307
    Nihal C. Rajapakse

15. Minimizing Postharvest Rots and Quality Loss in New Zealand Carrots 327
    L-H. Cheah and D. W. Brash

16. Minimal Processing of Fruits and Vegetables 345
    Catherine Barry-Ryan and David O'Beirne

    Index 359

Contributors

Daryl Joyce, Cranfield University, Silsoe Bedfordshire MK 45 4DT, UK.

Professor C.D. Tsadilas, National Agricultural Research Foundation, Institute of Soil Classification and Mapping, 1 Theophrastos street, 413 35 Larissa, Greece.

Dimitrios Savvas, Faculty of Agricultural Technology, Technological Education Institute (T.E.I.) of Epirus, P.O. Box 110, 47100 Arta, Greece.

Sally A. Bound, Tasmanian Institute of Agricultural Research 13 St John's Avenue, New Town, Tasmania 7008, Australia.

Gary J. Pickering, Senior Lecturer in Wine Science, Faculty of Science and Technology, Eastern Institute of Technology, Hawke'sBay, New Zealand.

Andreas Klieber, Senior Lecturer, The University of Adelaide Waite Campus, Department of Horticulture, Viticulture and Oenology, PMB 1, Glen Osmond SA 5064, Australia.

Salvador Pérez González, Facultad de Química/Area Agrícola, Universidad Autónoma de Querétaro, CU, Querétaro, Qro. Mexico CP 70100, México.

Ramdane Dris, Department of Applied Biology, P.O. Box 27 Fin-00014 University of Helsinki, Finland.

Gerd Fleischer, Stefan Agne, Institute of Economics and Communication, Faculty of Horticulture, University of Hannover, Herrenhäuser Straße 2, 30419 Hannover, Germany.

Professor José M. Olias, Dr. Carlos Sanz, Dr. Ana G. Pérez, Instituto de la Grasa, C.S.I.C. Dept. Physiology and Technology of Plant Products Padre Garcia Tejero 4, 41012-Seville Spain.

Giovanni Arras, CNR-Istituto per la Fisiologia della Maturazione e della Conservazione del Frutto delle Specie Arboree Mediterranee, Via Dei Mille, 48 - 07100 Sassari, Italy.

E. Sfakiotakis, Laboratory of Pomology, Department of Horticulture, School of Agriculture, Aristotle University, Thessaloniki 54006, Greece.

Nikolaos Niklis, Regional Center of Plant Protection and Quality Control Thessaloniki, 54626 Greece.

Maria Doulce Carlos Antunes, Universidade do Algarve, U.C.T.A., Campus de Cambelas, 8000 Faro, Portugal.

G. Stavroulakis, Higher Technical Educational Centre, Heraklion, Greece.

Weston Msikita, Ohio State University, Department of Plant Biology 2021 Coffey Rd., 310F Kottman Hall, Columbus, OH 43210 USA.

Henry T. Wilkinson, Professor Robert M. Skirvin, University of Illinois, Department of Natural Resources and Environmental Sciences, Urbana, IL. 61801, USA.

Nihal C. Rajapakse, Department of Horticulture, Clemson University, Clemson, SC 29634, USA.

L-H Cheah, D W Brash, New Zealand Institute for crop and Food Research Ltd Private Bag 11-600, Palmerston North, New Zealand.

Catherine Barry-Ryan, David O'Beirne, Food Science Research Centre, Department of Life Sciences, University of Limerick, Limerick, Ireland.

N. Barbayiannis, Aristotle University of Thessalonki, School of Agriculture, Soil Science Laboratory, 54006 Thessaloniki, Greece.

1

The Quality Cycle

Daryl Joyce
Cranfield University, Silsoe Bedfordshire MK45 4DT, UK

1. Preamble

Quality is a key issue common to all products of horticulture, whether fresh, minimally processed or fully processed. In virtually all cases, product quality is set or fixed at harvest. Postharvest operations are predominately aimed at maintaining harvest quality as well as possible for as long as practicable. Therefore, it is important that compilations such as the present text are published. For the most part, the chapters of this book document interactions between preharvest or postharvest factors and quality for a wide variety of horticultural products. They serve as valuable written records to facilitate scientific and technical progress towards presentation of the highest possible quality to consumers. In this introduction, horticultural produce quality is examined in general terms with respect to the drivers that determine quality expectations and the range of preharvest and postharvest factors that affect fresh produce quality.

2. Quality

Quality is defined in the Oxford Dictionary (1976) as 'degree of excellence, relative nature or kind or character'. A more functional and often touted definition is 'fitness for purpose'. For the most part, fresh produce quality is initially assessed by sight (Wills et al. 1998). Other quality attributes include taste, smell and texture. Each of these four quality attributes can be assessed subjectively (e.g. human eye) or objectively (e.g. machine vision). Nutritional value can also be regarded an important quality issue; for instance, fruits with high vitamin C contents. Microbial, chemical and/or physical food safety may also be considered within the ambit of quality; e.g. toxin-producing bacteria, pesticides residues and fragments of metal, respectively.

Generally speaking, production of horticultural produce exceeds the demand for it. This discrepancy leads to crop surpluses on markets in relatively more developed parts of the world. It has been said 'that there is enough food in the world for everyone's needs, but not enough for everyone's greed'. Whatever the moral implications may be, produce quality, as opposed to human nutrition, becomes the single-most important factor differentiating similar produce lines. Quality judged by external appearance is the basis for selection for the consumer. Visual appeal is the only basis for choice among similar lines of ornamental produce such as cut flowers. However, although attractiveness may be determined solely by appearance, purchase will be tempered by a value for money decision.

As noted above, produce quality cannot be improved after harvest. There are arguably some exceptions to this rule, such as waxing of apples and dyeing of oranges to enhance their appearance (Wills et al. 1998). However, there are relatively few exceptions to the general rule. Even the final quality of semi-processed and processed products such as minimally processed vegetables or wine respectively, is largely determined at harvest.

3. The Production and Marketing Cycle

The preceding discussion asserts that quality is fixed at harvest and that the judgement of quality is made in the marketplace. Recognition of the significance of this relationship has given rise to the 'market-led production' concept. That is, production aimed specifically at satisfying consumer needs. Accordingly, it is not unreasonable to pose the question, 'when is postharvest really preharvest?' If a simple cyclic model is accepted (Fig. 1) then the marketplace, which might intuitively be considered postharvest, becomes preharvest. The marketplace determines produce quality standards that production practices must attain and postharvest practices must strive to maintain.

Fig. 1. The quality cycle.

4. Drivers Affecting Quality Standards

Quality standards have largely been prescribed and monitored by government agencies. In various countries, for some commodities, this is still the case. However, because of the costs involved and in the context of the global economy, governments are moving away from a policing role in favour of industry self-regulation. A general exception is where there are either quarantine (e.g. insects, diseases) or food safety issues. Where industry self-regulation applies, it is anticipated that sales and marketing forces will exert feedforward control over production and postharvest (Fig. 1).

In many, if not most, countries, large retailers have emerged as the most powerful industry force determining quality standards. By virtue of their enormous buying and selling powers, supermarkets now prescribe product quality descriptions that must be met by their suppliers. Typically, suppliers adopt or develop the same or similar quality standards. Quality standards are normally applied back down the line of supply to the growers. Product descriptors or standards are often an important part of a company's quality assurance program. Perhaps unfortunately, quality standards often focus on the cosmetic features of produce (i.e. appearance) and frequently fail to address internal characteristics, like flavour.

The influence of retailers over produce quality is tempered by the attitudes of consumers. Retailers may target, say, the low versus the high end of the market in terms of price. Nonetheless, within and across such market segments, consumers exercise substantial control over the quality of produce presented for sale to them. For example, a consumer driven push for more flavour in tomatoes has resulted in tastier lines being presented for sale. This 'consumer power' is a function of competition among retailers. A recent and dramatic example of consumer influence over retailers has been the withdrawal, rightly or wrongly, of food containing genetically modified organisms from supermarket shelves in Europe and elsewhere.

Consumer awareness of fresh produce quality is generally strengthening. This has 'knock on' implications for quality standards, especially in the comparatively more developed countries. For example, interest in the nutritional value of fruits and vegetables is increasing. Consumers are also concerned about the environment, and not least in synthetic chemical inputs in production and postharvest management of horticultural produce. Increasing sales of organic produce is an example of developing consumer appreciation of intrinsic versus extrinsic quality parameters. Likewise, consumers are interested in ethical issues associated with the production, handling and marketing of their food. The perception of undue influence of large supermarket retailers is, for example, an issue likely to rise in public profile. Consumer concerns also include fiscal exploitation of producers and suppliers, and monopolization of markets at the expense of small retailers and of variety of choice. In each of these cases, consumer driven changes in produce quality expectations can be anticipated. For instance, consumers may accept comparatively higher levels of surface blemish on organic produce lines.

Consumers are also concerned about convenience, waste, fashion and choice. Sustained 'double digit percent' growth in sales (i.e. >10% per annum) of minimally processed (fresh cut, lightly processed) fruits and vegetables exemplifies the demand for convenience. Waste, especially that associated with packaging of horticultural produce, is a particularly vexing issue. On the one hand, consumers are generally concerned about high levels of waste and the community cost of its disposal. On the other, they are confronted with high use levels of disposable retail packaging. There is little doubt that this conundrum will become a higher profile issue. The results may include reduced retail or convenience packaging and accelerated development of biodegradable packaging materials. Changing fashions are particularly important in the ornamentals industry; for instance, bunched flowers are usually coordinated to reflect 'this years' or 'this seasons' colours. Similarly, new 'fashion varieties' of cut roses or potted petunias, for example, are released annually. Consumer demand for choice is reflected in the wide range of produce on offer and the duration of its availability. Regular consumption of fresh fruit was once largely the province of the privileged few. Now-a-days, thanks to technology and trade, apples are available for purchase year round. Similarly, exotic tropical fruits such as mangosteen, are now sold around the world. Quality is an issue in each of these drivers. For example, minimal processing of fruit may well increase convenience, but artificial plastic packaging is a poor substitute for the natural skin in terms of preserving the intrinsic quality of fresh produce.

The impact of molecular biology on horticultural produce quality will be profound. Tomatoes that have been genetically modified by antisense technology (Picton et al. 1995) to reduce polygalacturonase enzyme activity is an excellent example. Because of suppressed cell wall softening, these genotypes can be harvested later to give better taste in both fresh and processed products. Moreover, the fresh fruit have longer shelf-life through reduced disease susceptibility. Some manipulations may be more acceptable than other engineered changes; e.g. a similar molecular technology applied to the browning enzyme, polyphenoloxidase, may mask one (i.e. browning) but not all (e.g. off flavour) symptoms of a disorder. The current public debate over genetically modified organisms should probably be welcomed, since it is largely concerned with horticultural produce quality in a broad sense.

In the interests of brevity, preceding discussion has risked oversimplification of the issues. However, it is hoped that the discussion has served to illustrate direct effects of the marketplace on production, and to exemplify drivers associated with retailers and consumers. In short, social drivers have marked influences on quality through bringing about changes in production and postharvest practices. At least in some instances, changes with implications for produce quality are initiated by consumers, mediated by retailers and adopted by those concerned with preharvest and postharvest operations. A period of research and development often precedes adoption of changed practices. This phase generally involves scientists, technicians and extension personnel, and is reflected in the chapters that follow this introduction. In some topical cases such as food safety, changes may be enshrined in legislation effected by politicians.

5. Effectors of Quality

In a technical context, factors that determine produce quality can be divided into preharvest and postharvest. Preharvest factors, in turn, can be subdivided into genetic, environmental and management (Table 1). It is beyond the scope of this introductory chapter to discuss each of these factors in detail. The reader is referred to the following reviews for comprehensive treatments of the subject (Pattee, 1985; Beverley et al. 1993; Arpaia, 1994; Hofman and Smith, 1994). The development of supersweet sweetcorn varieties represents an excellent example of genetic effects on quality. Splitting of cherries because of rapid uptake of moisture during rainy periods typifies environmental factors. Nutrient management to achieve increased nitrogen to calcium ratios in pear fruit can enhance their resistance to postharvest disease. There is, of course, interaction between the three sets of preharvest factors (Table 1). For instance, in the case of greenhouse tomato production, the growth environment is modified substantially by management (e.g. computer controlled ventilation). More-over, certain tomato genotypes are more appropriate for greenhouse production than others. For instance, indeterminate versus determinate varieties. Research and development, such as that described throughout this book, is required to develop and optimize preharvest strategies for enhanced produce quality. For example, in a comparison of two types of fruit bags to reduce blemish on developing mango fruit, plastic fruit bags were shown to result in accelerated postharvest water loss, whereas paper bags had no adverse effect (Colour Plate 1.1) (Joyce et al. 1997).

Table 1. Some preharvest factors that affect fresh produce quality.

Genotype
Genus, species, cultivar differences; e.g. climacteric vs. non-climacteric cultivars

Environment
Temperature; e.g. chilling injury
Moisture—relative humidity, dew, precipitation; e.g. floral organ infection
Storms; e.g. hail damage
Wind; e.g. wind rub
Radiation—intensity, wavelength, daylength; e.g. blush

Management
Modulation and control of environmental factors—water, temperature, wind, radiation; e.g. greenhouse production
Mineral nutrition; e.g. calcium
Plant management—rootstocks, pruning, thinning, plant growth regulators; e.g. gibberellins

Pest and disease management
Mammals, insects, fungi, bacteria; e.g. biological control

Harvest
Maturity, method and time

As for preharvest, many postharvest factors determine quality. Principle biological issues are respiration, ethylene production, compositional changes, growth and development, transpiration, physiological breakdown, physical damage, and pests and diseases (Kader, 1992a). Measures to regulate biological deterioration processes include handling and storage at low temperature and high humidity, careful handling, and physical, chemical and biological pest and disease control measures (Table 2). These issues are discussed in detail in many general (e.g. Hardenburg et al. 1986; Kader, 1992b; Thompson, 1996; Wills et al. 1998) and somewhat more commodity specific (e.g. Weichmann, 1987; Nowak and Rudnicki, 1990; Seymour et al. 1996; Mitra, 1997) postharvest texts. Temperature is generally regarded to be the single-most important postharvest factor. In the case of cut grevillea flowers, lowering temperature from 22 to 0° C resulted in a 12-fold reduction in their rate of respiration Fig. 2 (Joyce et al. 2000).

Effective quality control through the pre-postharvest continuum can be approached through recognition that all processes from the decision to sow a crop to its display for sale comprise a single system. Accordingly, growers, intermediaries and retailers need to work together to ensure that the consumer is presented with horticultural produce of a suitable standard. Adoption of systems thinking and systems methodologies, such as descriptive or quantitative models, can assist this process (Shewfelt and Prussia, 1993). Likewise, product quality descriptions established

Table 2. Some postharvest factors that affect fresh produce quality.

Rate of metabolism
Low temperatures
Modified and controlled atmospheres
    Reduced concentrations of oxygen
    Elevated concentrations of carbon dioxide
Ethylene control
    Ventilation
    Inhibitors of production and binding
Rate of water loss
High humidity and moisture barriers

Degree of wounding
Packaging and handling to prevent bruises, cuts and abrasions

Pest (e.g. insects) and disease (e.g. fungi) control
Physical controls (e.g. hot water)
Chemical controls
    Fungicides (e.g. sulfur dioxide)
    Fungistats (e.g. benomyl)
Biological controls (e.g. microbial antagonists)

Miscellaneous treatments
Ethylene gas ripening/degreening
Application of sprout inhibitors
Vase solutions for cut flowers

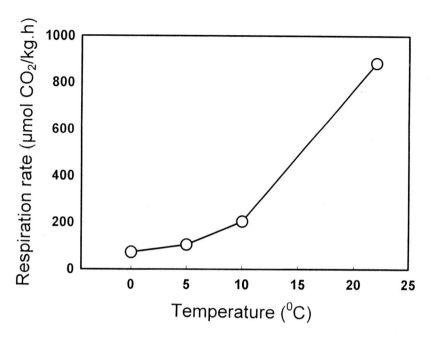

Fig. 2. Respiration rates (μmol. CO_2/kg.h) for *Grevillea* 'Sylvia' inflorescences at 0, 5, 10 and 22° C (Joyce et al. 2000).

through a process of consultation and shared ownership can assist greatly in assuring supply of quality produce. Quality assurance (QA) programs, both within and across organizations, constitute a particularly useful and practical approach to managing quality within a system (Wills et al. 1998). QA programs, in their various guises, including HACCP (hazard analysis and critical control points), are now widespread in the industry. They have contributed greatly to improvement of the quality of horticultural produce presented to consumers and, in a sense, serve to help complete the quality cycle (Fig. 1).

6. The Future for Quality Standards

Economic, ethical and environmental issues should continue to drive changes in quality standards. The gap between the rich and poor in societies is judged to be widening at a rapid rate. This economic disparity could result in a small proportion of the population purchasing extremely high quality produce, including expensive blemish-free organic lines, with little concern for price. At the other end of the bimodal spectrum, quality sacrifice can be anticipated in the interest of price. This scenario at least partly reflects present day government preoccupation with 'the economic imperative' at the cost of 'the social imperative'.

Changes in major horticultural producing areas are also likely to cause an impact on regional economies and quality of the produce. A general rise in incomes in developing countries may ultimately result in an economically more 'level playing field'. If so, horticultural production may again increase in today's high cost regions (e.g. UK, USA) and, conversely, decrease in production regions that are currently low cost (e.g. China, Spain). Near market production would result in improved quality through decreased transport duration and reduced handling. Despite the availability of rapid and high technology handling and transport systems, there is no doubt that import to major markets from production areas far away results in loss of both quality and quantity.

Ethical issues that may impact on produce quality include the anticipated consumer backlash against large retailers that was mentioned above. As evidenced by the increasingly successful 'cancel the third world debt' campaign championed by entertainers and religious groups, there is genuine public concern in developed countries over economic exploitation. Despite so-called 'ethical trading policies', large retailers achieve large profits in the face of rapidly diminishing farm returns. Intertwined with this issue, are other public concerns over undue power and influence of supermarkets and reduced freedom of choice of retailers. If these concerns lead to changes in consumer buying habits, quality expectations may also change. For instance, consumers tend to accept less cosmetically attractive produce from farmers markets, local greengrocers and roadside stalls.

Environmental issues likely to impact on fresh produce quality include the genetically modified organisms (GMOs) debate and pollution, including waste disposal. GMOs offer such significant opportunities for improving fresh produce quality (Patterson, 1998) that they are likely to gain wide acceptance. Education of the public about the true nature of GMOs and careful and transparent legislation for measures to prevent environmental risk should go a long way toward allaying concerns.

The problem of pollution is likely to prove less easy to resolve in the medium to long term. Pollution is intimately associated with cosmetic produce quality. For instance, fertilisers and pesticides are applied to achieve attractive produce as well as acceptable yields, and packaging materials are used to help protect harvested produce during storage and distribution. Consumers must make a quality sacrifice, perceived or real, if pollution is to be reduced. 'Human nature' suggests that although consumers may readily apportion accountability to others, such as producers or retailers, they may be less willing to accept responsibility themselves. Nonetheless, environmental pollution is such a critical issue that significant changes in pre- and postharvest practices are unavoidable. While such changes may reduce the extrinsic quality (e.g. appearance) of horticultural produce, they may well enhance its intrinsic quality (e.g. disease resistance) (Joyce and Johnson, 1999).

7. Conclusion

Within the general technical context of this book, the introduction asserts that preharvest production practices such as mineral nutrition, are the main effectors of horticultural quality. Subsequent postharvest practices contribute principally to quality maintenance. Both preharvest and postharvest practices are dictated to a large extent by social issues that affect market preferences. Changing consumer attitudes have a profound influence, largely through competition among large retailers, in the marketplace. These attitudes constitute drivers that help set quality standards. This feedforward regulation completes the quality cycle (Fig. 1). Thus, quality standards, and associated preharvest and postharvest practices, are largely market driven by the interaction of both social and technical factors.

References

Arpaia, M.L. 1994. Preharvest factors influencing postharvest quality of tropical and subtropical fruit. *Hort Science* **29**: 982-985.

Beverley, R.B., Latimer, J.G. and D.A. Smittle. 1993. Preharvest physiological and cultural effects on postharvest quality. Chapter 4. *In:* Postharvest Handling—A Systems Approach, R.L. Shewfelt and S.E. Prussia (Eds), Academic Press, London, 73-98.

Hardenburg, R.E., Watada, A.E. and C.Y. Wang. 1986. The Commercial Storage of Fruits, Vegetables, and Florist and Nursery Stocks. US Department of Agriculture, Agriculture Handbook No. 66 (revised), 130 pp.

Hofman, P.J. and L.G. Smith. 1994. Preharvest effects on postharvest quality of subtropical and tropical fruit. *Australian Centre for International Agricultural Research Proceedings* No. **50**: 261-268.

Joyce, D.C., Beasley, D.R. and A.J. Shorter. 1997. Effect of preharvest bagging on fruit calcium levels, and storage and ripening characteristics of 'Sensation' mangoes. *Australian Journal of Experimental Agriculture* **37**: 383-389.

Joyce, D.C. and G.I. Johnson. 1999. Prospects for exploitation of natural disease resistance in harvested horticultural crops. *Postharvest News and Information* **10**: 45N-48N.

Joyce, D.C., Meara S.A., Hetherington S.E., and P. Jones. 2000. Effects of cold storage on cut *Grevillea* 'Sylvia' inflorescences. *Postharvest Biology and Technology* **18**: 49-56.

Kader, A.A. 1992a. Postharvest biology and technology: An overview. Chapter 3. *In:* Postharvest Technology of Horticultural Crops, A.A. Kader (Ed.), University of California, Publication 3311; Oakland, CA, 15-20.

Kader, A.A. 1992b. Postharvest Technology of Horticultural Crops. University of California, Publication 3311; Oakland, CA, p. 296.

Mitra, S. 1997. *Postharvest Physiology and Storage of Tropical and Subtropical Fruits*. CAB International, Wallingford, UK, 423 pp.

Nowak, J. and R.M. Rudnicki. 1990. *Postharvest Handling and Storage of Cut Flowers, Florist Greens and Potted Plants*, Timber Press; Portland, OR, 210 pp.

Oxford Dictionary. 1976. The Concise Oxford Dictionary. 6th Edition. J.B. Sykes (Ed.), Oxford University Press, London.

Pattee, H.E. 1985. *Evaluation of Quality of Fruits and Vegetables*. The AVI Publishing Co, Westport, CT, 410 pp.

Patterson, B. 1998. *Fresh! Seeds from the Past and Food for Tomorrow*. Allen & Unwin; St Leonards, NSW, 170 pp.

Picton, S., Gray, J.E. and D. Grierson. 1995. Ethylene genes and fruit ripening. Chapter E4. *In:* Plant Hormones, P.J. Davies (Ed.), Kluwer Academic Publishers; Dordrecht, The Netherlands, 372-394.

Shewfelt, R.L. and S.E. Prussia. 1993. *Postharvest Handling—A Systems Approach*. Academic Press, London, 356 pp.

Seymour, G.B., Taylor, J.E. and G.A. Tucker. 1996. *Biochemistry of Fruit Ripening*. Chapman and Hall, London, 454 pp.

Thompson, A.K. 1996. *Postharvest Technology of Fruit and Vegetables*. Blackwell Science Ltd, Osney Mead, UK, 410 pp.
Weichmann, J. 1987. *Postharvest Physiology of Vegetables*. Marcel Dekker, Inc., New York, 597 pp.
Wills, R., McGlasson, B., Graham, D. and D. Joyce. 1998. *Postharvest—An Introduction to the Physiology and Handling of Fruit, Vegetables & Ornamentals*. 4th edition. CAB International; Wallingford, UK, 262 pp.

Colour Plate 1.1 Appearance of 'Sensation' mangoes during shelf-life evaluation for fruit that were not bagged (control), paper bagged or plastic bagged during development prior to harvest (Joyce et al. 1997).

2

Testing Soils for Determining Fertilizer Needs of Horticultural Crops

C.D. Tsadilas* and N. Barbayiannis
*National Agricultural Research Foundation, Institute of Soil Classification and Mapping, 1 Theophrastos street, 413 35 Larissa, Greece
Aristotle University of Thessaloniki, School of Agriculture, Soil Science Laboratory, 540 06 Thessaloniki, Greece*

Abbreviations

BCSR Basic Cation Saturation Ratio
CR   Cation Ratio
CEC   Cation Exchange Capacity

1. Introduction

Soil testing and plant analysis are important tools for tackling problems related to plant growth. Both of them can be used for diagnosis and prognosis but in different ways. Diagnosis and prognosis are two concepts that are usually not used with their precise meaning, resulting in disappointing results in efforts of estimating the nutritional status of plants and soils. An excellent analysis of this subject has been presented by Bell (1997).

Diagnosis is the estimation of the nutritional status of a plant during sampling while prognosis is the prediction of the possibility of a deficiency impairing plant growth at stages in the growth cycle after the sample is taken (Smith, 1986). Diagnosis is based on comparisons of nutrient concentrations in plant parts (mainly leaves) with standard values related to physiological requirements of the plants. Such critical concentrations have been established for several crops (Jones et al. 1991; Mills and Jones, 1996) so that diagnosis is possible for them. In diagnosis, mainly plant analysis and rarely soil analysis is used. Soil testing for diagnosis requires a knowledge of the minimum external nutrient requirements in soil solution for normal growth of roots. Efforts have been made so far for a small number of crops (Fox, 1981). Prognosis is also based on critical concentrations of some plant parts. This sometimes has led to confusion between these two concepts, which have been used with the same meaning (Bell, 1997). In contrast to diagnosis, prognosis may be based on either leaf or soil analysis. The main difference between prognosis and diagnosis lies in the time between sampling and appearance of the influence on plant growth. In the case of soil analysis, the 'time' is the whole growing season while in the case of plant analysis this 'time' is shorter. This reduces the possibility of a correct prediction in the case of soil analysis because more elapse enables more factors influencing plant growth to be involved (Bell, 1997).

Horticultural crops comprise crops with large differences between them. Vegetables are usually intensive crops requiring high nutrient quantities and water and have a very short life cycle which ranges from 8 to 12 weeks from plantation to harvesting date and they are harvested at an early ripening stage (Geraldson and Tyler, 1990). Crop yield and quality are strongly influenced by the nutrient concentration in the root environment, i.e. the intensity factor, and less by the nutrient forms associated with solid phases, i.e. the capacity factor. Furthermore, the root system of vegetable crops is usually concentrated in the surface soil layers and the estimation of soil fertility is easier in the effective depth. Thus soil testing is an effective tool for these crops. In contrast, fruit trees are perennial with higher development and a greater effective depth. Another important difference between vegetable and fruit trees in terms of using soil testing is the role of the stored nutrient pools in fruit trees that strongly affect their ability to satisfy nutrient plant needs for a considerable period. The purpose of this chapter is to present the usefulness of soil testing mainly for prognosis of nutritional status of horticultural crops.

2. Steps of Soil Testing

Soil testing is the physicochemical determination of a soil. However, for practical reasons this definition has been confined to comprise the determination of nutrients available to plant using fast chemical analysis, salinity, and toxicity of these elements. Furthermore, the concept comprises the evaluation and interpretation of the results as well as fertilizer recommendation. The steps followed in soil testing aimed at making fertilizer recommendations are extraction and chemical analysis, correlation and interpretation of the results, and making fertilizer recommendations (Peck and Soltanpour, 1990).

2.1. Soil Sampling

Sampling is one of the most crucial steps of soil testing. Soil has a significant variation in its properties, and hence selection of a representative sample of a tested area is difficult. James and Wells (1990) gave an excellent description of this problem. For obtaining a representative sample, variability sources must be well known. Soil variability observed either in a horizontal or a vertical direction is due to pedogenic processes and also due to human influence. Pedogenic processes cause variability mainly to A and B horizons referred on the properties, organic matter, pH, cation exchange capacity (CEC) and availability of nutrients. Human influence may cause a sharp change in soil properties mostly through tillage practices and crop management. For example, plowing and leveling may cause sharp changes in the properties of surface soil by bringing to the surface, parts of subsoil horizons that substantially differ from the surface horizons. Other human activities causing soil variability are the applied tillage system, fertilization system, type of fertilizers, application depth, etc. Variability caused by all these reasons are distinguished to micro-, meso- and macro-variability, i.e. to soil variation separated from 0 to 0.05 m, 0.05 to 2 m, and > 2 m (James and Wells, 1990). Soil sampling method must be adapted to all these variation scales which means that, to obtain a good sampling of a field, all the reasons causing variability must be well known.

2.2. Sampling Methods

In uniform fields, i.e. in fields with small macro- and meso-variation, composite soil sample collection from random places is advisable. In practice, this can be done following a zigzag course in the sampled field. The number of subsamples of the composite sample cannot be defined. It depends on the field size, but it is not strongly influenced by the increase of field size (Cameron et al. 1971). In non-uniform fields, where macro-variability is high, a non-random sampling process is recommended. By this process, both the estimation of the mean situation of the field as well as the localization of the places with extreme values are aimed (James and Wells, 1990). In cases with high meso-variation that usually occurs in band-applied fertilizer fields, a random sampling similar to that previously referred for composite samples is applied, but the sample number must be higher. In addition, the initial sample must be greater in volume and requires a special treatment for thorough mixing after reducing the bulk appropriately to the bulk necessary for soil analysis. Sub-sample numbers in this case should be four to five-fold compared to the number required in the usual sampling of random composite samples.

In all the above mentioned cases, it is very clear that in order to carry out a good sampling, a detailed description of the basic soil properties is required. Detailed soil maps are valuable for this purpose and should be utilized. Peck and Melsted (1973) in a relevant study using soil maps showed clearly the difficulties in soil sampling and the usefulness of soil maps.

2.3. Sampling Depth

Soil sampling depends on the mobility of the elements to be determined, root distribution, and rooting depth. Mobile nutrients like NO_3^- or SO_4^{2-} should be determined for the whole soil profile while for the less mobile nutrients like P, determination of the soil content in the plowing layer seems to be sufficient for most cases. Meisinger (1984) suggested that for NO_3, soils should be sampled to a depth of 180 cm. Root development depends on the species and variety cultivated, soil fertility, soil moisture, and soil structure. Soil structure and partial pressure of O_2 and CO_2 affect rooting depth and distribution especially in the case of dense (i.e. high bulk density) horizons close to the soil surface. In cases as that mentioned above, roots develop in the loose horizon but it has also been observed that there is in depth root development in cases of uniform high bulk density with soil depth (Fried and Broehart, 1967). Generally, enhanced root development is found where nutrients accumulate and sampling should aim in assessing nutrient supplying power of soils in the effective feeding zone. Under optimum soil conditions plants differ in their root development. Potatoes have a rooting system that develops horizontally while beets are deep-rooted with few side roots. Peas and beans have a larger part of side roots compared to beets. Onions have roots developing within 18 cm for the whole growing period (Brewster, 1994). Jackson and Stivers (1993) found that lettuce roots under actual management practices in grower's field are most abundant in the top 50 cm of the soil. In the same study it was found that irrigation system significantly affected root depth. Surface drip irrigation resulted in the highest root length and root biomass in the 0-15 cm soil layer and root distribution responded to the availability of water and NO_3–N. Oliveira et al. (1996) found the largest proportion of tomato (*Lycopersicum esculentum* Mill.) roots in the top 40 cm of the soil that rapidly decreased with depth because of a horizon with high density immediately below this depth. Gallardo et al. (1996) reported that wild species of lettuce (*Lactuca seriola*) has deeper root system being extended to a depth of 20-80 cm than cultivated lettuce (*Lactuca sativa*). Strawberries have a shallow root zone managing water and nutrients approximately in the upper 15 cm (Anonymous, 1994).

Root development of fruit crops depends on the rootstock used and the soil type. 'Orlando' tangelo trees on rough lemon or 'Palestine' sweet lime rootstocks in sandy well drained soils develop the main part of their roots to the depth of 75 cm while the 'Rusk' rootstock is shallow rooted. In soils with hardpans or high water table, the main volume of the roots remain within 15-45 cm (Davies and Albrigo, 1994). Kiwifruit develops a root system that can reach to the depth of 4 m (Davison, 1990). However, Smith and Clark (1989) and Tsadilas et al. (1997) found that boron concentration in kiwifruit leaves was strongly correlated with soil water extractable B at a depth of 0-60 cm and 0-50 cm, respectively, suggesting that this depth is critical for B supply to kiwifruit. In similar studies Neilsen et al. (1985) found strong relationships between soil boron at 0-90 cm depth and B concentration in peach leaves (*Prunus persica*, L.) and Tsadilas et al. (1994) for B concentration in olive leaves (*Olea europea*, L.) at a depth 0-90 cm. Kikuchi and Sato (1990) suggested that total root numbers and root numbers in the 0-50 cm and 50-100 cm soil layers vary with soil type. The depth of the root system for a high-yielding orchard of apple cv. Fuji on M 26 rootstocks growing in sandy loam or silt loam should be 60-100 cm. However, Koike and Tsukahara (1993) found that in a gravelly lowland soil with effective soil layer 50-60 cm deep, most of the roots developed within 30 cm of the soil surface. The majority of pear roots in several rootstocks was found in the 10-30 cm on podzolic, silty loam soil (Devvatov, 1993).

2.4. Sampling Time

Determination of the optimum sampling time is related to the seasonal variation of the nutrient levels in the soil. Time of sampling also depends on the nutrient form found in the soil, soil type, and species cultivated. Intensity factor shows greater variations within growing season than capacity factor. Changes of pH are greater in acid soils compared to neutral or alkaline soils. Soil P and K levels tend to reduce from spring to autumn. Sampling at the end of summer gives more accurate results for P, K, and lime requirements for soils testing low as compared to well supplied or neutral soils (Peck and Melsted, 1973).

2.5. Sample Handling

In every stage from sampling to the analysis, care should be taken to avoid any contamination from all possible sources (sampling tools, bags, etc). Composite soil samples are air dried, not above 30-35°C and by continuous subdivision the required quantity is sent for analysis (James and Wells, 1990).

3. Correlation—Calibration—Interpretation

Development of a soil test for a certain element includes the choice of the extractant, the correlation of the extracted values with the concentration in the plant tissue or the total uptake and calibration of the soil test values with a crop parameter, usually yield (Corey, 1987).

3.1. Correlation

Correlation studies between the element extracted and the concentration in plant tissue attempt to establish that the element extracted is also taken by the crop and a high correlation coefficient provides a strong evidence that the above hypothesis is true. Correlation studies are carried out first in the greenhouse and are later extended in the field. Sometimes correlation in the relationship is improved after inclusion of soil properties that affect the nutrient uptake like pH, organic matter content, clay content, etc (Reisenauer et al. 1973).

3.2. Calibration

Calibration studies attempt to identify the degree of deficiency or sufficiency of an element and the quantity of the element needed to be applied if it is deficient (Evans, 1987). The data are obtained from pot or field experiments and ideally should be applicable for the geographical area where recommendations are given. Calibration studies should also be repeated in order to include new research findings or introduction of new varieties.

3.3. Interpretation

Based on the correlation and calibration data, interpretation is the final stage that should produce a fertilization guide. Interpretation of a soil test value must be done by experts who have a good knowledge of the soils of the particular area and the management practices followed. Interpretation is also related to the adopted philosophy on fertilization in the area. One is the sufficiency concept according to which a low, medium or high range is identified. Crop response is expected or not expected depending on the range in which the soil test value falls. According to that concept, the crop is fertilized. The other is based on building up and maintaining a relatively high level of nutrient in the soil and fertilization aims to replace the amount removed by the crop. In this approach, the soil is fertilized (Olson et al. 1987).

4. Methods Assessing Available Nutrient Forms in the Soil

4.1. Nitrogen

Assessment ot soils for available N is based on measuring residual N and on estimation of the quantity that will be available to plants during the growing period. Total N in soils ranges from 0.08 to 0.4% with the major part available in an organic form (Bremner, 1965) and of this 1-3% is transformed to inorganic (Broadbent, 1984) depending on factors like soil water content, temperature, and aeration. Chemical and biological procedures have been proposed in an attempt to estimate N (availability index) that will be released from soil organic matter during the growing period. Biological indexes are considered realistic since they approach field soil conditions but are time consuming and are not applicable as routine soil tests. Keeney (1982) suggested the use of NH_4-N produced after incubation for seven days at 40°C as a biological index that can be used routinely. Chemical indexes involve extraction of soil with an acid or alkaline solution that extracts a part of the organic matter that will mineralize easily. Keeney (1982) proposed the use NH_4-N produced in 16 h at 120°C. The soil organic matter content has also been used and Buchholz et al. (1981) proposed the use of this soil property as presented in Table 1.

Residual N depends on the crop and the quantity of the residues remaining in the soil after the crop is removed. The best time to assess residual N is before the start of the growing period (Stanford, 1982). Of the two N forms NO_3-N and NH_4-N, the latter is dependent on soil conditions (pH, temperature, aeration) that affect nitrification and for

Table 1. Nitrogen rate adjustments based on soil texture, cation exchange capacity, organic matter, and time of major crop growth (Buchholz et al. 1981)

Soil texture	CEC cmol (+) kg^{-1}	OM %	Cool-season crops kg ha^{-1}	Warm-season crops kg ha^{-1}
Sand to		0.5	10	20
sandy loam	10	1.0	20	40
		1.5	30	60
Silty to loam		2.0	20	60
	10-18	3.0	30	80
		4.0	40	20
Clay loam to clay		2.0	10	20
	18	3.0	15	30
		4.0	20	40
		5.0	25	50

this reason most times it is not included in a soil testing program for available N (Dahnke and Johnson, 1990). For NO_3–N, despite the fact that in the past it was ignored as an N availability index, lately its inclusion in the soil analysis seems to be generalized. However, time and depth of sampling are critical factors if useful results are expected from the analysis. Ideally NO_3–N should be assessed for the effective depth of the crops' root development. Extraction of the soil with 2M KCl (1 : 10 soil : solution ratio) is most commonly used. NH_4–N and NO_3–N are determined in the extracts (Mulvaney, 1996). Very little data is available on the use of a soil test for N as a means of fertilization guide. There is however, data on the quantity of N required for specific target yield (Tables 2 and 3) and by using the simple relationship (Dahnke and Johnson, 1990):

N requirement – soil test N = Fertilizer N

the fertilization practices for horticultural crops can be improved.

4.2. Phosphorus

It was previously mentioned that the assessment of the concentration of a nutrient in the soil solution is more useful as a soil index for vegetables because the requirements of a nutrient is high for a short biological period. Attempts to identify the minimum P concentration in soil solution that can support maximum yield date back to Tidmore (1930) and Bingham (1949). They found that tomato (*Lycopersicum esculentum* Mill.) and Romaine lettuce (*Lactuca sativa* var. longifolia) required 0.03 and 0.4 ppm P, respectively in soil solution to obtain maximum yield. Phosphate concentrations in the soil solution that are necessary for maximum yield are called external P requirements. Fox (1981) describes the methodology for obtaining such a value. Separate soil samples are equilibrated at 25°C with solutions containing graduated quantities of $Ca(H_2PO_4)_2 \cdot H_2O$ in 1 : 10 soil to solution ratio for 6 days, shaking twice a day for 30 minutes.

Table 2. Nutrient demand of vegetable crops in relation to crop yield (Halliday and Trenkel, 1992)

Crop	Target yield ton/ha	Nutrient demand, ton/ha				
		N	P_2O_5	K_2O	Ca	Mg
Asparagus (*Asparagus officialis* L.)	6.4	15.4	4.3	19.3	13.9	1
Carrot (*Dacuus carota* L. var. *sativus* Hoff.)	30	90-120	30-45	150-300	—	—
Celery (*Apium graveolens* L. var. *dulce* (Mill.) Pers.)	20	130	50	180	60	4
Cucumber (*Cucumis sativus* L.)	300*	45-50*	20-25	80-100	30	13
French beans (*Phaseolus vulgaris* L.)	13	12.9	2.1	6.8	5	1.7
Lettuce (*Lactuca sativa* L.)**	—	1.5-3	1-1.5	4-7	1-1.5	0.2-0.4
Melon (*Cucunis melo* L.)***	20-30	5-12	1.5-2.5	5-20	7-10	2
Onion (*Allium cepa* L.)	30	0.116	0.044	0.144	0.131	0.02
Peppers (*Capsicum annuum* L.)	35-50	0.18-0.4	0.045-0.12	0.25-0.68	0.11-0.16	0.032-0.05
Spinach (*Spinacea oleracea* L.)	15-30	0.06-0.1	0.03-005	0.1-0.23	0.02-0.03	0.01-0.02
Tomato (*Lycopersicum esculentum* L.) for table outdoors	40-50	0.1-0.15	0.02-0.04	0.15-0.3	—	0.02-0.03
Tomato (*Lycopersicum esculentum* L.) under glass	100	0.2-0.6	0.1-0.2	0.6-1	—	—
Watermelon (*Citrulus lanatus* L.)**		1.7	1.3	2.7	—	0.7

* under glass, ** rates in kg/ton, *** outdoor crops

Table 3. Net uptake (removed in fruit and incorporated in framework) of fruit crops in relation to crop yield (Halliday and Trenkel, 1992)

Crop	Target yield ton/ha	Nutrient demand, ton/ha				
		N	P_2O_5	K_2O	Ca	Mg
Apple (Malus pumila L)(Red Delicious)	45	39.2	24	85.1	70.3	7.5
Pear (Pyrus communis L.)	25	21	6	43.3	70	6
Peach (Prunus persica L.)[1]	28	33.6	7.8	98.2	4.8	9.3
Orange (Citrus sinensis)[2]		1773	506	3194	1009	367
Mandarin (Citrusreticulata)[2]		1532	376	2465	706	184
Lemon (Citrus limon)		1638	366	2086	658	209
Grapefruit (Citrus paradisi)[2]		1058	298	2422	573	183
Avocado (Persea americana L. var. Fuert)	10	11.3	3.9	23.5	2.9	8.3
Avocado (Persea americana L. var. Fuert)	10	28	8	54.6	3.3	1.8
Pineapple (Ananas comosus L.)	100	123	34	308	—	—
Banana (Musa spp)[3]		1.7	0.45	6	0.3	0.4
Strawberry (Fragaria X ananassa)[4]		6-10	2.5-4	10	—	—
Grape (Vitis vinifera L.)	7-25	22-84	5-35	41-148	28-204	6-25
Kiwifruit (Actinidia deliciosa)	30	129	39	219	176	35
Pistachio (Pistacia vera L.)	1	30	12	15	3	—
Walnut (Juglas regia L.)[5]		28.3	8.9	6	0.8	2.7
Pecan (Carya illioinensis)	1.2 (nuts)	9.7	5.3	5.4	5.2	1

[1] Removed in fruit, [2] Grams per ton of fresh fruit, [3] kg/t whole bunches, [4] Nutrient removal per metric ton, [5] 0.9 t kernel

The graphical presentation of P concentration in the equilibrium solution versus P quantity sorbed (i.e. the difference between P added and P in equilibrium solution) is the P sorption isotherm. Using these isotherms we can estimate the P needed to be applied in the soil to obtain the P external requirements of a given crop. In Table 4 the yield and the respective external P requirements for maximum yield of several vegetable crops are presented. Field experiments showed that external P requirements were relatively stable for a wide range of soil conditions (Fox, 1981). Furthermore, it was found that P concentration in the equilibrium solution is strongly dependent on some soil properties. For example, Tsadilas et al. (1996) found that 61% of the variation of the P required to obtain a P concentration in the equilibrium solution equal to 0.2 ppm, that satisfies P requirements of most crops (Fox, 1981), was explained by clay content and free iron content. So, external P requirements in conjunction with P sorption curves seems to be extremely effective in estimating P fertilizer requirements of many crops, especially for vegetables.

Table 4. Yield of several vegetable crops in relation to P concentration in soil solution (Lorenz and Vittum, 1980)[*]

Crop	P in soil solution, ppm							
	0.003	0.006	0.012	0.025	0.05	0.1	0.2	0.4
	% of maximum yield							
Head Lettuce	1	2	6	14	26	52	81	100
Cucumber	20	32	45	58	72	83	97	–
Tomato			43	70	80	89	94	99
Chinese cabbage	27	44	58	70	81	90	96	100
Sweet potato	72	74	77	82	87	94	99	100
Head cabbage			87	91	96	99	100	100

[*] Fox (1981) proposed for 95% of maximum yield, P concentration in solution 0.30 ppm for head lettuce, 0.04 ppm for cabbage, and 0.20 ppm for tomato.

Estimating P bound with solid phase of the soil is another approach to assess soil P status (labile-P). In Table 5 several methods of extractable P are presented but only a few have been calibrated for horticultural crops. Van Diest (1963) found satisfactory correlation between P extracted with Bray 1, Olsen, and Mehlich 1 extractants and P uptake by tomato (r = 0.7, 0.8 and 0.73 respectively). In Table 6 the interpretation of available P for several vegetable crops is presented.

Thére is very little data relating to the use of soil P tests and their use in fruit crops. Soil analysis is used to determine the soil properties that will affect its growth when the fruit crop is established. The response of fruit crops to P addition is rare and was reported only in soils with high P retention capacity (Righetti et al. 1990). In general, P addition is not recommended and the probable response is attributed to carrier effects (Ca and S) rather than directly to P.

Table 5. Soil test extractants commonly used for determination of available P

Extractant	Solution	Reference
Olsen	0.5 M NaHCO$_3$ – pH 8.5	Olsen et al. 1954
Bray P1	0.03 M NH$_4$F + 0.025 M HCl	Bray and Kurtz, 1945
Bray P2	0.03 M NH$_4$F + 0.1 M HCl	Bray and Kurtz, 1945
Mehlich I	0.05 M HCl + 0.0125 M H$_2$SO$_4$	Mehlich, 1953
Mehlich II	0.015 M NH$_4$F + 0.2 M CH$_3$COOH 0.2 M NH$_4$Cl + 0.012 M HCl	Mehlich, 1978
Mehlich III	0.015 M NH$_4$F + 0.2 M CH$_3$COOH 0.25 M NH$_4$NO$_3$ + 0.013 M HNO$_3$	Mehlich, 1984
AB-DTPA	1 M NH$_4$HCO$_3$ + 0.005 M DTPA	Soltanpour and Schwab, 1977
Acetate methods	Sodium acetate-acetic acid 1 : 2, pH 4.8	Chapman, 1966
Acetate methods	Ammonium acetate-acetic acid 1 : 5, pH 4.8	Chapman, 1966
Fluoride methods	0.03 N ammonium fluoride-0.025 N hydrochloric acid 1 : 10, pH 3.5	Chapman, 1966

Table 6. Evaluation of some chemical P soil tests for vegetable crops (Chapman, 1966)

Method	Phosphorus level, ppm			Crop
	Low	Medium	High	
Sodium acetate - acetic acid pH 4.8; 1 : 2; 3 min.	10	11-25	26+	Lettuce
	10	11-50	51+	Certain vegetables
Sodium acetate - acetic acid pH 3.0; 1 : 2; 5 min.	12	13-50	51+	All vegetable crops
Sodium acetate - acetic acid pH 4.8; 1 : 4; 30 min.	2	3-4	5+	All vegetable crops
Ammonium acetate - acetic acid pH 4.8; 1 : 5; 30 min.	4	5-9	10+	All vegetable crops
0.03 N ammonium fluoride- 0.025 N hydrochloric acid pH 3.5; 1 : 10; 1 min.	6	7-20	21+	All vegetable crops
Sodium bicarbonate 0.5 M pH 8.5; 1 : 20; 30 min.	14	15-21	22+	All vegetable crops
Sodium bicarbonate 0.5 M pH 8.5; 1 : 20; 30 min.	2	3-4	5+	Beans
Electrodialysis method	10	11-20	21+	Asparagus, beans, beets, cucurbits
0.05 N boric acid 1 : 15; 20 min.	15	16-30	31+	Celery, eggplant, onions, peas, leafy vegetables

4.3. Potassium

Horticultural crops have a higher demand for K as compared to N (Tables 2, 3). The rate of K uptake exceeds the rate of organic matter accumulation resulting in a further stress of the soil's capacity to supply adequate K during certain stages of plant growth. The method most commonly used for available K involves measurement of exchangeable K (extraction with 1M ammonium acetate, pH 7.0 at 1 : 10 soil : solution ratio). It has been found that exchangeable K is related with K uptake by many crops (Haby et al. 1990). However, in addition, a number of different extractants have been tested and were found satisfactory for available K in soils. Several of these (AB-DTPA, Bray 1, Mehlich I and II and Olsen) are also used for other elements like P and B and are at times called universal extractants.

4.4. Calcium

Calcium in soils is classified as non-exchangeable, exchangeable, and soil solution Ca^{2+}. Non-exchangeable Ca includes mineral forms such as Ca-bearing aluminum silicates like feldspars, amphiboles, Ca phosphates, plagioclase etc. and Ca carbonates which is the dominant form in calcareous soils. Plants adsorb calcium as Ca^{2+} from the soil solution, the uptake is affected by calcium supply, soil pH, and CEC. Methods of testing soils for Ca include the following:

4.4.1. Determination of Total Carbonate

This Ca form is usually quantified by acid dissolution leading to production of CO_2 from which the total amount of carbonates [calcite—$CaCO_3$, and dolomite—$CaMg(CO_3)_2$] is calculated. The most common procedure includes decomposition of carbonates by treating the soils with 2N H_2SO_4 containing $FeSO_4$ as antioxidant to prevent release of CO_2 from organic matter. Total carbonate content (usually called equivalent calcium carbonate) is calculated by the amount of CO_2 produced. Apart from this procedure many other methods proposed by several investigators were reviewed by Loeppert and Suarez (1996).

Calcium carbonate influence on crop nutrition is indirect through soil pH and availability of P, B, and metallic elements like Fe, Mn, Zn, and Cu. Soil pH in calcareous soils range from 7.5 to about 8.5, and the total calcium carbonate content is not a good availability index for crop plants, although in some laboratories it is used as such an index. For example, Lahav and Kadman (1980) proposed a general guideline where the Guatemalian rootstock avocado (*Persea americana*, Mill.) had a maximum total carbonate of 4-5%, 20-25% for the Mexican rootstock and for the West Indian rootstock it was above 40%.

4.4.2. Active Carbonate

Another more reliable index for soil carbonate is the so-called active carbonate. Particle-size distribution, surface area and reactivity of soil carbonates are important properties influencing rizosphere processes. On calcium carbonate surface, many elements like P and B as well as trace metals and organic acids may be adsorbed. The carbonate reactivity may also substantially influence rizosphere processes, especially those depending on acidification like Fe (Loeppert and Suarez, 1996).

Among the several methods proposed for measuring carbonate activity, that suggested by Drouineau (1942) and improved by Boischot and Hebert (1947) is most widely used. This method involves reaction of the soil with 0.1 M ammonium oxalate followed by determination of the unreacted oxalate by titration with 0.1 M $KMnO_4$ (Loeppert and Suarez, 1996). An evaluation of this test for horticultural crops is presented in Table 7.

Table 7. Evaluation of active calcium carbonate values for horticultural crops

	Active carbonate content, %	Reference
Vegetable crops		Zuang, 1982
No or little chlorotic action	4 - 5	
Slight chlorotic action	6 - 7	
Chlorotic action	7 - 12	
Strong chlorotic action	12 - 20	
Excessive chlorotic action	> 20	
Fruit crops*		CRF-SCA, 1985
Pear on cognassier rootstock	> 9	
Peach	> 7	
Grape rootstocks**		Levandoux, 1961
Riparia glorie de Montpellier	6	
Rupestris du Lot	14	
Hybrids Riparia - Rupestris	11	
Hybrids Rupestris - Berlandieri	17	
Hybrids Riparia - Berlandieri	25	
Hybrids Vinifera - Berlandieri	> 40	

* Possible chlorosis, ** Prohibitive values

4.5. Magnesium

The most common extractant for determining available soil Mg is 1M ammonium acetate (pH 7.0) as for K and Ca. Exchangeable Mg is also estimated by using other extractants like dilute salt solutions, dilute acids or acidified salt solutions. Among them are Morgan's solution (0.52 M CH_3COOH, 0.62 M NH_4OH), double acid (Mehlich I – 0.025 M HCl, 0.0125 M H_2SO_4), 1M ammonium nitrate, calcium chloride, 0.05 M NaCl, and 1M sodium acetate (Haby et al. 1990).

Correlation studies of extractable soil Mg with crop response usually refer to surface soil Mg. However, significant exchangeable Mg moves into the subsoil to depths of about 50 cm and is concentrated in horizons with clay accumulation like argillic horizon in Alfisols or Ultisols. Furthermore, some of these soils may contain interstratified mica and vermiculite in which Mg can exist in the interlayer positions substituting for Al in the octahedrals. This Mg, although non-exchangeable, may be important for plant growth and should be taken into account when we evaluate soil Mg (Rice and Kamprath, 1968).

Two main methods are usually used for making fertilizer recommendations, exchangeable Mg and Mg per cent saturation of the soil CEC as is discussed below. An evaluation of exchangeable Mg is given in the Table 8 (Doll and Lucas, 1973).

Table 8. Evaluation of soil exchangeable Magnesium for horticultural crops (Doll and Lucas, 1973)

Concentration, mg Mg kg^{-1}	Crop response
0 - 25	Deficiency symptoms in most vegetables, fruit and glasshouse crops
26 - 50	Deficiency expected in fruit and glasshouse crops
51 - 100	Deficiency is not likely in vegetable crops
101 - 175	Mg is required for glasshouse crops tomato, cucumber, pepper
176 - 250	Mg is not suggested for horticultural crops

5. Cation Saturation Levels

Soil test interpretations for K, Ca, and Mg referred above were based on the philosophy of sufficiency level, i.e. the concept that there are definable levels of individual nutrients in the soil below which crops will respond to added fertilizers. However, for cations K, Ca, and Mg in addition to the previously described philosophy there is another concept, i.e. the basic cation saturation ratio (BCSR).

5.1. Calcium

According to this, maximum yields can be achieved by establishing an ideal ratio of Ca, Mg, and K in the soil (Eckert, 1987). Furthermore the cation ratio (CR) concept is used. A general guide for Ca in accordance with the BCSR concept is that many crops respond to Ca applications when the Ca saturation of soil CEC is below 25% (Haby et al. 1990). However, it may vary widely proportionally to the clay type. For example, kaolinite clays can satisfy Ca requirements of most plants at saturation levels of 40 to 50% while 2:1 clays such as montmorillonite require a saturation of 70% in order to release sufficient supply to plants. The range of 65-85% of the CEC was suggested as ideal Ca saturation (Haby et al. 1990).

5.2. Magnesium

Little data can be found in the literature relating crop growth to Mg saturation for horticultural crops. Martin and Page (1969) working on the influence of percentage base saturation on plant growth found that at 30, 50, and 100% a base saturation of 3 to 6% exchangeable Mg was associated with Mg deficiency and growth reduction of sweet orange (*Citrus sinensis* L.). Seven to thirteen per cent exchangeable Mg was found to be associated with Mg deficiency symptoms but no reduction of plant growth was observed.

In contrast to per cent Mg saturation, many studies have been carried out for horticultural crops referred on cation ratios. In general, the favorable ratio of K/Mg varies widely, from 0.8:1 to 20:1 (Haby et al. 1990). Magnesium deficiency symptoms appeared in apple (*Malus sylvestris* Mill) when the K/Mg ratio in the Morgan extract of the top 20 cm soil was 2:1 to 1.6:1 (Mulder, 1950). K to Mg ratio provided better estimate of Mg availability to citrus than available Mg in the top 30 cm soil. The same was true for carrots, for which it was found that K/Mg ratio was better related to the Mg content of the carrot leaves than exchangeable Mg in the top 0 to 46 cm soil (Charlesworth, 1967). In general, slow-growing crops may tolerate wider ratios of K/Mg in the soil than intensive crops like horticultural crops (Batey, 1967).

6. Micronutrients

6.1. Zinc, Copper, Manganese and Iron

Soil testing for metallic micronutrients is very difficult because of the following reasons (Viets and Lindsay, 1973): a) plants require very small quantities of these elements so that contamination problems usually arise, b) small changes in environmental conditions may either correct or induce deficiencies in soils, and c) there is a considerable lack of dependable criteria for estimating nutritional status. Very frequently plants may suffer from insufficiency of metallic micronutrients resulting in yield reduction but without showing deficiency symptoms. So, chemical analysis of leaves for these elements are superior to soil tests for diagnosing nutritional status. This is especially true for deciduous fruit trees and citrus. However, soil testing may provide valuable information which being co-estimated with plant analyses may lead to the best decision about the fertilizer needs.

Several methods are in use for estimating the available metallic micronutrients. They may be grouped according to the extractants used as: a) water and neutral salts extractable, b) weak and strong acids extractable, and c) chelating agents extractable.

The first group includes very slight extractants that extract very small quantities that are difficult to be evaluated. However, in some cases they were found useful in estimating available forms of these micronutrients. For example, Hoyt and Nyborg (1971) found a very good relationship between 0.01 M $CaCl_2$ extractable soil Mn and its concentration in turnip rape. Healy (1953) found that peach trees growing on a silt loam with less than 0.2 ppm of Mn extracted by NH_4OAc showed Mn deficiency symptoms. Sherman et al. (1951) found that concentration of extractable Mn 0.5 to 3 ppm in the top 40 cm corresponded to severe deficiency symptoms in cherry leaves. In greenhouse and field studies, Brown and Krantz (1961) found that soil Zn 0.5 ppm extracted by ammonium acetate-dithizone-carbon tetrachloride reagent was critical for vegetable crops. Drouineau and Mazoyer (1953) found that spinach developed in rows previously planted to peach trees showed Cu toxicity symptoms when soil Cu concentration extracted by ammonium acetate method was 98 to 130 ppm. The symptoms dissappeared when Cu concentration was reduced from 4.5 to 4.7 ppm.

The second group includes extractants from very slight acid solutions like HOAc plus 0.5 N KCl, 0.1 N HCl up to strong acids extracting the total amounts of the metallic micronutrients. In general, total amounts of a nutrient are not good indicators of their availability. However, sometimes it was observed the opposite. For example, Bould et al. (1953) found that soils containing more than 2 ppm of total Cu did not produce Cu deficiency symptoms in apple and pear trees while Knott (1933) reported that concentration of total Cu in soils (12 ppm) caused reduction in onion yield which was not observed in Cu concentration (31 ppm).

The third group is based on a more fundamental and theoretical base providing one of the most promising means of soil testing for metallic micronutrients. Two of the procedures of this category are the most important. The first one, using as extractant 0.02 or 0.05 M solution of chelating agent EDTA (ethylenediaminetetraacetic acid) has been used especially for Zn, Mn, and Cu (Viets and Lindsay, 1973). The other one developed by Lindsay and Norvell (1978) used 0.005 M solution DTPA (diethylenetriaminepentacetic acid) in 0.005 M $CaCl_2$ and 0.1 M TEA (triethanolamine) buffered at pH 7.30. Ten g soil is shaken for two hours, filtered and the concentrations of Zn, Fe, Mn, and Cu are determined in the supernatant by atomic absorption spectrometry. This test is considered one of the most promising tests available now. Unfortunately, there is no calibration data for both EDTA or DTPA tests for horticultural crops so far. However, a general guide for DTPA test for sensitivity to metallic micronutrient is given in Table 9 (Viets and Lindsay, 1973).

Table 9. Critical values (ppm d.w. soil) of DTPA-extractable metallic micronutrients for sensitive crops (Viets and Lindsay, 1973)

Nutrient	Deficient	Marginal	Adequate
Fe	< 2.5	2.5-4.5	> 4.5
Zn	< 0.5	0.5-1.0	> 1.0
Mn	< 1.0	—	> 1.0
Cu	< 0.2	—	> 0.2

6.2. Boron

Plants absorb boron (B) primarily from the soil solution, independently of the amount of B adsorbed in soil (Ryan et al. 1977). Thus, estimation of soil available B should involve soil solution B. However, soil solution B is a very small fraction of the total B uptake by most crops during the growing season within which soil solution B is replenished several times. So, a soil test for available B to plants must assess the potential of the soil to replenish soil solution B, i.e. the capacity factor (Aitken and McCallum,

1988). Capacity factor for B against losses by plant uptake was found to be dependent on several soil parameters such as texture, clay mineralogy, soil pH, organic matter and sesquioxides content (Gupta et al. 1985).

Several procedures have been proposed for estimating available soil B to plants. Most of them are shown in Table 10. One of the most widely used for several decades is that proposed by Berger and Truog (1939) involving extraction of B with boiling the soil in water for 5 min. However, there are still many limitations in using this procedure for assessing B deficiency and toxicity in soil coming mainly from the fact that hot water extracts B from different pools like organic, adsorbed inorganic and soluble inorganic pool (Gupta et al. 1985). So, for the use of hot water procedure to be effective, the interplay of the various B pools and their relationship to B uptake should be defined. In such an effort Jin et al. (1987) and Tsadilas et al. (1994) have fractionated soil B to soil solution B, non-specifically adsorbed B, specifically adsorbed B, B occluded in Mn oxyhydroxides, B associated with amorphous and crystalline Al and Fe oxyhydroxides, and B in silicate minerals and tried to find their relations to B uptake by plants. In 20 quite different soils from Greece, Tsadilas et al. (1994) found that hot water extractable B was significantly correlated with all B fractions except that associated with Mn oxyhydroxides which, however, contributed to plant B uptake. This means that hot water does not extract B from Mn oxyhydroxides and this is a possible explanation why hot water extraction fails to predict plant response to B in many cases. Furthermore, because of some serious technical difficulties involved in the hot water extraction procedure, several investigators tried to find alternatives to this method. Mahler et al. (1984) in order to avoid problems related to glassware proposed the use of sealed plastic pouches in place of the glass refluxing apparatus. Jeffrey and McCallum (1988) and Bingham (1982) proposed 0.01 M $CaCl_2$ solution instead of water in order to obtain clear suspensions free of colloidal matter. Jeffrey and McCallum (1988) and Odom (1980) proposed instead of 5 min, 10 min boiling time while Spouncer et al. (1992) suggested 30 min.

Several attempts were made to find effective procedures differing in extractants and extraction time for estimating available B in soils. Many of them have been tested for horticultural crops, especially vegetables. Some of the most effective methods for estimating available B to horticultural crops are presented in Table 11. For vegetables, hot water procedure seems to be the most effective as it was ascertained by several investigators (Table 11). The same was also true for fruit trees for which apart from the hot water method some other methods were also tested. For example, Tsadilas (1999) found very good correlation between B extracted by hot water, cold water, 0.05 M HCl, 0.05 M mannitol in 0.01 M $CaCl_2$, and hydroxylamine HCl and B uptake by olives. The critical levels for appearance B deficiency are shown in Table 11. For kiwifruit, except the hot water, 0.01 M HCl, 0.05 M mannitol in 0.01 M $CaCl_2$, and resin procedures gave good results in predicting B availability in high B soils.

Table 10. Soil Boron tests used for determination of available B

Extractant	Extraction time	Method of determination	Reference
Hot water	5 min	Quinalizarin	Berger and Truog, 1939
Hot water	5 min	Azomrthine-H	Gupta, 1979
Hot water	5 min	Curcumin and 2-hydroxy-4chlorobe-indophenol	Winker and Uppstrom, 1980
Hot water	10 min		Odom, 1980
0.01 M CaCl$_2$ + 0.05 M mannitol (pH 8.5)	1h	ICP-AES	Cartwright et al. 1983
0.01 M mannitol (pH 8.5)	1h	ICP-AES	Cartwright et al. 1983
0.01 M CaCl$_2$ (pH 8.5)	1h	ICP-AES	Cartwright et al. 1983
0.5 M NaHCO$_3$ (pH 8.5)	1h	ICP-AES	Cartwright et al. 1983
1.0 M NH$_4$OAc (pH 7.0)	1h	ICP-AES	Cartwright et al. 1983
1.0 M NH$_4$OAc (pH 4.8)	1h	ICP-AES	Cartwright et al. 1983
Hot water in seal plastics	7 min	Azomethine-H	Mahler et al. 1984
1 M NaHCO$_3$, 0.05 M DTPA	15 min	ICP-AES	Gestring and Soltanpour, 1987
Saturation past	—	ICP-AES	Gestring and Soltanpour, 1987
Hot water	5 min	ICP-AES	Gestring and Soltanpour, 1987
0.01 M mannitol-CaCl$_2$	16 h	ICP-AES	Gestring and Soltanpour, 1987
Hot 0.01 M CaCl$_2$	10 min	ICP-AES	Jeffrey and McCallum, 1988
Hot 0.01 M CaCl$_2$	5 min	Azomethine-H	Bingham, 1982
Cold water	24 h	ICP AES, Azomethine-H	Jin et al. 1987; Tsadilas et al. 1994
0.02 M CaC$_2$	24 h	ICP AES, Azomethine-H	Jin et al. 1987; Tsadilas et al. 1994
0.1 M NH$_2$.OH.HCl in 0.01 M HNO$_3$	30 min	ICP AES, Azomethine-H	Jin et al. 1987; Tsadilas et al. 1994
0.175 M ammonium oxalate (pH 3.25 in dark	4 h	ICP AES, Azomethine-H	Jin et al. 1987; Tsadilas et al. 1994
Na$_2$CO$_3$ fusion (950°C)	—	ICP AES, Azomethine-H	Jin et al. 1987; Tsadilas et al. 1994
0.05 M HCl	5 min	Curcumin, Azomethine	Ponnamperuma et al. 1981
			Tsadilas et al. 1997
Resin (Amberlite IRA 743)	—	Azomethine-H	Adams et al. 1991
			Tsadilas et al. 1994

Among these procedures the 0.05 M HCl is the most convenient and time saving (Tsadilas et al. 1997).

Table 11. Concentration of soil B extractable (ppm) by various methods in relation to B needs of horticultural crops

Crop	B extraction method	Low	Adequate	High	Reference
Vegetables					
Pea and bean	Hot water	—	< 0.1	—	Berger, 1949
Bean		0.4-0.5	> 0.5	—	Bell, 1997
Strawberry	Hot water		< 0.1		Berger, 1949
				1.6	Haydon, 1981
			0.15-0.25		Blatt, 1982
Lettuce	Hot water		0.1-0.5		Berger, 1949
Carrot	Hot water		0.1-0.5	—	Berger, 1949
		< 0.4	—	—	Gupta and Cutliffe, 1985
Onion	Hot water		0.1-0.5		Berger, 1949
Turnip	Hot water		> 0.5		Berger, 1949
Asparagus	Hot water		> 0.5		Berger, 1949
Radish	Hot water		> 0.5		Berger, 1949
Celery	Hot water		> 0.5		Berger, 1949
Rutabaga	Hot water		> 0.5	—	Berger, 1949
		0.65-0.9	> 0.9	—	Bell, 1997
		0-1.3	> 1.3-1.8	> 3.1	Gupta and Munro, 1969
Cauliflower	Hot water	< 0.15	—	—	Bell, 1997
Broccoli	Hot water	0.28-0.34	—	—	Bell, 1997
Brussel sprouts	Hot water	0.28-0.34	—	—	Bell, 1997
Radish	Hot water	< 1.1	—	—	Adams et al. 1991
Fruits					
Peach	Hot water		0.1-0.5		Berger, 1949
Pear	Hot water		0.1-0.5		Berger, 1949
Cherry	Hot water		0.1-0.5		Berger, 1949
Olive	Hot water		0.1-0.5		Berger, 1949
	Hot water				Tsadilas, 1999
	Cold water	< 0.33			Tsadilas, 1999
	0.01 M HCl	> 0.17			Tsadilas, 1999
	0.05 M Mannitol in	< 0.05			
	0.01 M $CaCl_2$				Tsadilas, 1999
	Hydroxylamine HCl	< 0.41			Tsadilas, 1999
		< 0.14			
Pecan	Hot water		0.1-05		Berger, 1949
Apple	Hot water		> 0.5		Berger, 1949
		0.17			Chabannes and Lucas, 1965
			0.13-2.48	0.24-3.08	
			(0-15 cm)	(0-15 cm)	Neilsen et al. 1985
			0.10-0.88	0.29-1.37	
			(30-45 cm)	(15-40 cm)	
		< 1.0			Yamazaki et al. 1973
Kiwifruit	Hot water			> 0.5	Smith and Clark, 1989
Kiwifruit	Hot water			> 0.51	Tsadilas et al. 1997
Kiwifruit	0.05 M Mannitol in 0.01 M $CaCl_2$			> 0.80	Tsadilas et al. 1997

(Contd.)

Table 11. (*Contd.*)

Crop	B extraction method	Boron level			Reference
		Low	Adequate	High	
Kiwifruit	0.05 M HCl			> 0.18	Tsadilas et al. 1997
Kiwifruit	Resin			> 2.0	Tsadilas et al. 1997
Mungo	Hot water	< 0.53	—	—	Bell, 1997

While soil B tests properly calibrated can provide information about the necessity for B addition, it is difficult to relate test values to the quantity of B required. Such information is usually obtained from greenhouse or field experiments, a process that is slow and quite expensive. For this reason, attempts were made to estimate B requirements to increase its level in soil to a desired level. For estimating the B quantity needed to increase B concentration in soil solution up to external B requirements of a crop, Shunway and Jones (1972) proposed a procedure similar to that of Fox and Kamprath (1970) for estimating P. A concentration of 0.1 ppm B is considered satisfactory for most crops (Shunway and Jones, 1972). Using the sorption isotherm technique referred above, they found that 0.5 ppm B in the equilibrium solution should be the approximate upper limit of B concentration where the relationship between B added and B in equilibrium solution is linear. This is a very simple and easy technique to estimate B fertilizer needs for obtaining a desirable B concentration in soil solution. However, more data on external B requirements of plants are needed.

References

Adams, J.A., Hamzah, Z. and R.S. Swift. 1991. Availability and uptake of boron in a group of pedogenetically-related Canerbury, New Zealand Soils. *Aust. J. Soil Res.* **29**: 415-423.

Aitken, R.L. and A.J. McCallum. 1988. Boron toxicity in soil solution. *Aust. J. Soil Res.* **25**: 263-273.

Anonymous. 1994. Integrated pest management for strawberries. Strawberry growth and development. Univ. of Calif., DANR. Publ. 3351. p. 142.

Batey, T. 1967. The ratio of K to Mg in the soil in relation to plant growth. *In:* Soil Potassium and Magnesium Tech. Bull. 14. Her Majesty's Stationery Office, London, pp. 143-146.

Bell, R.W. 1997. Diagnosis and prediction of boron deficiency for plant production. *In:* Boron in Soils and Plants: Reviews, B. Dell, P.H. Brown and R.W. Bell (Eds.), Kluwer Academic Publ. London, pp. 149-168.

Berger, K.C. 1949. Boron in soils and crops. *In:* Advances in Agronomy, A.G. Norman (Ed.), Academic Press Inc., NY, (1): 321-351.

Berger, K.C. and E. Truog. 1939. Boron determination in soils and plants. *Ind. Eng. Chem. Educ.* **11**: 540-544.

Bingham, F.T. 1949. Soil test for phosphorus. *Calif. Agric.* **3(8)**: 11-14.

Bingham, F.T. 1982. Boron. *In:* Methods of Soil Analysis Part 2. 2nd Edition, Agronomy Mongr. 9, A.L Page A.L. Page, R.H. Miller, and D.R. Keeney (Eds), ASA, SSSA, Madison, WI, pp. 431-447.

Blatt, C.R. 1982. Effects of two boron sources each applied at three rates to the strawberry cv. Midway on soil and leaf boron levels and fruit yields. *Commun. Soil Sci. Plant Anal.* **13(1)**: 39-47.

Boischot, P. and J. Hebert. 1947. Determination of available calcium in soil by the ammonium oxalate method and its use to determine the readily assimable calcium of liming materials. Ann. Agron. **17**: 521-525.

Bould, C., Nicholas, D.J.D., Potter, J.M.S., Tolhurst, J.A.H. and T. Wallace. 1953. Zinc and copper deficiency of fruit trees. *Annual Rept. Agr. Hort. Res. Sta.*, Long Ashton, Bristol (England), **1949**: 45-49.

Bray, R.H. and L.T. Kurtz. 1945. Determination of total, organic and available forms of phosphorus in soils. *Soil Sci.* **59**: 39-45.

Bremner, J.M. 1965. Organic nitrogen in soils. *In*: Methods of Soil Analysis. Part 2, Agronomy Monogr. 9. C.A. Black, D.D. Evans, J.L. White, L.E. Ensminger, F.E. Clark, and R.C. Dinauer (Eds), ASA, Madison, WI, pp. 1324-1345.

Brewster, J.L. 1994. Onions and other vegetable alliums. CAB International, Cambridge, p. 236.

Broadbent, F.E. 1984. Plant use of soil nitrogen. *In*: Nitrogen in Crop Production, R.D. Hauck (Ed.), ASA, CSSA, SSSA, Madison, WI, pp. 171-182.

Brown, A.L. and B.A. Krantz. 1961. Zinc deficiency diagnosis through soil analysis. *Calif. Agric.* **15(11)**: 15.

Buchholz, D.D., Brown, J.R., Hamson, R.G., Wheaton, H.N. and J.D. Garret. 1981. Soil test interpretations and recommendations Handbook. Univ. of Missouri, Columbia.

Cameron, D.R., Nybert M. and J.A. Toogood. 1971. Accuracy of field sampling for soil tests. *Can. J. Soil Sci.* **51**: 165-175.

Cartwright, B., Tiller, K.G., Zarcinas, B.A. and L.R. Spouner. 1983. The chemical assessment of the boron status of soils. *Aust. J. Soil Res.* **84**: 133-144.

Chabannes, J. and D. Lucas. 1965. Severe boron deficiency in an orchard of Richared apples on a sandy alluvial soil in the Loire. Borax Consol. Ltd, Charlisle Place London, S.W.I.

Chapman, H.D. 1966. Diagnostic criteria for plants and soils. *Univ. of Calif. Div. Of Agr. Sci.* pp. 793.

Charlesworth, R.R. 1967. The effect of applied magnesium on the uptake of magnesium by and the yield of arable crops. *In*: Soil Potassium and Magnesium Tech. Bull. 14. Her Majesty's Stationery Office, London, pp. 11-124.

Corey, R.B. 1987. Soil test procedures: Correlation. *In*: Soil Testing: Sampling, Correlation, Calibration, and Interpretation, J.R. Brown (Ed.), SSSA Special Publication No 21, Madison WI, pp. 15-22.

CRF-SCA, Commission Romande des Fumures-Sous Comission Abboricole. 1985. La fertilisation des arbres fruitiers, directives pour la Suisse romande et le Tessin. Revue Swisse Vitic. *Arboric - Hortic.* **17(5)**: 321-331.

Dahnke, W.C. and G.V. Johnson. 1990. Testing soils for available nitrogen. *In: Soil Testing and Plant Analysis*, R.L. Westerman (Ed.), SSSA, Inc. Madison, WI, pp. 127-139.

Davies, F.S. and L.G. Albrigo. 1994. Citrus. CAB International, Cambridge, pp. 254.

Davison, R.M. 1990. The physiology of the kiwifruit vine. *In*: Kiwi Fruit Science and Management, I.J. Warrington and G.C. Weston (Eds), Ray Richards Publ. New Zealand Society for Horticultural Science pp. 127-154.

Devvatov, A.S. 1993. Comparison of root development of pear trees on seedling and Quince A and C rootstocks. *Fruit Varieties Journal.* **47(4)**: 193-197.

Doll, E.C. and R.E. Lucas. 1973. Testing soil for potassium, calcium, and magnesium. *In:* Soil Testing and Plant Analysis L.M. Walsh and J.D. Beaton (Eds), SSSA, Inc. Madison, WI, pp. 133-151.

Drouineau, G. 1942. Dosage rapide du calcaire actif de sols. *Ann. Agron.* Vol. 12.

Drouineau, G. and R. Mazoyer. 1953. Copper toxicity and soil evaluation under the influence of fungicides. *Compt. Rend. Acad. Agr. France*, **39**: 390-392.

Eckert, D.J. 1987. Soil Test Interpretations: Basic cation saturation ratios and sufficiency levels. *In*: Soil Testing: Sampling, Correlation, Calibration, and Interpretation, J.R. Brown (Ed.), SSSA Special Publication No 21, Madison, WI, pp. 53-64.

Evans, C.E. 1987. Soil test calibration. In: Soil Testing: Sampling, Correlation, Calibration, and Interpretation, J.R. Brown (Ed.), SSSA Special Publication No 21, Madison, WI, pp. 22-29.

Fox, R.L. 1981. External phosphorus requirements of Crops. In: Chemistry in the Soil Environment. M. Stelly, D.M. Kral and M.K. Cousin (Eds), ASA, SSSA, Madison, WI, pp. 223-239.

Fox, R.L. and E.J. Kamprath. 1970. Phosphate sorption isotherm of evaluating the phosphate requirement of soils. Soil Sci. Soc. Am. Proc. **34**: 902-906.

Fried, M. and H. Broehart. 1967. The soil-plant system in relation to inorganic nutrition. Academic Press, pp. 358.

Gallardo, M., Jackson, L.E. and R.B. Thompson. 1996. Shoot and root physiology responses to localized zones of soil moisture in cultivated and wild lettuce. Plant Cell and Environment **19(10)**: 1169-1178.

Geraldson, C.M. and K.B. Tyler. 1990. Plant analysis as an aid in fertilizing vegetable crops. In: Soil Testing and Plant Analysis, R.L. Westerman (Ed.), SSSA, Inc. Madison, WI, pp. 549-562.

Gestring, W.D. and P.N. Soltanpour. 1987. Comparison of soil test for assessing boron toxicity to alfalfa. Soil Sci. Soc. Amer. J. **51**: 1214-1219.

Gupta, U.C. 1979. Some factors affecting the determination of hot-water-soluble boron from podzol soils using azomethine-H. Can. J. Soil Sci. **59**: 241-247.

Gupta, U.C. and D.C. Munro. 1969. The boron content of tissue and roots of rutabagas and of soil associated with brown heart condition. Soil Sci. Soc. Amer. Proc. **33**: 424-426.

Gupta, U.C. and J.A. Cutcliffe. 1985. Boron nutrition of carrots and table beets grown in a boron deficient soil. Commun. Soil Sci. Plant Anal. **16(5)**: 509-516.

Gupta, U.C., Jame, Y.W., Campbell, C.A., Leyshon, A.J. and W. Nicholaichuk. 1985. Boron toxicity and deficiency: A review. Can. J. Soil Sci. **65**: 381-409.

Haby, V.A., Russelle M.P. and E.O. Skogley. 1990. Testing soils for potassium, calcium, and magnesium. In: Soil Testing and Plant Analysis, R.L. Westerman (Ed.), SSSA, Inc. Madison, WI, pp. 181-227.

Halliday, D.J. and M.E. Trenkel (Eds). 1992. IFA World Fertilizer Manual. IFA, pp. 632.

Haydon, G.F. 1981. Boron toxicity of strawberry. Commun. Soil Sci. Plant Anal. **12(11)**: 1085-1091.

Healy, W.B. 1953. Treatment of a lime-induced manganese deficiency in peach trees. New Zealand Journ. Sci. Tech. **34A**: 386-396.

Hoyt, P.B. and M. Nyborg. 1971. Toxic metals in acid Soil: II. Estimation of plant-available manganese. Soil Sci. Soc. Amer. Proc. **35**: 241-244.

Jackson, L.E. and L.J. Stivers. 1993. Root distribution of lettuce under commercial production: Implications for crop uptake of nitrogen. Biological Agriculture and Horticulture **9(3)**: 273-293.

James, D.W. and K.L. Wells. 1990. Soil sample collection and handling: Technique based on source and degree of field variability. In: Soil Testing and Plant Analysis, R.L. Westerman (Ed.), SSSA, Inc. Madison, WI, pp. 25-44.

Jeffrey, A.J. and L.E. McCallum. 1988. Investigation of a hot 0.01 M $CaCl_2$ soil boron extraction procedures followed by ICP-AES analysis. Commun. Soil Sci. Plant Anal. **19(6)**: 663-673.

Jin, J., Martens, D.C. and L.W. Zelancy. 1987. Distribution and plant availability of soil boron fractions. Soil Sci. Soc. Amer. J. **51**: 1228-1231.

Jones (Jr)., Benton, J., Wolf, B.,and H.A. Mills. 1991. Plant analysis handbook, a practical sampling, preparation, analysis, and interpretation Guide. Micro-Macro Publ., Inc. Athens, Georgia, 213 pp.

Keeney, D.R. 1982. Nitrogen-inorganic forms. In: Methods of Soil Analysis. Part 2. 2nd Edition. Agronomy Mongr. 9, A.L Page all authors. (Eds), ASA, SSSA, Madison, WI, pp. 643-698.

Kikuchi, T. and K. Sato. 1990. Studies on the yield of mature high-density apple orchards in Aomori Prefecture. 2. Root systems of Fuji trees on M.26 rootstock. Bull. Of the Fac. Of Agric. Hirosaki Univ. 1990, No 52: 19-30 (CAB Abstracts).

Knott, J.E. 1933. The effect of certain mineral elements on the color and thickness of onion scales. *New York Agr. Expt. Sta. Bull.* 552.

Koike, H. and K. Tsukahara. 1993. Studies on root system and growth of Fuji apple trees on Dwarfing interstock and rootstocks. *Journal of the Japanese Society for Horticultural Science* **62(1):** 49-54.

Lahav, E. and A. Kadman. 1980. Avocado fertilization. IPI-Bulletin No. 6, Intern. Potash Institute, Switzerland.

Levandoux, L. 1961. La vigne et sa culture. Presses Univ. de France. pp. 128.

Lindsay, W.L. and W.A. Norvell. 1978. Developmemt of a DTPA soil test for zinc, iron, manganese, and copper. *Soil Sci. Soc. Am. J.* **42:** 421-428.

Loeppert, R.H. and D.L. Suarez. 1996. Carbonate and gypsum. *In:* Methods of Soil Analysis, Part 3, Chemical Methods J.M. Bartels (Ed.), SSSA, ASA, Madison, WI, pp. 437-474.

Lorenz, O.A. and M.T. Vittum. 1980. Phosphorus nutrition of vegetable crops and sugarbeets. *In:* The Role of Phosphorus in Agriculture, R.C. Dinauer (Ed.), ASA, CSSA, SSSA, Madison, WI, pp. 737-762.

Mahler, R.L., Naylor, D.V. and M.K. Fredickson. 1984. Hot water extraction of boron from soils using sealed plastic pouches. Commun. *Soil Sci. Plant Anal.* **15(5):** 493-506.

Martin, J.P. and A.L. Page. 1969. Influence of exchangeable Ca and Mg and of percentage base saturation on growth of plants. *Soil Sci.* **107:** 39-46.

Mehlich, A. 1953. Determination of P, Ca, Mg, K, Na, and NH_4 by the North Carolina Soil Testing Laboratory. Mimeographed, Raleigh, North Carolina.

Mehlich, A. 1978. New extractant for soil test evaluation of phosphorus, potassium, magnesium, calcium, sodium, manganese, and zinc. *Commun. Soil Sci. Plant Anal.* **9:** 455-476.

Mehlich, A. 1984. Mehlich 3 soil test extractants: A modification of Mehlich 2 extractant. *Commun. Soil Sci. Plant Anal.* 15:1409-1416.

Meisinger, J.J. 1984. Evaluating plant available nitrogen in soil-crop systems. *In:* Nitrogen in Crop Production, ASA, Madison, WI. pp. 391-416.

Mills, H.A. and J.B. Jones (Jr). 1996. Plant analysis handbook. Micro-Macro Publ., Inc. USA, pp. 422.

Mulder, D. 1950. Mg-deficiency in fruit trees on sandy and clay soils in Holland. *Plant Soil* **2:** 145-157.

Mulvaney, R.L. 1996. Nitrogen—inorganic forms. *In:* Methods of Soil Analysis, Part 3—Chemical Methods, J.M. Bartels (Ed.), SSSA, ASA, Madison, WI, pp. 1123-1184.

Neilsen, G.H., Yorston, J., Lierop, W.V. and P.B. Hoyt. 1985. Relationship between leaf and soil boron and boron toxicity of peaches in British Columbia. *Can. J. Soil Sci.* **65:** 213-217.

Odom, J. 1980. Kinetics of the hot water soluble boron Soil Test. *Commun. Soil Sci. Plant Anal.* **11(7):** 759-765.

Oliveira, M.R.G., Calado, A.M. and C.A.M. Portas. 1996. Tomato root distribution under drip irrigation. *J. Amer. Soc. Hort. Sci.* **121(4):** 644-648.

Olsen, S.R., Cole, C.V., Watanable, F.S. and L.A. Dean. 1954. Estimation of available phosphorus in soils by extraction with sodium bicarbonate. USDA Circ. 939.

Olson, R.A., Anderson, F.N., Frank, K.D., Grabouski, P.H., Rehm, G.W. and C.A. Adriano. 1987. Soil testing intrpretations: Sufficiency vs. build-up and maintenance. *In: Soil Testing: Sampling, Correlation, Calibration, and Interpretation,* J.R. Brown (Ed.), SSSA Special Publication No 21, Madison, WI, pp. 41-52.

Peck, T.R. and S.W. Melsted. 1973. Field sampling for soil testing. *In:* Soil Testing and Plant Analysis, L.M. Walsh and J.D. Beaton (Eds), SSSA, Inc. Madison, WI, pp. 57-75.

Peck, T.R. and P.N. Soltanpour. 1990. The principles of soil testing. *In:* Soil Testing and Plant Analysis, R.L. Westerman (Ed.), SSSA, Inc. Madison, WI, pp. 3-9.

Ponnamperuma, F.N., Cayton, M.T. and R.S. Lantin. 1981. Dilute hydrochloric acid as an extractant for available zinc, copper, and boron in rice soils. *Plant and Soil* **61:** 297-310.

Reisenauer, H.M., Walsh, L.M. and R.G. Hoeft. 1973. Testing soils for sulfur, boron, molybdenum, and chlorine. *In:* Soil Testing and Plant Analysis L.M. Walsh and J.D. Beaton (Eds), SSSA, Inc. Madison WI, pp. 173-200.

Rice, B. and E.J. Kamprath. 1968. Availability of exchangeable and non-exchangeable Mg in sandy coastal plain soils. *Soil Sci. Soc. Am. Proc.* **32**: 386-388.

Righetti, T.L., Wilder, K.L. and G.A. Cummings. 1990. Plant analysis as an aid in fertilizing orchards. *In:* Soil Testing and Plant Analysis, R.L. Westerman (Ed.), SSSA, Inc. Madison WI, pp. 563-601.

Ryan, J., Mayamoto, S. and J.L. Stroehleim. 1977. Relation of solute and sorbed boron to the boron hazard in irrigation water. *Plant Soil* **47**: 253-256.

Sherman, G.D., McHargue, J.S. and W.S. Hodgkiss. 1951. Determination of active manganese in soil. *Soil Sci.* **54**: 253-257.

Shunway, J.S. and J.P. Jones. 1972. Boron sorption isotherm: a method to estimate boron fertilizer requirements. Commun. *Soil Sci. Plant Anal.* **3(6)**: 477-485.

Smith, F.W. 1986. Interpretaion of plant analysis: concepts and principles. *In:* Plant Analysis An Interpretation Manual, D.J. Reuter and J.B. Robinson (Eds), Inkata Press, Melbourne, pp. 1-12.

Smith, G.S. and C.J. Clark. 1989. Effect of excess boron on yield and post-harvest storage of kiwifruit. *Scientia Horticulturae* **38**: 105-1989.

Soltanpour, P.N. and A.P. Schwab. 1977. A new soil-test for simultaneous extraction of macro- and micronutrients in alkaline soils. Commun. *Soil Sci. Plant Anal.* **8**: 195-207.

Spouncer, L.R., Nable R.O., and B. Cartwright. 1992. A procedure for the determination of soluble boron in soils ranging widely in boron concentrations, sodicity, and pH. *Commun. Soil Sci. Plant Anal.* **23**: 441-453.

Stanford, G. 1982. Assessment of soil nitrogen availability. *In:* Nitrogen in Agricultural Soils. Agronomy Monogr. 22. F.J. Stevenson (Ed.), ASA, CSSA, SSSA, Madison, WI. pp. 651-688.

Tidmore, J.W. 1930. The phosphorus content of the soil solution and its relation to plant growth. *Agron. J.* **22**: 481-488.

Tsadilas, C.D. 1999. Boron deficiency in olive trees in Greece in relation to soil boron concentration. *Acta Horticulturae*, **474(1)**: 293-296.

Tsadilas, C.D., Yassoglou, N., Kosmas, C.S. and Ch. Kallianou. 1994. The availability of soil boron fractions to olive and barley and their relationships to soil properties. *Plant and Soil* **162**: 211-217.

Tsadilas, C.D., Samaras, V. and D. Dimoyiannis. 1996. Phosphate sorption by red Mediterranean soils from Greece. *Commun. Soil Sci. Plant Anal.* **27(9&10)**: 2279-2293.

Tsadilas, C.D., Dimoyiannis, D. and V. Samaras. 1997. Methods of assessing boron availability to kiwifruit plants growing on high boron soils. *Commun. Soil Sci. Plant Anal.* **28**(11&12): 973-987.

Winker, B. and L. Uppstrom. 1980. Determination of boron in plants and soils with a rapid modification of the curcumin method utilizing different 1,3-Diols to eliminate interference. *Commun. Soil Sci. Plant Anal.* **17(7)**: 697-714.

Viets (Jr), F.G. and W.L. Lindsay. 1973. Testing soils for zinc, copper, manganese, and iron. *In:* Soil Testing and Plant Analysis L.M. Walsh and J.D. Beaton (Eds), SSSA, Inc. Madison, WI, pp. 153-172.

van Diest, A. 1963. Soil Test Correlation Studies on New Jersey Soils. 1. Comparison of Seven Methods for Measuring Labile Inorganic Soil Phosphorus. *Soil Sci.* **96**: 261-266.

Yamazaki, T., Nuzuma, T. and T. Taguchi. 1973. Studies on the soil fertility of apple orchards. IX. The relationship between boron deficiency and boron content in soils and leaves. *Hort. Abs.* **45**: No. 5602.

Zuang, H. 1982. La fertilisation des cultures legumieres. C.T.I.F.L. Paris. pp. 391.

3

Nutritional Management of Vegetables and Ornamental Plants in Hydroponics

Dimitrios Savvas

Faculty of Agricultural Technology, Technological Education Institute (TEI) of Epirus, PO Box 110, 47100 Arta, Greece

Abbreviations

EC	electrical conductivity
STV	salinity threshold value
SYD	salinity yield decrease
BER	blossom-end rot
DTPA	diethylene-Triamine-Penta-Acetic acid
EDDHA	ethylene-Diamine-Di(2-Hydroxyphenyl)-Acetic acid
CCS	concentration of cations in the target nutrient solution
CAS	concentration of anions in the target nutrient solution
CCW	concentration of cations in the tap water
CAW	concentration of anions in the tap water
CAF	addition of cations through fertilizers
AAF	addition of anions through fertilizers

1. Introduction

Some researchers, based on a strictly linguistic interpretation of the term hydroponics, reserve this definition exclusively for soilless growing methods that do not employ any material other than nutrient solution as a rooting medium (Cooper, 1979). However, most scientists use the term hydroponics as a synonym to 'soilless culture' to describe all types of plants growing without the use of soil as a rooting medium (Steiner, 1976). Nevertheless, in all types of soilless-growing plants, the soil is substituted—in its function to supply nutrients to the plants—by an artificial nutrient solution which is provided to the crop by the irrigation system. This technique resembles liquid fertilization of plants growing in the soil, which is known as 'fertigation'. Liquid fertilization of soil crops is aimed at supplementing the nutritional requirements of the plants for particular nutrients (mainly nitrogen and potassium), whose availability in the soil is often insufficient. In contrast, in hydroponics, the only source of nutrients for the plant is the nutrient solution. This is particularly true when no substrate or an inert substrate is used as a rooting medium. However, it is also valid for crops grown in porous materials which exhibit a relatively low exchange capacity, due to the limited volume of substrate per plant that is usually applied in hydroponics. Since in hydroponics the only source of nutrient supply to the roots is virtually the nutrient solution, the latter should contain, in sufficient quantities, all the nutrients which are essential for plant growth and development.

To formulate the composition of a nutrient solution for a certain crop, experimental results concerning the nutritional requirements of the particular plant species should be available. Such data is also essential to check and adjust the nutritional status of the plants during the cropping period. However, even when an optimal nutrient solution composition is available, to achieve high yields, it is essential to correctly prepare and supply the desired nutrient solution by employing proper equipment. Moreover, the nutrient solution should be provided to the plants by following an efficient irrigation scheduling regarding the total amount of water supplied to the crop and the frequency of watering application. Last but not least, if the nutrient solution is recycled, the runoff solution should be replenished with the appropriate quantities of nutrients and water in each watering application in order to be reused. To achieve it, different recycling techniques (Savvas and Manos, 1999) and strategies (Raviv et al. 1998) may be applied.

In the present paper, a survey on the nutrition of hydroponically grown crops via the nutrient solution is given. Special emphasis has been laid on the topics mentioned above, as they are considered to be the key factors for success in commercial hydroponics.

2. Historical Review of Research on Hydroponical Plant Nutrition

Since Knop (1859) and Sachs (1859, 1861) published their classic research work on cultivation of plants in nutrient solutions, our knowledge about nutrition management of soilless cultivated plants has increased enormously. Among the numerous publications referring to the composition of nutrient solutions up to the end of the sixties, the most worth to mention are the reports of Gericke (1937,1938), who introduced the widely used term hydroponics to describe all methods of commercial plant cultivation without the use of soil, as well as those of Hoagland and Arnon (1950) and Hewitt (1966). Detailed literature reviews on research work about the composition of nutrient solutions for soilless culture which was conducted during those years have been given by Cooper (1979) and Jones (1982). The results of all these investigations, but especially the research work that had been carried out in the California Agriculture Experimental Station for over two decades, were used by Hoagland and Arnon (1950) to formulate a nutrient solution composition which is suitable for most cultivated plants. Since various hydroponic techniques such as sand culture, water culture, etc. were often used in plant nutrition studies, Hoagland and Arnon (1950), besides a balanced solution, also developed nutrient solution formulations which were deficient in one or more essential nutrients. These formulations were prescribed to induce deficiency symptoms in the plants for research purposes. The nutrient solutions proposed by Hoagland and Arnon (1950) are widely used in plant nutrition studies even today, either in the original or in modified forms.

Until the end of the sixties, hydroponics was a cultivation technique mainly used for research purposes. The main hydroponic systems used in such investigations were sand, gravel or water culture. However, in the seventies, a continually increasing interest on the application of soilless cultivation methods in commercial vegetable and ornamental plant production was developed. This trend was most marked, particularly in The Netherlands, United Kingdom and some of the Scandinavian countries. Whereas in the United Kingdom, the Nutrient Film Technique (Cooper, 1975, 1979; Graves, 1983) was the first soilless culture system which found practical application on a large scale, in the Netherlands and in Scandinavia, the cultivation on rockwool and other inert substrates was dominant (Verwer, 1976, 1978; Ottosson, 1977; Verwer and Welleman, 1980).

Increasing interest in commercial application of hydroponic techniques brought about intensive research activity focussing on the composition of nutrient solutions for soilless culture. These efforts, supported by the development of modern analytical techniques and equipment, resulted in the formulation of new nutrient solution compositions, adapted to the specific requirements of most horticultural species grown under glass, as those suggested by Sonneveld and Straver (1994), De Kreij et al. (1997), Resh (1997) and Hanan (1998). Thus, the currently recommended nutrient solution formulations are more balanced and specific when compared to the composition suggested by Hoagland and Arnon (1950). This enables more efficient nutrient management of most horticultural crops in commercial hydroponics.

3. Nutrients and Fertilizers needed to prepare a Nutrient Solution

In soilless culture, all essential plant nutrients are supplied via the nutrient solution, with the exception of carbon, which is taken up from the air as carbon dioxide. As a rule, inorganic fertilizers are used to provide all essential nutrients in nutrient solutions. Iron is an exception to this rule, since it is added in chelated forms to enhance its availability to the plants (Jacobson, 1951; Wreesman, 1996). In most cases, the fertilizers used to prepare nutrient solutions are highly soluble inorganic salts. However, some inorganic acids, particularly nitric, phosphoric and boric acid, are also used. In Table 1, the fertilizers usually used in the constitution of nutrient solutions, accompanied by their chemical names, chemical formulae, nutrient content percentages, molecular weights and solubility thresholds are cited. In commercial hydroponics, proper amounts of the fertilizers needed to prepare the nutrient solution are mixed with water in tanks to form concentrated stock solutions. Thus, when the soilless cultivated plants are to be watered, the stock solutions are diluted with the irrigation water in proper ratios through automatic fertilizer injection systems, to form a fresh nutrient solution, which is supplied to the crop.

4. Soilless Versus Soil Crop Nutrition

4.1. Differences in Nutrient Management

Most horticultural substrates used for soilless cultivation of plants including rockwool (Blaabjerg, 1983), perlite (Gizas et al. 1999), expanded clay (Verdonck et al. 1984), and polyurethane foam (Benoit and Ceustermans, 1995) are chemically inert. However, there are also substrates used in hydroponics, such as peat (Verdonck et al. 1981), vermiculite (Verdonck et al. 1981) and zeolite (Challinor et al. 1996), which exhibit considerable cation exchange capacity. Thus, the growth of the root

Table 1. A brief description of the water soluble fertilizers commonly used in hydroponics

Fertilizer	Chemical formula	Percentage nutrient	Molecular weight (g)	Solubility (kg L^{-1}, 0°C)
Ammonium nitrate	NH_4NO_3	N: 35	80.0	1.18
Calcium nitrate	$5[Ca(NO_3)_2 2H_2O]NH_4NO_3$	N: 15.5, Ca: 19	1080.5	1.02
Potassium nitrate	KNO_3	N: 13, K: 38	101.1	0.13
Magnesium nitrate	$Mg(NO_3)_2 \cdot 6H_2O$	N: 11, Mg: 9	256.3	2.79 (20°C)
Nitric acid	HNO_3	N: 22	63.0	—
Monoammonium phosphate	$NH_4H_2PO_4$	N: 12, P: 27	115.0	0.23
Monopotassium phosphate	KH_2PO_4	P: 23, K: 28	136.1	1.67
Phosphoric acid	H_3PO_4	P: 32	98.0	—
Potassium sulphate	K_2SO_4	K: 45, S: 18	174.3	0.12
Magnesium sulphate	$MgSO_4 \cdot 7H_2O$	Mg: 9.7, S: 13	246.3	0.26
Potassium bicarbonate	$KHCO_3$	K: 39	100.1	1.12
Iron chelates	various types	Fe: 6-13	—	—
Manganese sulphate	$MnSO_4 \cdot H_2O$	Mn: 32	169.0	1.05
Zinc sulphate	$ZnSO_4 \cdot 7H_2O$	Zn: 23	287.5	0.62
Copper sulphate	$CuSO_4 \cdot 5H_2O$	Cu: 25	249.7	0.32
Borax	$Na_2B_4O_7 \cdot 10H_2O$	B: 11	381.2	0.016
Boric acid	H_3BO_3	B: 17.5	61.8	0.050
Sodium octaborate	$Na_2B_8O_{13} \cdot 4H_2O$	B: 20.5	412.4	0.045
Ammonium heptamolybdate	$(NH_4)_6Mo_7O_{24} \cdot 4H_2O$	Mo: 54	1235.3	0.43
Sodium molybdate	$Na_2MoO_4 \cdot 2H_2O$	Mo: 40	241.9	0.56

system into an inert medium cannot be considered a universal characteristic for all soilless-grown crops. Moreover, in the last few decades, the supply of a nutrient solution to the plant instead of pure irrigation water in order to optimize the crop nutrition has become a routine operation, not only in hydroponics but also in the soil-grown crops. Hence, the substantial differences between the crops grown in the soil and those grown hydroponically are limited to the volume of the rooting medium per plant and to the uniformity of the rooting medium. As pointed out by Sonneveld (1981), the volume of rooting medium per square meter which is available to a tomato crop grown in the border soil of a greenhouse is approximately 500 L, whilst the corresponding values for peat, and rockwool are 25 and 14 L, respectively. As a result, the volume of nutrient solution which is available to the tomato plants amounts to 150, 12 and 10 L m^{-2} in a soil, peat and rockwool crop, respectively. In case of an NFT tomato crop, this value is even lower; in particular, only 4 L m^{-2} of nutrient solution are available to the plants (Sonneveld, 1981). On the other hand, it is not possible to drastically raise the nutrient concentrations in the nutrient solutions supplied to the crops in hydroponics, since an increase in the total salt concentration would lead to the appearance of growth and yield restrictions due to salinity (Sonneveld, 1989). Consequently, the nutrient reserves which are available in the root environment of plants grown hydroponically are dramatically lower than those corresponding to plants which are grown in the border soil, regardless of the soilless culture system employed.

A superficial consideration of the above data may lead to the conclusion that hydroponics is inferior to cultivation in soil since the nutrient and water reserves in the former are dramatically lower than in the latter. In fact, the opposite is true. In modern greenhouse horticulture, the frequency of nutrient and water (nutrient solution) application may be easily raised to several times per day by using fully automated equipment (Hanan, 1998; Nelson, 1998). Thus, the availability of extensive nutrient and water reserves in the root environment of greenhouse grown plants is of minor importance. In contrast, if the nutrient reserves in the root environment are restricted, the nutrition of the plant may be controlled more efficiently through the composition of the nutrient solution supplied to the plants. Therefore, in most cases, completely inactive substrates are preferred to those which exhibit a substantial cation exchange capacity. The influence of the cation exchange capacity of the substrate on the composition of the nutrient solution which is retained in the substrate pores has been demonstrated by Gizas et al. (1999). These investigators measured the macronutrient concentrations in water extracts from perlite and pumice prior and after application of a nutrient solution, and found that the macrocation ratios in the nutrient solution retained by the substrate changed more markedly in pumice than in perlite, compared to those prevailing in the supplied solution. This was ascribed to the fact that perlite has a very low cation exchange capacity whilst pumice is chemically more active.

As mentioned above, the second specific characteristic of soilless culture is the uniformity of the rooting medium. If the roots grow in pure nutrient solution, as in the case of Nutrient Film Technique (Graves, 1983), the uniformity of the rooting medium is obvious. However, the aggregate substrates are also uniform porous materials which in most cases are industrially standardized and, therefore, their physical and chemical properties are well known. Thus, the supply of nutrients and water to the crop may be more accurately adapted to the plant requirements and controlled, if experimental data concerning the physical and chemical properties of the particular substrate are known.

4.2. Differences in Product Quality

Some consumers are skeptical of hydroponically produced vegetables. This attitude is mainly based on the assumption that the soilless grown plants are fed with 'chemicals', in contrast to the plants grown in soil which acquire 'natural substances' for their nutrition. However, this belief obviously contrasts with the principles of the science of plant nutrition. It is well known that higher plants need only inorganic substances, mainly in an ionic form, to satisfy their nutritional requirements (Marschner, 1995). Thus, plants take up NO_3^- and NH_4^+ but not organic substances to supply their cells with nitrogen, regardless of the content of availability of organic N in the soil. However, the inorganic ions do not have any memory concerning their origin when they are used in the plant and human metabolism. Consequently, with respect to the quality of the edible vegetable products, it is completely irrelevant whether the nitrogen contained in the plant tissues stems from organic substances of the soil or from inorganic fertilizers. The only factor influencing the vegetable quality is the quantity of absorbed nitrogen and the way in which it is utilized in plant metabolism, with respect to the nitrate-nitrogen content in the edible plant tissues. However, both these factors are better managed in hydroponics (Santamaria et al. 1996), as the nutrient supply is more efficiently controlled through the composition of the nutrient solution due to the small volume of the rooting medium applied in soilless culture. Thus, as reported by Benoit and Ceustermans (1995) reducing the nitrate-nitrogen content in the nutrient solution supplied to lettuce during the last week prior to harvesting considerably lowered the NO_3^- content in the leaves of the plants, without significant yield losses. Similar responses have also been reported by Wendt (1982) in kohlrabi and celery. Moreover, since in hydroponics the substrates are free from pathogens when they are initially supplied to the grower, the pressure from soil-borne diseases is much weaker than in soil grown crops. As a result, the need to use soil disinfecting chemicals is considerably reduced in hydroponics, with obvious advantages to the quality of the vegetables produced. Furthermore, the taste of some fruit vegetables such as tomato, melon, etc. may be substantially improved in hydroponics by manipulating the total salt and nutrient concentration in the supplied nutrient solution.

5. The Composition of the Nutrient Solution

5.1. General

The composition of a nutrient solution is completely defined, if the total salt concentration, the pH, the individual micronutrient concentrations and the ratios between the macronutrients are given, and the mineral composition of the tap water used is known (Savvas and Adamidis, 1999).

Usually, the amounts of nutrients taken up by the plants per liter water consumption in hydroponics tend to diverge considerably from the corresponding concentrations prevailing in the nutrient solution (Sonneveld, 1981; Savvas and Lenz, 1995). Any attempt to relate the nutrient concentrations in the supplied solution to the ratio of nutrient to water uptake by the plants meet with two problems. First, the ratios of nutrient to water uptake are not constant during the cropping period; usually they fluctuate widely, even within short time intervals, depending on the weather conditions, the time of day, the developmental stage of the plant and the current nutritional requirements of the individual plants (Sonneveld, 1981; Adams and Massey, 1984; Van Goor et al. 1988; Savvas and Lenz, 1995). Second, the supply of some bivalent nutrient ions to the crop, which are not taken up easily by the plants (e.g. Ca^{2+}, Mg^{2+}, Fe^{2+}, SO_4^{2-}), is reduced and may become insufficient when attempting to lower the concentrations of these ions in the nutrient solution in order to achieve levels corresponding to their uptake ratios (Sonneveld, 1981; De Kreij, 1995).

Since, usually, the amount of nutrients and other elements taken up by the plants per liter water absorbed tend to diverge from the corresponding concentrations in the nutrient solution, the composition of the latter changes progressively in the root environment after its supply to the crop. Obviously, the plant growth is directly influenced only by the composition of the nutrient solution which is present in the root environment. However, the composition of the nutrient solution supplied to the plants is the main factor determining the ionic concentrations in the root zone (Sonneveld, 1981). Thus, the plant growth is also indirectly influenced by the composition of the nutrient solution supplied to the crop. This issue will be discussed in more detail in section 5.5. Nevertheless, when recommendations concerning the composition of a nutrient solution are given, it should be clarified whether the suggested data refer to the nutrient solution supplied to the plants or to that in the root zone (Sonneveld and Welles, 1984).

5.2. Total Nutrient Concentration

5.2.1. Definition

In commercial hydroponics, the nutrient solutions are prepared by using irrigation water which, in most cases, contains substantial amounts of inorganic ions. Therefore, usually, besides the added fertilizers, the nutrient solutions also contain some undesired salts (mainly NaCl) in concentrations depending on the quality of the water used. Consequently, when measuring the total ionic concentration in a nutrient solution for hydroponics, the value obtained may be partly due to ballast salts and not only due to the added nutrients.

In older studies, the total salt concentration in the hydroponical nutrient solutions was expressed mainly as osmotic pressure (Steiner, 1961, 1966) and less frequently as total weight of salts per liter of solution (Pardossi et al. 1987). However, in the last decades, in most references concerning soilless culture, the total salt concentration is expressed in terms of electrical conductivity (EC in dS m^{-1}) because this quantity can be readily and accurately measured even in the glasshouse, using portable instruments.

It can be concluded from data presented by the US Salinity Laboratory Staff (1954), that the EC of single salt solutions is almost linearly related to the corresponding equivalent salt concentration (meq L^{-1}) in the range of EC which is usually applied in commercial hydroponics. However, the relationships obtained for each salt are not identical. Nevertheless, the differences in the composition of most nutrient solutions used in commercial soilless culture are not very extensive (De Rijck and Schrevens, 1998c; Savvas and Adamidis, 1999). Therefore, for most balanced nutrient solutions, a single linear relationship may be used in commercial practice to convert the electrical conductivity E, in dS m^{-1} to total salt concentration C (meq L^{-1}) and vice versa (Savvas and Adamidis, 1999):

$$C = 9.819E - 1.462 \tag{1}$$

The electrical conductivity of the nutrient solutions used to obtain (1) ranged from 0.8 to 4.0 dS m^{-1}. Consequently, (1) is valid only in this particular EC range. Certainly, a small discrepancy between the actual total salt concentration and the value estimated through (1) is likely to occur in some cases. However, in commercial hydroponics such small discrepancies are acceptable.

When a certain level of electrical conductivity is considered to be optimal for a particular crop in hydroponics, it should be noted whether this value refers to the nutrient solution supplied to the plants or to that in the root environment.

5.2.2. Yield Response to the Electrical Conductivity in Hydroponics

5.2.2.1. The relationship between EC and yield. The total salt concentration is considered to be one of the most important properties of the nutrient solutions used in soilless culture. The nutrient solutions are prepared by adding all the essential plant nutrients to the irrigation water at ratios that are supposed to meet the plant requirements. Thus, their total salt concentration constitute a coarse estimate of their nutritional value for the plants. If a nutrient solution has too low a total salt concentration, the supply of some nutrients to the crop may be inadequate. Similarly, when a nutrient solution with a too high salt concentration is supplied to a crop, the plants are exposed to salinity. However, the yield response of the plants to the total salt concentration of the nutrient solution may vary widely among the different species. Therefore, for each cultivated plant species, the terms 'too low' and 'too high' need to be quantitatively defined on the basis of experimental results.

The growth and yield response of hydroponically grown plants to the total salt concentration of the nutrient solution may be described by a generalized model as shown in Fig. 1. According to this model, if the total salt concentration (EC) is lower than a particular value, a, an increase of the EC to a value not exceeding a enhances the yield of the crop. If the EC ranges between a and t, where t is the upper EC value for optimal growth which is known as salinity threshold value (STV) (Maas and Hoffman, 1977; Sonneveld and Van der Burg, 1991), the yield of the crop remains constant. However, any further increase of the electrical conductivity above t is accompanied by yield depressions which, in most cases, are linearly proportional to the increase of the EC above t. Detailed reviews on the responses of vegetables (excluding the fruit vegetables) to high salinity and on the salinity-mineral nutrient relations in horticultural crops grown on both soil and soilless media have recently been given by Shannon and Grieve (1999) and Grattan and Grieve (1999), respectively.

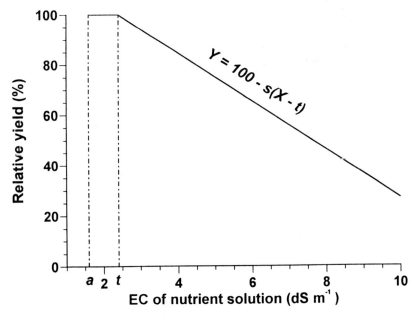

Fig. 1. A generalized model showing the influence of the nutrient solution EC in the root zone on the yield of most vegetables and ornamental plants when all nutrients are present in adequate concentrations in the solution.

Results concerning the yield responses of some fruit vegetables grown hydroponically under semicommercial conditions to increasing nutrient solution EC have been presented by Sonneveld and Van der Burg (1991) and Savvas and Lenz (1994b). Furthermore, Sonneveld et al. (1999) presented a comprehensive synopsis of experimental results concerning the effects of increasing nutrient solution EC on growth and yield of some of the most important cut flower species.

5.2.2.2. Influence of too low EC on yield. As pointed out by Graves (1983), a too low EC may result in inadequate nutrient supply and, subsequently, in growth retardation due to nutrient deficiencies. There is few published data dealing with the growth and yield response to nutrient solution EC in the range $0-a$ dS m^{-1} (Moustafa and Morgan, 1983; Shinohara and Suzuki, 1988; De Kreij and Van Os, 1988; De Kreij, 1999). Nevertheless, the little available data indicate that the declined growth and yield observed in this range of EC usually arise from a complex of nutritional deficiencies of individual nutrients, which have different lower threshold values of adequacy. Therefore, when the EC rises in the range $0-a$ dS m^{-1} the total salt concentration of the nutrient solution and the yield of the crop are not expected to increase proportionally to the EC.

5.2.2.3. Influence of too high EC on yield. In contrast to the range of too low EC values, a low or a moderate raise of EC above t decreases the yield mainly due to osmotic salt effects which are not influenced by the nutrient solution composition. Therefore, when the EC increases above t, in most cases the relative yield decrease can be estimated by employing a straight line equation, as suggested by Maas and Hoffman (1977). The model for this relationship is described by the equation

$$Y = 100 - s(X - t) \qquad (2)$$

where Y is the expected relative yield (% of the highest yield which is obtained when the EC is within $a-t$), X is a variable indicating the EC of the nutrient solution in dS m^{-1} and s is the slope which indicates the so called salinity yield decrease (SYD). The term SYD is used to express the relative yield (%) decrease per salinity unit (dS m^{-1}) increase above t. The above relationship (2) is valid only if $X > t$.

Although the model proposed by Maas and Hoffman (1977) to quantify the yield response of cultivated plants to salinity was initially applied to data obtained from soil grown crops, it proved to fit well also in hydroponics. For instance, the relationship between the total fruit yield of soilless cultivated eggplants and the EC of the nutrient solution, as determined by Savvas and Lenz (1994b), proved to fit well to the model of Maas and Hoffman (1977) as graphically shown in Fig. 2. The data of Fig. 2 were obtained by continuously supplying the plants with nutrient solution during the day and, therefore, the values of STV and SYD found refer to the EC prevailing in the root environment.

The results presented in Fig. 2 have been obtained by adding NaCl to the nutrient solution to increase the EC from 2.1 up to 8.1 dS m^{-1}. However, in two other experiments involving soilless cultivation of eggplants, when the EC was increased up to 4.7 dS m^{-1} (Savvas and Lenz, 2000a) and 6.1 dS m^{-1} (Savvas and Lenz, 2000b) by adding, in both cases, either NaCl or additional nutrients to a basic nutrient solution, the yield was depressed at the same extent, irrespective of the salts used to induce salinity. Similar results concerning the yield response to increased nutrient solution EC due to addition of NaCl, as compared to nutrient-induced salinity of the same EC level, were also found by Sonneveld and Van der Burg (1991) in tomato, cucumber and sweet pepper with EC values up to 5.2 dS m^{-1}, by Adams (1991) with tomato grown in rockwool in the range 3–8 dS m^{-1} and by Sonneveld et al. (1999) with gerbera, carnation, rose, aster and lilly exposed to EC values up to 5.2 dS m^{-1}. Thus, at least for the above mentioned crops, the STV and the SYD obtained in hydroponics by increasing the EC up to a particular level seem to be independent of whether the total salt concentration is raised by

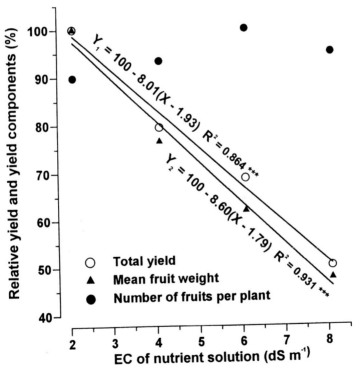

Fig. 2. Relationship between the relative fruit yield and yield components of soilless cultivated eggplants and the EC of the nutrient solution in the root environment. The relative number of fruits was not significantly different in the four EC treatments applied (2.1, 4.1, 6.1, and 8.1 dS m^{-1}). In the regression equations, Y_1 indicates the relative total yield whilst Y_2 indicates the relative mean fruit weight which are obtained at X dS m^{-1} (adapted from Savvas, 1992).

adding NaCl or additional nutrients, provided that the basic nutrient solution has a balanced composition. However, there are also crops such as bouvardia, which exhibit a specific sensitivity to NaCl salinity (Sonneveld et al. 1999).

The STV and the corresponding values of SYD which have been found for several vegetable and ornamental plants in experiments carried out in hydroponics are cited in Table 2. Considering these data it seems that, with the exception of carnation, the flower crops are more sensitive to salinity, since, as a rule, their salinity threshold values are lower than those of most vegetables. All data of Table 2 refer to the EC of the nutrient solution in the root environment with the exception of those referring to lettuce which are based on the EC of the supplied solution. The data of Table 2 referring to the influence of EC on the growth of lettuce indicate a surprisingly high salt tolerance for this plant species, which is not in agreement with previous results, as summarized by Maas and Hoffman (1977). Nevertheless, when only one particular salinity source (NaCl or nutrients) has been tested, the results obtained should be evaluated with cautiousness, particularly if the occurrence of ion specific effects in the range of EC under consideration has not been excluded in other experiments.

Table 2. Salinity threshold values (STV) and the corresponding values of salinity yield decrease (SYD) for some hydroponically grown vegetables and flowers as determined by various investigators.

Crop	STV ($dS\ m^{-1}$)	SYD (%)	Literature source
Anthurium	0.8-0.9	7.8-13.2	Sonneveld and Voogt, 1993
Bouvardia	2.1	16.8	Sonneveld et al. 1999
Carnation	4.3	3.9	Sonneveld et al. 1999
Cucumber	2.3-3.5	5.6-5.8	Sonneveld and Van der Burg, 1991
Eggplant	1.93	8.0	Savvas and Lenz, 1994b
Gerbera	1.5	9.8	Sonneveld et al. 1999
Lettuce	4.6	4.5	Shannon et al. 1983
Lily	1.6	4.6-9.6	Sonneveld et al. 1999
Roses	2.1	5.3	Sonneveld et al. 1999
Sweet pepper	2.8	7.6	Sonneveld and Van der Burg, 1991
Tomato	2.5-2.9	2.3-7.2	Sonneveld and Van der Burg, 1991

Note: When, for a particular horticultural crop, STV and SYD from different experiments or different cultivars are available, and the values are not identical, a range and not a constant value is given.

The quantification of salinity effects on growth and yield is valid only for crops supplied with nutrient solutions having adequate nutrient concentrations. If a nutrient solution is deficient in some nutrients but its EC has been raised to values which are within the optimal range (a–t) due to the presence or extra addition of one particular salt (e.g. NaCl, K_2SO_4, $Ca(NO_3)_2$, etc.) nutritional disorders and growth restriction will occur.

This has been observed by De Kreij (1999) in an experiment with sweet peppers which were supplied with a nutrient solution having optimal EC (2 dS m^{-1}) due to the presence of 10 mM NaCl whilst the macronutrient concentrations were approximately half as much as the usual levels corresponding to that EC value. Consequently, the EC of a nutrient solution alone is an insufficient measure to assess the nutritional value of the solution for a particular crop, even if the responses of this crop to salinity are known.

Another aspect concerning nutrient solution salinity which should be pointed out is the dependence of the results upon the climatic conditions prevailing when the experiment is conducted. As a result, experiments of different investigators or even different experiments of the same researcher concerning exposure of the same plant species to nutrient solution salinity may result in divergent STV and SYD. This is also obvious in the data given in Table 2. It is well known that climatic conditions such as air humidity (Sonneveld and Welles, 1988) and light intensity (Helal and Mengel, 1981) exhibit a strong interaction with salinity. Thus, as a rule, under high light intensity as well as under low air humidity the detrimental effects of nutrient solution salinity are more pronounced. However, as reported by Sonneveld and Welles (1988) a too high regime of air humidity may also deteriorate the adverse effects of an increased EC in the root environment. This response was ascribed to the fact that under too high air humidity the calcium transport to the growing leaves of the plants is affected and this acts synergistically with salinity.

If the salt concentration varies considerably in different parts of the root environment, the plants respond mainly to the lower EC levels (Sonneveld and Voogt, 1990; Sonneveld and De Kreij, 1999). Consequently, even when some salt accumulation occurs in particular sections of the rooting medium, the plants can still grow satisfactorily, provided that the EC is maintained within the target range in other parts of the root environment as, for instance, in the substrate segment below the irrigation nozzle.

The influence of increased nutrient solution EC on the growth of different plant parts varies among different horticultural plants. This is, for instance, obvious when considering the results of Caro et al. (1991) who worked with various normal fruited and cherry tomato cultivars and found quite different STV and SYD values for leaf dry weight, stem dry weight, and plant height. There is a general consensus in the experimental results of many investigators that moderate levels of salinity affect the above-ground growth more markedly than the roots (Meiri and Polljakoff-Mayber, 1970; Shannon et al. 1983; Seemann and Critchley, 1985; Savvas and Lenz, 1994b; Van Ieperen, 1996; Pardossi et al. 1999). In some plant species, the vegetative growth is affected less severely than the fruit growth by an increase in the nutrient solution salinity. Thus, the fresh fruit growth of eggplant was more severely affected by salinity than the vegetative growth (Savvas and Lenz, 1994b). A similar result was obtained also in melon (Mavrogianopoulos et al. 1999). In contrast, in tomato, the vegetative plant parts seem to be more sensitive to increased nutrient solution EC than the fruit growth, especially if the comparison is made in terms of dry matter production (Van Ieperen, 1996). Moreover, in tomato, eggplant, and melon, moderate levels of salinity affect mainly the mean weight of the fresh fruits (Adams and Ho, 1989; Adams, 1991; Nichols et al. 1994; Savvas and Lenz, 1994b, 2000a; Willumsen et al. 1996; Van Ieperen, 1996; Mavrogianopoulos et al. 1999). In contrast, the number of fruits per plant is reduced only at relatively high levels of nutrient solution EC, when the vegetative growth is also restricted, as has been found by Adams and Ho (1989), Adams (1991), and Van Ieperen (1996) in tomatoes and by Mavrogianopoulos et al. (1999) in melon. As shown in Fig. 2, in eggplants, even an EC of 8.1 dS m^{-1} has no significant influence on the number of fruits per plant. In case of peppers, both the mean weight and the number of fruits per plant are affected by nutrient solution salinity, whilst in cucumber a moderate increase of EC up to 5.2 dS m^{-1} seems to diminish only the number of fruits per plant (Sonneveld and Van der Burg, 1991). On the other hand, in almost all fruit vegetables the dry matter content is enhanced with increasing nutrient solution salinity (Ehret and Ho, 1986a; Sonneveld and Welles, 1988; Adams and Ho, 1989; Gaugh and Hobson, 1990; Savvas and Lenz, 1994b; Willumsen et al. 1996; Van Ieperen, 1996; Petersen et al. 1998). Thus, it seems reasonable to conclude that the initial effect of increasing the nutrient solution EC above the threshold values is a reduction in fruit water accumulation due to adaptation of the plant to a lower internal water potential in order to maintain its ability to acquire water (Johnson et al. 1992; Savvas and Lenz, 2000a). In agreement with this statement, the yield of smaller fruits (cherry tomatoes or additional fruits in eggplants) responds less sensitively to moderate salinity levels in hydroponics (Gaugh and Hobson, 1990; Caro et al. 1991; Savvas and Lenz, 1994b, 2000a) since the small fruits have a lower bulk volume per fruit surface unit and are, therefore, more capable of attracting water.

In the case of ornamentals, there is no consistency in the response of different cut-flower species to salinity with respect to the yield components. Thus, Sonneveld et al. (1999) found that gerbera and roses respond to salinity by reducing the number of flowers rather than the mean flower weight, whilst in carnations and bouvardia, the opposite is true. In case of *Anthurium andreanum*, a small increase of nutrient solution salinity seems to decrease predominantly the flower weight (Sonneveld and Voogt, 1993). However, a further increase of EC also depressed the number of flowers per m^2. Nevertheless, at high salinity levels in the nutrient solution, the dry matter content of some flowers like gerbera also increased (De Kreij and Van Os, 1988).

In hydroponics, the salinity may be easily controlled via the composition of the supplied nutrient solution due to the decreased volume of the root zone. Thus, since the reproductive and vegetative organs exhibit different responses to increased EC, some researches tested the possibility of exposing the plants to moderate salinity levels in order to reduce excessive vegetative vigour (Hall, 1983; Gosselin et al. 1988; Adams, 1991; Savvas and Lenz, 1994b, 2000b). However, there was only one case in which a marginally higher yield was obtained at moderate nutrient solution salinity ranging between 4.6-5.5 dS m^{-1} (Adams, 1988). Consequently, when growing fruit vegetables hydroponically, the application of a controlled salinity during the harvesting period in order to enhance the yield cannot be suggested. Nevertheless, as can be concluded from some experimental results presented by Massey et al. (1984) for tomatoes and Savvas and Lenz (2000b) for eggplants, it is possible to achieve a better control of the early, excessive vegetative growth by exerting a moderate salt stress only during the vegetative growing stage prior to the start of fruit setting. Moreover, in some fruit vegetables like tomato, and perhaps also cucumber, the early vegetative growth can be more efficiently controlled by additionally increasing the EC of the nutrient solution supplied to the young seedlings prior to planting, to levels up to 3.6 dS m^{-1} (Boertje, 1986).

5.2.3. Quality Response to the Electrical Conductivity

As can be concluded from the data of Table 2, for most hydroponically cultivated plants the STV ranges between 1.5-3.0 dS m^{-1}. However, for most fruit vegetables, the recommended EC in the root environment is somewhat higher than the respective STV given in Table 2 and ranges from 2.5 to 3.5 dS m^{-1} (Sonneveld and Straver, 1994; De Kreij et al. 1997). This strategy is aimed at improving the quality of the vegetables produced in hydroponics. If the EC of the supplied nutrient solution is only slightly higher than the experimentally calculated STV, the yield losses occurring in crops which are moderately sensitive to salinity may in fact be smaller than the usually expected experimental errors and, therefore, even experimentally not detectable. Thus, in experiments with eggplants, in which the EC of the nutrient solution was increased from 2.1 dS m^{-1} to 2.9, 3.7 and 4.7 dS m^{-1} by adding NaCl to the basic nutrient solution, no significant yield reduction could be found at 2.9 dS m^{-1} (Savvas, 1992). However, in another experiment with eggplants which was conducted in the next year, a STV of 1.93 dS m^{-1} was obtained, as also shown in Fig. 2 (Savvas, 1992; Savvas and Lenz, 1994b). Thus, when growing vegetables such as tomato and melon which are moderately sensitive to salinity, the EC in the root environment may be increased to levels slightly higher than the theoretically calculated STV in order to improve the fruit quality without the appearance of yield losses.

In contrast to vegetables, the standard EC values recommended for the nutrient solutions in the root environment of flower crops seem to approximate more closely to the experimentally determined STV, since no benefit in terms of flower quality are expected by an increased EC (De Kreij and Van Os, 1988; Sonneveld, 1989; Sonneveld et al. 1999). According to results of various trials, as summarized by De Kreij (1995), in many cases a low EC in the root environment of ornamental plants seems to improve the vase-life of the cut flowers.

To increase the EC above the standard level whilst ensuring adequate concentrations for all nutrients several options are available. In most cases, the EC is increased by adding either NaCl (Adams, 1988, 1991; Sonneveld and Van der Burg, 1991; Savvas and Lenz, 1994b; Sonneveld et al. 1999) or a mixture of major nutrients (Ehret and Ho, 1986a; De Kreij and Van Os, 1988; Adams and Ho, 1989; Adams, 1991; Van Ieperen, 1996; Petersen et al. 1998; Savvas and Lenz, 2000a, b) or both (Sonneveld and Welles, 1988; Gaugh and Hobson, 1990) to the basic nutrient solution. However, the use of only a single fertilizer (K_2SO_4) to supplement the EC of the nutrient solution has also been tested (Papadopoulos et al. 1999).

The favourable effects of salinity on the quality of tomato are an increased dry matter content and higher sugar and titratable acid contents in the fruit juice (Ehret and Ho, 1986a; Adams, 1988; Adams and Ho, 1989; Gaugh and Hobson, 1990; Sonneveld and Van der Burg, 1991). In other studies, the shelf life of tomato was also prolonged by nutrient solution salinity (Sonneveld and Van der Burg, 1991) whilst the incidence of physiological disorders like blotchy ripening (Sonneveld and Welles, 1984), gold specks (Sonneveld and Voogt, 1990) and russeting (Sonneveld and Van der Burg, 1991) were decreased. Gold specks are considered symptoms of excess Ca in the tomato fruits (De Kreij et al. 1992). Thus, the favourable effect of nutrient solution salinity on gold specks is attributed to the decreased translocation of Ca into the fruits of tomatoes which is observed when the plants are exposed to high nutrient solution EC (Ehret and Ho, 1986b). Furthermore, increasing the EC of the nutrient solution results in higher concentrations of vitamin C and total carotene in the fresh tomato fruit and increased fruit firmness (Petersen et al. 1998). However, recalculation of the above data on mg per fruit or mg per 100 g dry fruit revealed equal or even decreasing concentrations with increasing salinity, in agreement with previous results of Adams and Ho (1989). Only the sugar content of dry tomato fruit seems to increase slightly with increasing salinity (Petersen et al. 1998). As suggested by Adams and Ho (1989) the increased concentrations of titratable acids and sugars in the fruit juice are merely due to the reduced water content (increased dry matter content) of the fruit with increasing nutrient solution EC. Presumably, the increased vitamin C and total carotene concentrations found by Petersen et al. (1998) in the fruit juice of salt stressed tomatoes in hydroponics are also due to the lower water content of the fruit. Nevertheless, in terms of consumer quality, the determining factors when assessing the response of tomato to nutrient solution salinity are the concentrations of the above compounds in the fruit juice. It can, therefore, be concluded that moderate levels of salinity improve the fruit quality of tomato.

Based on the above data one could suppose that the percentage of tomato fruits graded Class 1 would be enhanced by increasing nutrient solution EC. Indeed, some researchers reported an increased proportion of fruits graded Class 1 when the nutrient solution salinity was increased above the standard recommended values (Adams and Ho, 1989; Adams, 1991). However, the nutrient solution salinity also increases the incidence of the calcium related physiological disorder, blossom-end rot (BER) in tomato fruits (Ehret and Ho, 1986b; Sonneveld and Van der Burg, 1991; Adams and Ho, 1992; Willumsen et al. 1996; Van Ieperen, 1996). As a result, in some cases, the favourable effects of salinity on the percentage of Class 1 fruits may be counteracted by a marked incidence of BER, as has been reported by Willumsen et al. (1996).

The influence of nutrient solution EC on the quality of other vegetable species, including pepper, cucumber, eggplants, beans, lettuce, celery, etc. is mainly dependent on the inherent salt tolerance of the particular species. However, the specific characteristics and the utility of the edible part of the vegetable are also important for the estimation of the quality response to the nutrient solution EC. For instance, eggplants are eaten cooked and, therefore, the sugar and the acid contents in their juice are of minor importance for the assessment of their quality by the consumer. On the other hand, the fruit size is an important criterion when grading eggplant fruits. Therefore, in experiments with eggplants the percentage of fruits graded Class 1 was always lower when the plants were exposed even to moderate salinity, since an increased EC results in a higher percentage of undersized fruits (Savvas, 1992; Savvas and Lenz, 2000a). Furthermore, in some cases, an increased incidence of the physiological disorder internal fruit rot which is observed when the eggplants are exposed to nutrient solution salinity (Savvas and Lenz, 1994a) may also contribute to the decline of the Class 1 fruit percentage (Savvas, 1992). In Color Plate 3.1, eggplant fruits affected by internal fruit rot are shown.

Compared to tomato, relatively little information can be found in the literature about the influence of total salt concentration in the nutrient solution on the quality of other fruit or leafy vegetables grown hydroponically. Thus, Mizrahi and Pasternak (1985) found that melon fruits from plants exposed to salinity scored higher in taste than the controls. De Kreij (1995) postulates that an increased nutrient solution EC improves the taste of pepper fruits and decreases the incidence of cuticle cracking, 'white flecks' and 'green spot' but the shelf life of the fruits is also shortened. In a study with NaCl or nutrient-induced salinity in the range 2.5-5.2 dS m^{-1}, Sonneveld and Van der Burg (1991) found that increasing the EC resulted in an increased incidence of blossom-end rot in sweet pepper whilst the appearance of russeting and green spot diminished. The increase of EC by adding NaCl to the nutrient solution had a more pronounced effect than a nutrient-induced salinity on BER in peppers. Moreover, the percentage of deformed pepper fruits was not influenced by the increased nutrient solution EC, whilst the effect on the vitamin C content was not consistent, regardless of the salts used to induce salinity. In another series of experiments with peppers, De Kreij (1999) found that within the range of low nutrient solution EC (1-3 dS m^{-1}) the incidence of blossom-end rot is primarily influenced by the cation ratios in the root environment and not by the total salt concentration. With respect to cucumber, Sonneveld and Van der Burg (1991) found that increasing the nutrient solution EC may enhance the color and the shelf life of the fruits, although the results concerning the latter quality characteristic were not always consistent. In the case of leafy vegetables, EC affects the quality mainly to the extent that it influences the incidence of physiological disorders. Thus, raising the EC of the nutrient solution, the incidence and the severity of 'blackheart' is significantly decreased in celery thus increasing the percentage of good quality product, although Ca also seems to be involved in the occurrence of this physiological disorder (Pardossi et al. 1999). Moreover, a high EC level in the root environment may prevent the incidence of glassiness in lettuce (Maaswinkel and Welles, 1986). Nevertheless, some internal quality characteristics of leafy vegetables such as vitamin C and sugar contents of lettuce, may also be influenced by the nutrient solution EC (Shinohara and Suzuki, 1988). However, the impact of such effects on the taste of these vegetables is questionable. Thus, in another study, Mizrahi and Pasternak (1985) did not find any significant difference in the taste of iceberg lettuce which had been grown under saline conditions compared to control plants.

5.3. The pH of the Nutrient Solution

Two questions are primarily raised, with respect to the management of pH in hydroponics. The first question is related to the range of pH in the root environment which is optimal for the hydroponically grown crops. The second question, addresses the methods of achieving and maintaining the optimal pH in the root environment of plants grown without soil.

5.3.1. The optimal pH range in hydroponics

It is well known that the main effects of pH on the growth and development of the plants arise from its influence on the nutrient availability in the root environment, regardless of whether the plants are grown in the soil (Marschner, 1995) or hydroponically (Islam et al. 1980; Jones, 1982; Graves, 1983; Schwarz, 1995). Thus, from a physiological point of view, it seems that the pH values recommended for crops grown in the soil would be optimal also for soilless-grown plants. However, this consideration is only partly true. Most soils contain various substances which are not taken up by the plants but they may indirectly influence the nutrient availability. For instance, most soils contain natural chelating agents and humic substances which strongly influence the uptake of micronutrients, especially that of Fe, Mn, Zn, and Cu. Moreover, the changes of pH in the soil are normally gradual and not very extensive during the cropping period due to the large volume of rooting medium per plant, and also due to ion exchange processes between the soil colloids and the soil solution. In contrast, in soilless culture, large changes of pH may occur even within hours in the root environment owing to the low volume of substrate per plant and to the lack of buffering capacity, particularly in chemically inert substrates. Therefore, in soilless grown crops, the pH in the root environment should be maintained within a narrow range around a target value.

Usually, the pH in the root zone which seems to be optimal for most crops grown hydroponically ranges between 5.5-6.0. Values between 5.0-5.5 and 6.5-7.0 may not cause problems in most crops (Islam et al. 1980; Graves, 1983). However, in soilless culture, when maintaining marginal values of the optimum pH range, the risk to exceed or drop below them during some periods of the day is increased due to the reasons stated above. Most plants, when they are supplied with nutrient solutions with a pH higher than 7 or lower than 4.8-5, may show growth restrictions (Islam et al. 1980; Jones, 1982). Nevertheless, there are also plants such as cut chrysanthemums which perform better at a nutrient solution pH as low as 4 (De Kreij and Van der Hoeven, 1996). This response is attributed to increased susceptibility of this plant species to chlorosis induced by Fe and Mn deficiency.

Values of pH above 7.0 may quickly result in inadequate uptake of P, Fe, and Mn, and sometimes even in Cu and Zn deficiency symptoms. The appearance of P deficiency at pH values above 6.5-7.0 is attributed to the increasing transformation of $H_2PO_3^-$ into HPO_3^{2-} (De Rijck and Schrevens, 1997a) since the latter ion is not readily taken up by the plants (Hendrix, 1967). Furthermore, the precipitation of calcium phosphate at pH values above 6.2 is another reason to maintain the pH below this level in the root environment of plants grown hydroponically (Sonneveld, 1989; De Rijck and Schrevens, 1998b). The appearance of Fe, Mn, Zn, and Cu deficiencies at pH values above 6.5-7.0 is associated with increased precipitation of these nutrients. In case of manganese, this process is mainly induced by an increased activity of Mn oxidizing bacteria which is observed at rising pH levels in the nutrient solution (Sonneveld and Voogt, 1980). Reduced leaf Cu and Zn concentrations due to increased pH have been reported by De Kreij and Van der Hoeven (1996). Moreover, Savvas and Manos (1999) found a considerably reduced Zn concentration in the drain solution of a rose culture which was ascribed to the very high pH (7.05) prevailing in the root environment. Iron is the micronutrient showing the lowest solubility at a high pH. In solution cultures, the free iron ions precipitate even at pH values lower than 6.5, mainly as iron phosphate (De Rijck and Schrevens, 1998c). Therefore, Fe should always be added in the form of Fe-chelates in hydroponics, preferably as Fe-DTPA or Fe-EDDHA (Wreesman, 1996).

In soilless culture, the growth of most plants is impaired at pH values lower than 4.5-5.0 in the nutrient solution. If pH values lower than 4.0 are involved, this response is mainly attributed to direct H^+ injury to the roots (Islam et al. 1980). Moreover, various forms of insoluble oxides and hydroxides of Mn and Al which are present in the substrate may become soluble at pH lower than 5. As a result, the concentrations of these ions in the nutrient solution may drastically increase, thus resulting in the appearance of Mn and Al toxicities in the plants. In addition, at pH 4 the uptake of Ca, Mg and K by the plants seems to be considerably restricted as compared to the uptake of these ions at pH 5.5 (Islam et al. 1980). Thus, at relatively low pH levels, the plant growth may also be impaired by inadequate Ca, Mg and K supply. The suppressive effect of low pH levels on the nutrient cation uptake is partly attributed to the presence of NH_4^+ at relatively high concentrations (see also section 5.5). Besides the yield, a reduced cation uptake at low pH may also impair the quality of fruit vegetables (Willumsen, 1980).

Some researchers have reported improved plant growth when humic substances are added to the nutrient solution in hydroponics (David et al. 1994; Siminis et al. 1998). However, De Kreij and Van der Hoeven (1996) provided evidence that this response is related to the effect of humic substances on the pH of the nutrient solution. In particular, it was demonstrated that the addition of humic substances to the nutrient solution enables a more consequent maintenance of a low pH in the root zone.

5.3.2. Adjustment of the target pH

To adjust the target pH in the nutrient solution supplied to the plants, and to maintain the desired pH level in the root zone by simultaneously maintaining optimal nutrient levels, some knowledge about the interactions between pH and nutrient concentrations is required. In fact, besides H^+ and OH^-, there are only three ions in the nutrient solutions which are involved in pH changes. These ions are $H_2PO_4^-$, NH_4^+, and HCO_3^-. In the desired pH range, these ions are in chemical equilibrium with HPO_4^{2-}, NH_3 and H_2CO_3 ($H_2O + CO_2$), respectively. All other cations and anions included in a nutrient solution form strong bases or acids respectively, and are therefore, not hydrolyzed in the pH range applied in hydroponics (De Rijck and Schrevens, 1997b).

The pH has a strong impact on the relative proportions of the phosphate forms which are present in a nutrient solution. As pH drops to values lower than 4.5, an increasing proportion of the phosphate contained in a nutrient solution forms non-dissociated phosphoric acid (H_3PO_4). At pH 4.6 all phosphate ions of the solution are present in form of $H_2PO_4^-$ whilst at pH 6.5 about 20% and at pH 9 nearly 100% of phosphate ions dissociate into HPO_4^{2-}. Furthermore, as pH rises above 10, increasing amounts of HPO_4^{2-} dissociate further into PO_4^{3-} (Steiner, 1961; De Rijck and Schrevens, 1997a). The above data indicate that in nutrient solutions having pH values within the desired range (5-6.2) almost all phosphate ions are present in form of $H_2PO_4^-$ and only an extremely low proportion may be in form of HPO_4^{2-}, particularly at pH higher than 6.0.

The addition of phosphate ions influences the pH of the nutrient solution. If P is added as phosphoric acid, the latter dissociates, thus reducing the pH of the solution. If phosphorus is added as KH_2PO_4 the pH is again reduced, due to partial dissociation of $H_2PO_4^-$ into HPO_4^{2-}, which renders hydrogen ions. Nevertheless, $H_2PO_4^-$ is a much weaker acid than H_3PO_4. However, when tap water is used to prepare a nutrient solution, even the addition of the entire desired amount of phosphate in the form of phosphoric acid may be insufficient to lower the pH to the target levels (5-6). This subject will be discussed in more detail below, because it is related to the HCO_3^- concentration in the tap water.

The concentration of NH_4^+ in a nutrient solution is also influenced by the solution pH. When ammonium is added to a nutrient solution, almost the entire quantity will be present in form of NH_4^+, provided that the pH is lower than 7.5 (De Rijck and Schrevens, 1997a). If the pH rises above this level, NH_4^+ is gradually replaced by NH_3 due to hydrolysis as follows:

$$NH_4^+ + H_2O \leftrightarrow H_3O^+ + NH_3 \quad K_a = 10^{-9.2} \quad (3)$$

However, since the pH values aimed at in hydroponics never exceed 6.5, virtually all the ammonium ions added when preparing a nutrient solution are present in form of NH_4^+.

Considering the above data, it becomes obvious that the immediate effect of adding NH_4^+ to a nutrient solution on its pH is very weak provided that the pH of the latter is below 7-7.5. However, the NH_4^+ in the nutrient solutions is converted into NO_2^- by the autotrophic bacterium *Nitrosomonas* sp. and subsequently quickly into NO_3^- by another bacterium (*Nitrobacter* sp.) through the nitrification process which can be summarized as follows:

$$NH_4^+ + 2O_2 + H_2O \rightarrow NO_3^- + 2H_3O^+ \quad (4)$$

This process takes place gradually in the rooting medium and results in a reduction of the pH due to formation of H_3O^+. Thus, NH_4^+ is the only nutrient ion which has a long-term influence on the pH of the nutrient solution in the root environment owing to a slow release of hydrogen ions. This property is very important for the maintenance of a target pH in the root environment, especially in substrate cultures where no continuous nutrient solution recirculation is applied.

In most cases, the concentration of HCO_3^- in the tap water used to prepare nutrient solutions is substantial. However, even if rain water is used, the HCO_3^- concentration is usually higher than 0.001 meq L^{-1} due to a partial dissociation of dissolved CO_2. Nevertheless, a chemical equilibrium between carbonate, bicarbonate, and carbonic acid (in fact, dissolved CO_2) is established in the water and the nutrient solutions:

$$HCO_3^- + H_3O^+ \leftrightarrow H_2CO_3 + H_2O \quad K_{a1} = 10^{-6.3} \quad (5)$$

$$HCO_3^- + H_2O \leftrightarrow H_3O^+ + CO_3^{2-} \quad K_{a2} = 10^{-10.3} \tag{6}$$

The fraction HCO_3^- can be calculated as a function of pH (H_3O^+ concentration) through the equation (De Rijck and Schrevens, 1997a):

$$\frac{[HCO_3^-]}{[CO_3^{2-}] + [HCO_3^-] + [H_2CO_3]} = \frac{K_{a1}}{[H_3O^+]B} \tag{7}$$

where [·] denotes concentration, $B = 1 + K_{a1}[H_3O^+]^{-1} + K_{a1}K_{a2}[H_3O^+]^{-2}$ and all concentrations are expressed on equivalent basis. The fractions H_2CO_3 and CO_3^{2-} can be also calculated as functions of pH. Based on these equations it can be shown that at pH 8 virtually no CO_3^{2-} or H_2CO_3 but only HCO_3^- is present in a nutrient solution. If the pH rises above 8, the bicarbonate is gradually replaced by carbonate. Similarly, if pH decreases progressively to values lower than 8, the HCO_3^- is gradually replaced by H_2CO_3 (dissolved CO_2) until pH 4.5 where only dissolved CO_2 is present in the solution.

Since K_{a1} is much higher than K_{a2}, from (5) it is obvious that the presence of HCO_3^- at relatively high concentrations in a nutrient solution is associated with a decreased H_3O^+ concentration, accompanied by formation of CO_2 and, therefore, with an increased pH. Since, in most cases, the tap water contains substantial amounts of bicarbonate, its pH is expected to be relatively high. Indeed, in most cases the pH of the tap water exceeds 7 and in some cases it may be higher than 8. Thus, when preparing a nutrient solution by using tap water, it is necessary to add sufficient amounts of an acid (H_3O^+) in order to decrease the pH. However, according to (5), this will result in replacement of HCO_3^- by CO_2 until the desired pH level is established.

The concentration of HCO_3^- which corresponds to the desired H_3O^+ concentration (pH) in the nutrient solution according to (5) may be calculated using (7). In particular, at first the concentrations of HCO_3^- and H_3O^+ (pH) measured in the tap water are substituted in (7). Thus, it is possible to calculate the sum of the concentrations of carbonic acid, bicarbonate and carbonate ions. Then, since the latter is known, (7) can be used again to calculate the concentration of HCO_3^- corresponding to the target pH of the nutrient solution (Savvas and Adamidis, 1999).

Due to the much lower HCO_3^- concentration in the rain water, its pH usually does not exceed the 5.5-6.5 range. Thus, if rain water is used to prepare a nutrient solution, only small amounts, if any, of an acid are required to achieve the desired pH.

The above considerations indicate that the regulation of pH in commercially used nutrient solutions is in fact a problem of adjusting the HCO_3^- concentration. As already mentioned, (5) is the chemical equilibrium equation which indicates the alterations occurring in the HCO_3^- concentration when an acid is added to the nutrient solution. The concentration of HCO_3^- which is established in the nutrient solution when the desired pH has been achieved may be denoted as C_b (meq L^{-1}), whilst the bicarbonate concentration in the tap water is denoted as $[HCO_3^-]_w$ (meq L^{-1}). Then, the acid required to decrease the bicarbonate concentration from $[HCO_3^-]_w$ to C_b is roughly equal to the difference $[HCO_3^-]_w - C_b$, since (5) indicates that 1 meq of acid (referring to the anion released at the usual nutrient solution pH) is required to convert 1 meq HCO_3^- into CO_2.

For a certain HCO_3^- concentration in the tap water, the target concentration of the anion released when the acid added to adjust pH dissociates, determines whether this acid may be used to control the pH or not. For instance, it is assumed that the HCO_3^- concentration in tap water amounts to 5 meq L^{-1}, whilst the desired $H_2PO_4^-$ concentration and the target pH in the nutrient solution are 1.5 meq L^{-1} and 5.7 respectively. Furthermore, it is assumed that taking the pH of the tap water into account and using (7) a C_b of 0.5 meq L^{-1} HCO_3^- in the nutrient solution is calculated. Then, 4.5 meq L^{-1} of acid should be added to achieve the target pH. However it is not possible to add more than 1.5 meq L^{-1} of H_3PO_4 (equivalent concentration referring to $H_2PO_4^-$) without violating the target value for the phosphate concentration in the nutrient solution. Consequently, another acid should be used. If it is further assumed that the maximum desired SO_4^{2-} concentration in the nutrient solution is 3 meq L^{-1}, sulphuric acid can also not be used to adjust the pH to the desired value. Usually, the desired sulphate concentrations in the nutrient solutions do not exceed a threshold of 5 meq L^{-1}, whereas the target $H_2PO_4^-$ concentrations range between 1.0-1.5 meq L^{-1} (Sonneveld and Straver, 1994; De Kreij et al. 1997; De Rijck and Schrevens, 1998c). This is the main reason why these two acids are usually not used to adjust the pH of the nutrient solution. In the past, some researchers have used phosphoric or sulphuric acid to control the pH of the nutrient solution (Cooper, 1979). In most cases, either the yield and the fruit quality were impaired or the irrigation system was blocked, the latter being mainly an effect of calcium phosphate precipitation. Presumably, neither the HCO_3^- concentration in the tap water nor the target concentrations of $H_2PO_4^-$ and SO_4^{2-} in the nutrient solution were taken into account when phosphoric or sulphuric acid, respectively, were used to adjust pH, thus resulting in overdosing of either H_2PO_4 or SO_4^2 in the nutrient solution.

An additional reason to avoid the use of H_2SO_4 to adjust the pH in the nutrient solutions used in hydroponics is the fact that this acid is highly corrosive and, therefore, too dangerous to be used in routine operation in the glasshouses. Moreover, as the addition of Cl^- to the nutrient solution is not desired, HCl is usually not used to adjust the pH in hydroponics. However, in crops like tomato, a small addition of Cl^- may be desirable if the Cl^- and the total salt concentration in the tap water are relatively low. In such cases, if the HCO_3^- concentration in the tap water is also low, HCl may be used to adjust the pH of the nutrient solution. Indeed, as reported by Papadopoulos and Pararajasingham (1998) under such conditions the use of HCl to control the nutrient solution pH in hydroponically grown tomatoes may result in improved fruit quality and higher marketable yield.

Nevertheless, in most cases nitric acid is used to adjust the pH of the nutrient solution, since the desired NO_3^- levels in the nutrient solution are usually above 10 meq L^{-1}. High HCO_3^- concentrations in the tap water are essentially accompanied by equally high concentrations of cations, particularly Ca^{2+} and Mg^{2+}. Thus, when preparing a nutrient solution using tap water with a high HCO_3^- concentration, an increased addition of NO_3^- in the form of HNO_3, in order to control pH is compensated by a decreased supply of NO_3^- in the form of $Ca(NO_3)_2$. If a high HCO_3^- concentration in the tap water is also accompanied by a high Mg^{2+} concentration, less, or sometimes even no Mg^{2+} is added in the form of $MgSO_4$. Then, the necessary SO_4^{2-} is added in the form of K_2SO_4, thus resulting in a reduced addition of NO_3^- in the form of KNO_3 as well. Consequently, even if the HCO_3^- concentration in the tap water is relatively high, there is no risk to add too much NO_3^- to the nutrient solution when HNO_3 is used to adjust the pH.

5.3.3. Maintenance of pH in the Root Zone

The composition of the nutrient solution in the root zone changes gradually, mainly due to selective ion uptake by the plants in accordance with their nutrient requirements (Steiner, 1980; Morard and Benavides, 1990). In periods of intensive growth and sufficient light intensity, the anion uptake usually exceeds that of cations, owing to an elevated nitrate absorption and utilization in the plant metabolism. In terms of electrochemical potential, a higher anion uptake ratio which exceeds that of cations is compensated by release of HCO_3^- and OH^- by the plant roots (Ben-Zioni et al. 1971; Graves, 1983). As a result, the pH of the nutrient solution in the root environment increases. However, under conditions of low light intensity, this situation may be reversed. In particular, under poor light conditions, the nitrate reductase activity declines, thus resulting in a depression of nitrate utilization in the plant. This situation leads to a decreased nitrate uptake ratio. Consequently, the difference between the anion and the cation uptake ratio is reduced and in some cases may become negative. Again, in terms of electrochemical potential, a more rapid cation uptake than the anion uptake is compensated by release of H^+ by the roots (Graves, 1983). Hence, under poor light conditions, the pH in the root zone of hydroponically grown plants does not tend to increase rapidly whilst in some cases it may decrease.

In exceptional cases where the pH of the nutrient solution in the root zone drops below the optimal range, KOH (Graves, 1983), $KHCO_3$ (Sonneveld, 1989) or K_2CO_3 (De Kreij and Van der Hoeven, 1996) may be used to adjust it. As pointed out by Sonneveld (1989) if it is necessary to increase the pH in the root environment, a separate solution containing only concentrated $KHCO_3$ should be supplied to the plants. If $KHCO_3$ is added directly to the supplied nutrient solution, the pH of the latter becomes very high, resulting in phosphate and carbonate precipitation.

Obviously, the use of a separate stock solution is also suggested when KOH or K_2CO_3 is used to increase the pH in the root environment.

In most cases, the control of pH in the root environment of soilless cultivated plants requires measures aimed at preventing the occurrence of a too high rather than a too low pH. To achieve it, the primary operation is the supply of a nutrient solution having a pH somewhat lower than the target pH in the root environment. Thus, if a pH of 6.0 is desired in the root environment, the pH of the supplied solution should be set at 5.5. If this is not sufficient and the pH increases above an acceptable threshold value (e.g. 6.3) the pH of the supplied solution might be reduced even more (up to 5.0). Alternatively, if the percentage of drainage solution is relatively low, the frequency and/or the water dosage at each irrigation cycle could be increased. However, often these measures are not sufficient to control the pH in the root environment. The addition of some nitrogen in the form of ammonium in the supplied nutrient solution may be a more efficient operation. Indeed, as has been previously illustrated, the NH_4^+ has a long-term depressing effect on the pH in the root zone, due to a slow release of hydrogen ions which are produced during the nitrification process. Normally, the inclusion of 5-15% of total nitrogen of a nutrient solution in the form of NH_4^+-N is considered to be sufficient to provide adequate control of pH in the root environment, if the solution supplied to the plants has a pH ranging from 5 to 6. The exact ratio is dependent on the specific susceptibility of the particular crop to ammonium supply via the nutrient solution.

As stated by De Kreij and Van der Hoeven (1996), besides ammonium, humic substances may also be used to provide a better control of the nutrient solution pH in the root environment. According to the same authors, organic pH buffers such as methyl-sulfonyl-methane, might also be used to control the pH in the root environment of plants grown hydroponically. Nevertheless, the use of the above substances did not enable a better pH control in the roots of a chrysanthemum crop and did not result in higher yields as compared to the treatment with inclusion of ammonium in the nutrient solution. Thus, the final decision on the means used to achieve a better control of pH in the root zone seems to depend mainly on cost estimations.

5.4. Nutrient Concentrations

If the nutrient concentrations in a nutrient solution are expressed in terms of meq L^{-1}, then the total cation concentration (C_c) is essentially equal to the total anion concentration (C_a):

$$C_c = C_a \tag{8}$$

The establishment of optimal nutrient concentrations for hydroponic nutrient solutions has been, since the last century, the objective of many experiments related to plant nutrition. The data gained were mainly based on the determination of the quantities of nutrients which disappeared from the nutrient solution due to plant uptake in pure water culture, in relation to the water consumption during a particular time interval. Moreover, data concerning the total amount of each particular nutrient per plant in combination with the total water consumption per plant were also used as indicators for optimal nutrient concentrations in the nutrient solutions. In addition, the nutrient concentrations usually prevailing in the natural soil solutions could in some cases give an additional indication on the possible optimal nutrient concentrations in artificial nutrient solutions. Finally, the performance of the plant at a particular nutrient concentration which seemed to be suitable according to the previously mentioned working methods was checked. Based on data gained with such methods of experimentation, Hoagland and Arnon (1950) proposed the nutrient concentrations given in Table 3 for the cultivation of almost all plant species in nutrient solutions. In the same table, recently recommended nutrient concentrations (Sonneveld and Straver, 1994) for the soilless cultivation of two representative plant species, one vegetable and one ornamental, are also cited. A comparison of the individual micronutrient concentrations proposed by Hoagland and Arnon (1950) with those suggested by Sonneveld and Straver (1994), but also with other recent recommendations published by De Kreij et al. (1997), reveals substantial differences only in the Zn, Mo and, to some extent, in the Cu levels. Hoagland and Arnon (1950) suggested a relatively low Zn concentration because they noticed that considerable amounts of Zn are released to the nutrient solution from galvanized pipes which were used at that time to transfer the irrigation water. In contrast, the macronutrient concentrations range at the same order of magnitude in all literature sources cited, although Sonneveld and Straver (1994) and De Kreij et al. (1997) have specialized the proposed nutrient concentrations according to the cultivated plant species.

Table 3. Composition of standard nutrient solutions.

Macro-nutrient	m mol L^{-1}			Micro-nutrient	m mol L^{-1}		
	Hoagland & Arnon	Sonneveld & Straver, cucumber	Sonneveld & Straver, roses		Hoagland & Arnon	Sonneveld & Straver, cucumber	Sonneveld & Straver, roses
NO_3^-	14.0	16.00	11.00	Fe	25.00	15.00	25.00
$H_2PO_4^-$	1.0	1.25	1.25	Mn	9.10	10.00	5.00
SO_4^{2-}	2.0	1.375	1.25	Zn	0.75	5.00	3.50
K^+	6.0	8.00	4.50	Cu	0.30	0.75	0.75
NH_4^+	1.0	1.25	1.50	B	46.30	25.00	20.00
Ca^{2+}	4.0	4.00	3.25	Mo	0.10	0.50	0.50
Mg^{2+}	2.0	1.375	1.125				

Note: The data given by Hoagland and Arnon (1950) has been converted by the author into millimolar (macronutrients) or micromolar (micronutrients) concentrations.

Several investigators found out that, especially in case of actively absorbed macronutrients such as N, P, and K, even extremely low concentrations may give good results, provided that these low concentrations can be maintained rigorously. Thus, Massey and Winsor (1980a) did not find any significant differences in the yield of NFT grown tomatoes when the plants were grown at nitrogen concentrations of 10, 20, 40, 80, 160, and 320 mg L^{-1} NO_3^--N. Similarly, Massey and Winsor (1980b) found that varying the P concentration in the range 5-200 mg L^{-1} had no influence on the yield of tomatoes grown in NFT. Furthermore, Adams and Grimmett (1986) did not find significant yield differences in tomatoes grown in recirculating nutrient solution at 20, 50, 150 and 400 mg L^{-1} K, whilst at 10 mg L^{-1} K a small yield reduction (7%) was observed. In all these experiments, the lower levels of nutrient concentrations were maintained constant using small peristaltic pumps controlled by interval timers and a time switch.

The above results indicate that the concentrations of the actively absorbed nutrients might be maintained at very low levels without any significant yield losses. However, in commercial practice it is difficult and, therefore, unsafe to maintain such low concentrations which are markedly below the respective nutrient to water uptake ratios. Nutrient concentrations ranging in the order of magnitude indicated in Table 3 seem to give the best results for most horticultural crops because they provide a safety zone between the minimal acceptable and the actual nutrient levels. As stated by Sonneveld (1981) and Graves (1983), best results are obtained when the nutrient concentrations correspond approximately to the nutrient to water uptake ratio. Under such conditions, the plants do not have to consume energy to take up or actively exclude any nutrient ions, whose concentrations are lower or higher than their nutrient to water uptake ratios, respectively (Steiner, 1980). However, as pointed out by several investigators (Adams, 1980; Adams and Massey, 1984; Van Goor et al. 1988; Savvas and Lenz, 1995; Sonneveld, 1995) the nutrient to water uptake ratios fluctuate widely, in response to different climatic conditions, even within the same day. Therefore, it is not possible to constitute a nutrient solution having a nutrient concentration which would be continuously in accordance with the nutrient to water uptake ratio. On the other hand, due to a very low volume of nutrient solution per plant in hydroponics, changes in the nutrient to water uptake ratio might quickly result in large alterations of the ionic concentrations in the root zone. Indeed, due to a more intensive plant uptake during particular time intervals, some nutrients may become depleted whilst some others may accumulate. Therefore, most investigators suggest higher nutrient concentrations than the expected mean nutrient to water uptake ratios, in order to ensure adequate supply with all nutrients even when the actual nutrient to water uptake ratios for some nutrients are strongly divergent from the expected mean uptake. Hence, the nutrient concentrations suggested in several literature sources (Hoagland and Arnon, 1950; Hewitt, 1966; Sonneveld and Straver, 1994; De Kreij et al. 1997; De Rijck and Schrevens, 1998c) for the supplied nutrient solution as well as for the solution in the root zone are higher than the nutrient to water uptake ratios found by various investigators (Adams, 1980; Schippers, 1980; Sonneveld, 1981; Letey et al. 1982; Gislerod and Adams, 1983; Savvas and Lenz, 1995). In open systems, the quantities of nutrients corresponding to the surplus concentration are leached in the drain solution. In closed systems, the nutrients added to replenish the drain solution before it is reused should be as much as needed to maintain the target concentrations in the supplied solution. This can be achieved if, for each individual nutrient, the mean ratio of nutrient to water added to the drain solution when it is replenished, is equal to the respective mean uptake ratio.

As illustrated above, the monovalent nutrient ions, particularly K^+, NO_3^- and $H_2PO_4^-$, are taken up very efficiently, even at concentrations markedly lower than the respective nutrient to water uptake ratio. Therefore, these ions may quickly become depleted in the root zone as reported, for instance, by Schippers (1980) if their concentrations are lower than the corresponding nutrient to water uptake ratios. To ensure an adequate supply of these nutrients in hydroponics, it is sufficient to increase their concentrations in the nutrient solution to levels slightly higher than their nutrient to water uptake ratios. However, the uptake of bivalent nutrient ions such as Ca^{2+}, Mg^{2+}, and SO_4^{2-} is more difficult than that of monovalent ions (Voogt, 1988a; De Kreij, 1995; Sonneveld, 1995). Therefore, in case of bivalent nutrient ions, a considerably higher nutrient concentration than the desired nutrient to water uptake ratio is required. This necessity is well illustrated by some data reported by Sonneveld (1981) who found that K concentrations of 4.8, 6.6, and 10.2 mM resulted in the uptake of 4.6, 5.3, and 5.7 mmol K per liter water absorbed whilst at 6.1, 8.0 and 10.9 mM Ca in the solution 1.6, 2.0 and 2.3 mmol Ca were absorbed per liter water taken up by the plants. The difference between the Mg concentration in the nutrient solution and the observed Mg/water (mmol L^{-1}) uptake ratio was even larger. Results reported by Savvas and Lenz (1995) concerning water and nutrient uptake at particular nutrient concentrations support the above findings.

If the nutrient concentrations in the supplied nutrient solution are balanced, based on the principles stated above, controlling the EC of the nutrient solution in the root environment is a good tool to avoid the occurrence of deficient or toxic macronutrient concentrations in the root zone. Nevertheless, a chemical analysis in a representative substrate or nutrient solution sample taken from the root environment at regular intervals (e.g. every month), especially in closed hydroponic systems, could enable a better and more safe nutritional management of the crop. However, the control of the EC in the root environment does not provide any information about the micronutrient concentrations. Therefore, special attention should be paid to select proper target micronutrient concentrations for both the supplied nutrient solution and that in the root environment. Generally, the pH maintained in the root environment is probably more important for the micronutrient availability than the macronutrient concentrations per se in the supplied nutrient solution. However, especially for some microelements, the concentration in the supplied nutrient solution is crucial. Boron has a narrow range of optimal concentrations (Graves, 1983). Therefore, boron concentrations higher or lower than this narrow optimal range may readily result in B toxicity or deficiency respectively. Moreover, as pointed out by De Kreij (1995) the requirements of various horticultural species for B differ strongly. The same author points out that especially roses have a low boron requirement and are susceptible to B toxicity. Nevertheless, the nutritional management for this element must be more precise in all crops.

In an experiment with cucumbers grown on rockwool at different micronutrient levels, it was concluded that Mn, Zn and B toxicities may quickly appear if their concentrations in the supplied nutrient solution are raised to some extent, whilst Fe, Cu, and Mo toxicities are less likely to occur even at much higher concentrations than those given in Table 3 (Sonneveld and De Bes, 1984). In the above study, Fe, Zn, Cu, and B deficiencies were observed only at 0, 2, 0.05 and 5 but not at 1, 4, 0.35 and 16 μM Fe, Zn, Cu and B in the nutrient solution, respectively. Even no addition of Mn and inclusion of only 0.1 μM of Mo did not result in Mn or Mo deficiencies. In another experiment with cucumbers in peat, the yield was not reduced when no Mn, Mo and Zn were added to the root environment, probably because minor, but sufficient quantities of these micronutrients could be released from the substrate (Adams et al. 1989). In general, the minimal micronutrient concentrations which are sufficient for optimal plant growth are so low, that in many substrates the plant requirements may be satisfied by minor amounts released from the rooting medium. Even rockwool seems to release small amounts of Fe and B, especially at the beginning of the cultivation (Ottosson, 1977). Nevertheless, the release of micronutrients from the substrate is also dependent on the pH of the nutrient solution in the root zone (Adams et al. 1989).

Another nutrient which may be added to nutrient solutions in some cases is chloride (De Kreij, 1995). The maintenance of Cl concentrations at approximately 5 mM seems to reduce the nitrate content in lettuce due to a partial replacement of NO_3^- by Cl^- in the cell vacuoles (Van der Boon, 1988). Moreover, the inclusion of 8-10 mM Cl^- in the nutrient solution is suggested in order to improve the quality of tomato fruits and to reduce the incidence of blossom end rot (Nukaya et al. 1991; De Kreij et al. 1997).

Besides the essential nutrients, silicon may be beneficial for the growth of certain plants in hydroponics. In particular, it has been shown that Si concentrations of 1.7 mM may ameliorate or even prevent the occurrence of powdery mildew in cucumber, melon, zucchini squash and roses

(Menzies et al. 1992). Miyake and Takahashi (1978, 1983) imply that Si may be an essential micronutrient for tomato and cucumber. Liang (1999) reports an increased salt tolerance in barley grown hydroponically when 1 mM Si is added to the nutrient solution. As reported by Voogt (cited by De Kreij, 1995) a concentration of 1 mM Si in the nutrient solution increased yield by 5%. De Kreij (1995) suggests Si concentrations of 0.6 mM for cucumber and 1.5 mM for roses. In most cases, potassium silicate is the preferable fertilizer used to add Si in hydroponical nutrient solutions.

5.5. Nutrient Ratios

The ratios between the ions included in a nutrient solution may be defined either in terms of chemical units (meq/meq or mmol/mmol) or on a weight basis (w/w). Thus, the K : Mg ratio in a nutrient solution which contains 6 mM K and 2 mM Mg is 3 on a molar basis, 1.5 on an equivalent basis and 4.83 on a w/w basis.

A nutrient ratio may involve either two or more nutrients. If all the cations or anions of a nutrient solution are involved, then the nutrient ratio may be expressed in terms of relative proportions of cations or anions in the solution. For instance, in a nutrient solution containing 8 meq L^{-1} Ca, 1.5 meq L^{-1} Mg, 7 meq L^{-1} K, 1 meq L^{-1} NH_4^+ and 2 meq L^{-1} Na, the relative cation proportions are 0.41, 0.077, 0.36, 0.051 and 0.10, respectively. The micronutrient cation concentrations have been neglected because they are much lower than those of macronutrients (see Table 3).

For a particular total salt concentration, the possible relative cation or anion proportions in a nutrient solution are theoretically unlimited. Nevertheless, the achievement of some relative cation or anion proportions in a nutrient solution may be not feasible due to precipitation owing to the low solubility of some of the salts that may be formed by combining all anions with all cations included in the solution (Steiner, 1961). However, although the chemically feasible relative anion and cation proportions still vary within a wide range, those applied in commercially used nutrient solutions are restricted within rather narrow ranges (De Rijck and Schrevens, 1998c). This is well demonstrated by the standard deviations calculated for the relative nutrient proportions cited in Table 4. The data of Table 4 have been obtained by analyzing 24 representative fresh nutrient solutions prepared according to various literature recommendations and prescribed to supply various commercial vegetable and ornamental crops (Savvas and Adamidis, 1999). The fact that the differences in the nutrient ratios which seem to be optimal for most horticultural plants are small is not surprising, since the plants are selective in their ability to absorb nutrients (Steiner, 1980; Morard and Benavides, 1990) and the proportions of nutrients contained in the whole plant tissue of different species also vary within relatively narrow ranges. Nevertheless, different plant species have specific preferences within the ranges indicated in Table 4, since in many experiments the crops seemed to respond to varia-

Table 4. Means ($\bar{x}$) and standard deviations (s) of relative macroanion and macrocation proportions (%) from 24 nutrient solutions used to supply various vegetables and ornamental plants (adapted from Savvas and Adamidis, 1999)

Cation	$\bar{x}$ (%)	s	Anion	$\bar{x}$ (%)	s
Ca^{2+}	38.35	5.55	SO_4^{2-}	18.67	5.87
Mg^{2+}	16.10	4.65	NO_3^-	65.06	7.91
K^+	34.18	5.74	$H_2PO_4^-$	5.96	1.09
NH_4^+	4.69	1.95	HCO_3^-	3.67	2.84
Na^+	6.69	4.29	Cl^-	6.64	4.75

tions in the relative ion proportions (Sonneveld and Voogt, 1985; Sonneveld, 1987; Voogt, 1988a, b; Nukaya et al. 1991, 1995; Willumsen et al. 1996; De Kreij, 1999). Thus, the determination of the most favourable nutrient ratio for each species is of major importance. Schrevens and Cornell (1993) and De Rijck and Schrevens (1999) propose the application of multifactorial experimental designs based on the mixture theory in the study of the effects of nutrient ratios on plant growth and yield. This concept demonstrates well the complexity of the interactions between the nutrients which constitute a nutrient solution, although usually, such experimental designs may be not feasible because they require a large number of different treatments.

Most experiments concerning effects of nutrient ratios in nutrient solutions deal with nutrient cation ratios (mainly K : Ca : Mg or K : Ca), nutrient anion ratios, the N : K (or K : N) ratio, or the ratio of NH_4^+ to total nitrogen. The ratio between the macronutrient cations K, Ca and Mg has probably attracted the interest of most investigators working with nutrient proportions in hydroponics. Steiner (1966) stated that the cation ratios prevailing in a nutrient solution are more crucial for the growth of tomato than the anion proportions. Based on a series of experiments with cucumbers, eggplants and sweet pepper, Sonneveld and Voogt (1985) found that the K : Ca : Mg ratio in the nutrient solution may influence the growth and yield but the responses of each of the above fruit vegetables are different. Thus, eggplant seems to be susceptible to a low Mg proportion whilst cucumber prefers a relatively higher calcium proportion and sweet pepper is less sensitive to variations in the nutrient cation ratios. Moreover, it was shown that the uptake of nutrient cations is markedly influenced by the nutrient cation ratios prevailing in the root environment. According to Voogt (1988b), tomato seems to prefer a molar K : Ca ratio as high as 63 : 37 in the root environment which can be obtained by supplying these nutrients at a molar K : Ca ratio of 78 : 22 to the recirculating nutrient solution. Sonneveld (1987) stresses the importance of adequate Mg supply in tomato crops which can be ensured if the Mg proportion in the root environment is at least 0.16 on a molor basis. In a rockwool crop, this could be achieved by maintaining the Mg proportionin the supplied nutrient solution at least at 0.09. With respect to the yield, lettuce does not respond to K : Ca variations within the range 0.8-5.6 in the supplied solution (molar basis) but raising the K : Ca ratio seemed to increase the incidence of tipburn and rot (Voogt, 1988a). Willumsen et al. (1996) stated that the cation activity ratios rather than the ratios of the cation concentrations should be used as a basis to assess the responses of plants to different cation ratios in the root environment. Indeed, the concentrations of some nutrients are not identical to their ionic activities, since small proportions of these nutrients, particularly Ca, Mg, SO_4, and H_2PO_4 may be present in the form of soluble salt complexes in the nutrient solution (De Rijck and Schrevens, 1998a). In the case of tomato, Willumsen et al. (1996) found that the increase of K and Mg activity ratios and the reduction of the activity of Ca in the root environment favours the occurrence of blossom-end rot. Nukaya et al. (1995) found that raising the K:Ca ratio in the nutrient solution increases the incidence of BER and decreases that of gold specks. De Kreij (1999) found that in sweet peppers BER is promoted by an increased K:Ca ratio in the root zone.

Relatively little information can be found in the literature about the effects of anion concentration ratios on the performance of horticultural crops in hydroponics. In general, the proportion of P to the total nutrient anions should be maintained at very low levels (usually less than 0.1), otherwise precipitation of phosphoric salts may occur (De Rijck and Schrevens, 1998b). Therefore, Steiner (1966), in an experiment with tomato, explored only the NO_3^- : SO_4^{2-} ratio. He found that NO_3^- : SO_4^{2-} ratios as high as 0.86 may result in a small yield decrease compared to lower ratios ranging from 0.67 to 0.36. Nukaya et al. (1991), working at a constant nutrient solution EC of 3.5 dS m^{-1} did not find any significant differences in tomato fruit yield when the NO_3^- : SO_4^{2-} : Cl^- ratio (molar basis) in the recirculating nutrient solution varied between (0.20-0.55) : (0.17-0.34) : (0.10-0.45), respectively, although the treatments with the highest sulphate proportions tended to slightly increase the yield. However, raising the proportion of Cl^- resulted in an increased incidence of gold specks which also affected the shelf life of the tomato fruits, whereas the occurrence of BER was diminished, probably because Cl^- enhances the uptake of Ca^{2+} (Nukaya et al. 1991).

Research on the N : K (or K : N) ratio in the nutrient solution in hydroponics has been carried out only in a limited number of experiments. Van Goor et al. (1988) have already pointed out that 'the effect of the N : K ratio in the nutrient solution deserves further study', since both elements are linked in the carbohydrate metabolism. According to data published by Adams (1980), the N : K uptake ratio in cucumber fluctuates considerably during the day. Thus, during the night and early morning the N : K uptake ratio (w/w) tends to be 1.0, but a gradual reduction up to 0.6 is observed until midday, when the light intensity reaches its maximum. Adams (1980) estimated an average N : K uptake ratio (w/w) of 0.81 (K : N = 1.26). On an equivalent basis the average N : K uptake ratio, found by Adams (1980) for cucumbers, amounts to 2.26. A similar trend, with respect to the variations observed during the day was also observed in tomato by Adams and Massey (1984). However, the minimal and maximal N : K ratios observed in the course of the day in tomato by Adams and Massey (1984) were approximately 0.55 and 0.75 (w/w), respectively. The mean daily N : K ratios (w/w) observed prior to fruit setting in the first trus were 0.86 and 0.81 in February and August, respectively (2.40 and 2.25 on equivalent basis). However, when the fruit load on the plants began to increase rapidly, the N : K ratio at which these nutrients disappeared from the solution dropped to 0.4 (1.12 in meq/meq) followed by a slight increase to 0.5 (1.40 in meq/meq) after a few weeks. Comparable results have also been found for tomato by Voogt and Sonneveld as cited by Van Goor et al. (1988). Moreover, Van Goor et al. (1988) report that in lettuce the N : K uptake ratio (mol and equivalent basis) is initially 0.77 but increases rapidly to 2.0 during the growing period. Based on data presented by Tapia and Dabed (1984) it seems that in sweet pepper, the ratio of cumulative N : K uptake (equivalent basis) is maintained approximately at 2.8 until anthesis of the first flower. This ratio decreases to a level of about 2.15 during the period prior to the start of fruit setting, increases again and exceeds 2.8 when full fruit set is observed and remains almost constant to a value close to 3.3 during full fruit production.

Some of the nitrogen contained in nutrient solutions may be lost due to denitrification (Daum and Schenk, 1998). Therefore, the amounts of nitrogen removed from the nutrient solution during the growth of plants in either type of water culture should not be considered as net N uptake. Nevertheless, these data are used to establish guidelines about the optimal nutrient levels in the nutrient solutions. Thus, they can still be utilized, since the depletion of a nutrient from the solution rather than the net uptake by the plant should be taken into account for this purpose.

The proportion of NH_4^+-N to total nitrogen is not relevant for the supply of N to the nutrient solution, since both NH_4^+ and NO_3^- are N sources. However, it is very important for the regulation of pH in the root environment (see also section 5.3.2.). Moreover, in soilless culture, a too low or too high ammonium nitrogen to total nitrogen ratio may cause serious problems to the crops. Many authors have observed NH_4^+ toxicity in soilless grown crops, when a relatively high ammonium proportion or only NH_4^+-N was added as N source (Elia et al. 1996; Chance et al. 1999). Presumably, this response should be attributed to excessive acidification of the rooting medium caused by the nitrification of NH_4^+ (see section 5.3.2.). However, an inadequacy of carbon skeletons in the root cells which are needed to detoxify ammonia by formation of amino acids and related compounds may be also responsible for the inhibitory effects of high NH_4^+ concentrations in the nutrient solutions (Mengel and Kirkby, 1987). Moreover, at high NH_4^+ concentrations, depletion of oxygen in the root environment due to enhanced O_2 consumption in the process of NH_4^+-nitrification may also be involved in the appearance of ammonium toxicity in solution cultures. In addition, relatively high NH_4^+ concentrations in the rooting medium suppress the uptake of other cations, especially Ca, Mg, and K (Graves 1983; Heuer, 1991; Elia et al. 1996; Chance et al. 1999). On the other hand, when a relatively low proportion of the target total nitrogen concentration in a nutrient solution is supplied in form of NH_4^+-N, the plants grow better and give higher yields as compared to complete absence of ammonium (Kirkby and Mengel, 1967; Feigin et al. 1979; Errebhi and Wilcox, 1990; Feigin, 1990; Stensvand and Gislerod, 1992; Elia et al. 1996; Chance et al. 1999). A low NH_4^+ concentration, in addition to its contribution to the control of pH in the root zone, enhances the availability of organic N in the root cells. Indeed, in the roots, the NO_3^--N needs to be reduced to NH_4^+ before it can be used in metabolism and this involves the presence of nitrate reductase and carbohydrates which should be transported from the leaves.

Feigin et al. (1984) suggest an ammonium proportion of 0.25 in the nutrient solutions supplied to roses. Elia et al. (1996) propose an even higher proportion (0.3) for the NH_4^+-N in nutrient solutions for eggplants, although they found that eggplants are less tolerant to increased NH_4^+ concentrations than many other vegetables. However, according to Sonneveld (1995), the proportion of NH_4^+-N to total N in the nutrient solutions supplied to soilless grown crops should range between 0.07-0.14.

Overall, the target nutrient ratios in the nutrient solution supplied to a particular crop are different than those aimed at in the root environment. This is due to the varying ability of the plants to take up different ions owing to the involvement of different absorption mechanisms in each case. In particular, for bivalent nutrients, the ratios of nutrient concentrations to the respective absorption concentrations (nutrient absorption per liter water taken up by the plants) are in most cases much higher than 1. If the Ca, Mg, and SO_4^{2-} concentrations in the supplied nutrient solution are reduced in order to lower the above ratio to values close to 1, thus avoiding accumulation of these ions in the root environment, their uptake may become insufficient for the plants. In contrast, the acquisition of monovalent ions which are taken up actively by the plants is substantially reduced only when their concentrations in the nutrient solution are diminished to values markedly lower than their absorption concentrations. This topic has already been discussed in section 5.4. Nevertheless, if the nutrient losses due to precipitation or immobilization in the root environment are neglected, it can be easily proved that in free drainage systems the mean concentration of a nutrient in the root zone during a particular time interval is given by the equation

$$C_{di} = (C_{si} - C_{ai}(1-d))/d \qquad (9)$$

where C_{di} is the mean concentration of the i nutrient in the root environment and in the leachate during that time; C_{si} is the concentration of the i

nutrient in the supplied nutrient solution; C_{ai} is the mean absorption concentration of the i nutrient during that time; and d is the mean proportion of drainage to the supplied solution (0-1) during that time. Thus, if during a particular time interval the proportion of drainage solution (d) is 0.2, the C_{sCa} and the C_{aCa} are 150 and 100 mg L^{-1}, respectively, and the C_{sK} and the C_{aK} are 300 and 280 mg L^{-1}, respectively, using (9) it can be calculated that C_{dCa} and C_{dK} are equal to 350 and 380 mg L^{-1}, respectively. Consequently, whilst the K : Ca ratio in the supplied nutrient solution (C_{sK}/C_{sCa}) is 300 : 150 = 2 (w/w), the K : Ca ratio in the root environment (C_{dK}/C_{dCa}) is 380 : 350 = 1.09 (w/w).

6. Nutrient Solution Calculation

In commercial hydroponics, the preparation of a nutrient solution is performed by diluting stock solutions of fertilizers with irrigation water at a particular dilution ratio (e.g. 1 : 100) at any time that the plants should be irrigated (Cooper, 1979; Graves, 1983; Sonneveld, 1989). The dilution ratio may be adjusted either directly or indirectly by entering a target EC to the controlling system. The main advantages obtained by employing stock solutions are the use of much smaller tanks and the reduced labour requirements for the preparation of a particular amount of nutrient solution.

According to the standard practice, the fertilizers needed to prepare a nutrient solution are distributed over at least two stock solutions, usually defined as A and B. This practice is aimed at separating the calcium ions from the phosphate and sulphate ions in order to prevent precipitation of calcium sulphate and calcium phosphate which is expected to occur if these ions are mixed in concentrated solutions (De Rijck and Schrevens, 1998b). A separate stock solution containing only an acid (usually HNO_3) for the regulation of pH is also necessary. Moreover, in some cases an additional tank containing a base (usually potassium bicarbonate) is also used to control too low pH levels in the root environment of the crop (see section 5.3.2). However, in large and modern commercial hydroponic units, a separate stock solution tank is usually used for each single fertilizer. Thus, a fertilizer mixing unit of this type operates by employing about 7-10 stock solutions for the macronutrients and, usually, 5-6 stock solutions for the micronutrients. However, in some units all micronutrients may be placed in one stock solution. The use of single fertilizer stock solutions enables full automation of the nutrient solution management, especially when frequent adjustments in the composition of the solution supplied to the crop are required (Savvas and Adamidis, 1999). Nevertheless, to prepare a fresh nutrient solution having a desired composition, it is essential to correctly calculate the amount of fertilizers that should be added per volume of irrigation water, taking into account the ratio in which the stock solutions will be diluted.

Some authors suggest nutrient solution formulae in terms of fertilizer amounts to be added to a particular volume of water (e.g. Arnon and

Hoagland, 1950). Such formulae can be realized even by a trained technician who has only a limited chemical background. Thus, when one has to prepare a universal nutrient solution in order to grow plants in a water culture and is not interested in specific target characteristics (e.g. a specific total salt concentration, or certain nutrient ratios) the use of such a formula is very convenient. However, this is possible only when deionized or rain or at least very soft water is used to prepare the nutrient solution, otherwise the mineral composition of the water employed should be taken into account. Moreover, as has been already discussed in previous sections, to obtain high yields and good quality in commercial crops grown hydroponically, the nutrient solution supplied to the plants must be specific for the particular crop, the growth stage, the climatic conditions, the substrate or hydroponic system used, etc. Obviously, the nutritional management of a soilless cultivated crop according to this concept is not compatible with the use of a standard formula suggested in the literature, especially when the water used to prepare the nutrient solution contains substantial amounts of inorganic ions (Ca^{2+}, Mg^{2+}, Na^+, Cl^-, HCO_3^-, etc.). Therefore, various investigators proposed several methods of calculating a nutrient solution satisfying particular requirements which are given as target values such as EC, nutrient concentrations, relative proportions of nutrients, etc. (Steiner, 1961, 1984; Cooper, 1979; Sonneveld, 1982; Schwarz, 1995; Resh, 1997; Hannan, 1998). However, all these methods require some understanding of chemistry and are more or less arduous, because routine calculations have to be repeated in each particular case. Therefore, Savvas and Adamidis (1999) proposed a standardized method to calculate nutrient solutions corresponding to any desired characteristic by taking into account the mineral composition of the water used. This method, due to the complete standardization of the calculations, can be used in a computer algorithm which enables fully automated formulation of a nutrient solution composition and calculation of the fertilizers needed to prepare it. Moreover, if single fertilizer stock solutions are used and a suitable computer controlled system for the preparation and the supply of the nutrient solution is available, this algorithm enables automated preparation of a nutrient solution merely by defining the target characteristics. The target values can change as frequently as desired without removing the currently used stock solutions in order to change the fertilizer concentrations in them. This method is presented below.

As pointed out by Savvas and Adamidis (1999), the composition of a nutrient solution is completely defined if target values for the following solution characteristics are given: (i) EC of the nutrient solution, (ii) pH of the nutrient solution, (iii) the K : Ca : Mg ratio, (iv) the N : K ratio, (v) the ratio of P to total nutrient anions, (vi) concentration ratio of ammonium nitrogen to total nitrogen and (vii) micronutrient concentrations. In this concept, priority is given to the N : K ratio rather than to the nutrient anion proportions, for the reasons outlined in section 5.5. If target values for the above characteristics are given and the mineral composition of the water used to prepare the nutrient solution is precisely known, it is possible to calculate the target concentrations of the nutrient solution as follows:

1. Using (1), the target nutrient solution EC (E_t in dS m^{-1}) is converted into total salt concentration (C_t), expressed in meq L^{-1} of diluted salts (see section 5.2.1).
2. The bicarbonate concentration (C_b in meq L^{-1}) which is established in the nutrient solution after adjustment of the desired pH is calculated using (7) as described in section 5.3.2.
3. The target macroelement concentrations in the nutrient solution are estimated using the equations in Table 5. In the equations of Table 5, the target K : Ca : Mg ratio is defined as X : Y : Z, the desired N : K as R, the target ratio of P to total nutrient anions as P_r, the ratio of ammonium nitrogen to total nitrogen as N_r, and the concentrations (meq L^{-1}) of Na$^+$ and Cl$^-$ in the tap water as $[Na^+]_w$ and $[Cl^-]_w$, respectively, whilst $[\cdot]_t$ denotes the target nutrient solution concentration (meq L^{-1}) of the ion in the argument. The calculations are based on the assumption that no Na$^+$ and Cl$^-$ are added when constituting the nutrient solution. Any Na$^+$ and Cl$^-$ input owing to fertilizer impurities is neglected. Consequently, the concentrations of these ions in the nutrient solution are identical to those measured in the tap water. Moreover, since the target pH of the nutrient solutions supplied to commercial hydroponic crops is almost always higher than 5, the target H$^+$ concentration (in fact H$_3$O$^+$) is negligible (less than 10^{-2} meq L^{-1}).
4. The dosages of the nutrients (meq L^{-1}), which should be added via the fertilizers to the irrigation water to prepare a nutrient solution satisfying the given requirements, are calculated using the equation

$$[\cdot]_f = [\cdot]_t - [\cdot]_w \qquad (10)$$

where $[\cdot]$ denotes a value in meq L^{-1} of the ion in the argument and the subscripts f, t, and w indicate dosage of nutrients to be added through fertilizers, target concentration, and concentration in tap

Table 5. The formulae used to estimate target macroelement concentrations in nutrient solutions based on given target values for EC, pH, and macronutrient ratios.

Cations	Anions
$[K^+]_t = \dfrac{X(C_t - [Na^+]_w)}{X + Y + Z + N_r RX}$	$[NO_3^-]_t = R[K^+]_t - [NH_4^+]_t$
$[Ca^{2+}]_t = [K^+]_t Y X^{-1}$	$[H_2PO_4^-]_t = P_r(C_t - C_b - [Cl^-]_w)$
$[Mg^{2+}]_t = [K^+]_t Z X^{-1}$	$[SO_4^{2-}]_t = C_t - [NO_3^-]_t - [H_2PO_4^-]_t - C_b - [Cl^-]_w$
$[NH_4^+]_t = N_r R[K^+]_t$	$[HCO_3^-]_t = C_b$
$[Na^+]_t = [Na^+]_w$	$[Cl^-]_t = [Cl^-]_w$
$[H^+]_t = 0$ (less than 10^{-2} meq L^{-1})	

water, respectively. Especially for $[HCO_3^-]_f$ and $[H^+]_f$ (8) is not valid, since HCO_3^- is never added when constituting nutrient solutions (see section 5.3.3) and the addition of acids (H^+) results in a reduction of the initial HCO_3^- concentration whilst most of the added hydrogen ions are consumed in the neutralization of the bicarbonates. Therefore,

$$[HCO_3^-]_f = 0 \quad \text{and} \quad [H^+]_f = [HCO_3^-]_w - C_b \tag{11}$$

5. When all the dosages of nutrients to be added through fertilizers ($[\cdot]_f$) are known, the required dosages of fertilizers ($[rs]$ in meq L^{-1}; r = Ca, Mg, K, NH$_4$, H; s = SO$_4$, NO$_3$, H$_2$PO$_4$) can be readily calculated using the equations given in Table 6.

The calculations included in the steps 1-5 can be performed more conveniently by using a table to introduce all the results in the sequence cited above, following an arrangement as shown in Table 7. In Table 8 a numerical example is given, concerning the estimation of the fertilizer dosages needed to prepare a nutrient solution for roses.

Table 6. The equations used to estimate the dosages of fertilizers (meq L^{-1}) required to achieve the target values of EC, pH and macronutrient ratios in nutrient solutions

---

$[Ca(NO_3)_2] = [Ca^{2+}]_f$

$[MgSO_4] = [Mg^{2+}]_f$ if $[SO_4^{2-}]_f > [Mg^{2+}]_f$; $[MgSO_4] = [SO_4^{2-}]_f$ if $[SO_4^{2-}]_f < [Mg^{2+}]_f$

$[Mg(NO_3)_2] = 0$ if $[SO_4^{2-}]_f > [Mg^{2+}]_f$; $[Mg(NO_3)_2] = [Mg^{2+}]_f - [MgSO_4]$ if $[SO_4^{2-}]_f < [Mg^{2+}]_f$

$[K_2SO_4] = [SO_4^{2-}]_f - [Mg^{2+}]_f$ if $[SO_4^{2-}]_f > [Mg^{2+}]_f$; $[K_2SO_4] = 0$ if $[SO_4^{2-}]_f < [Mg^{2+}]_f$

$[KH_2PO_4] = [H_2PO^+]_f$ if P is added as KH$_2$PO$_4$; $[KH_2PO_4] = 0$ if P is added as H$_3$PO$_4$

$[H_3PO_4] = 0$ if P is added as KH$_2$PO$_4$; $[H_3PO_4] = [H_2PO_4^-]_f$ if P is added as H$_3$PO$_4$

$[KNO_3] = [K^+]_f - [K_2SO_4] - [KH_2PO_4]$

$[NH_4NO_3] = [NH_4^+]_f$

$[HNO_3] = [H]_f - [H_3PO_4]$

---

6. Based on the previously calculated dosages of fertilizers ($[rs]$), the amounts of fertilizers (W_{rs} in kg) needed to prepare certain quantities (V_{rs} in m^3) of stock solutions are calculated using the following formula:

$$W_{rs\prime} = 10^{-3}[rs]\, E_{rs}\, V_{rs}\, A_{rs\prime} \tag{12}$$

r = Ca, Mg, K, NH$_4$, H; s = SO$_4$, NO$_3$, H$_2$PO$_4$

where E_{rs} is the equivalent weight of the rs^{th} fertilizer and A_{rs} denotes the dilution ratio of the stock solution containing the rs^{th} fertilizer. As shown in Table 1, the chemical formula of the commercially used calcium nitrate is 5[Ca(NO$_3$)$_2$.2H$_2$O]NH$_4$NO$_3$. Therefore, when using (12) to calculate the weight of calcium nitrate

Table 7. A tabular representation of the calculations required to estimate the dosages of macronutrient fertilizers that should be added to tap water to obtain a nutrient solution with the target EC, pH and macronutrient ratios (adapted from Savvas and Adamidis, 1999)

Anions/ cations	CCS	CCW	CAF	SO_4^{2-}	NO_3^-	$H_2PO_4^-$	HCO_3^-	Cl^-
CAS	$\Sigma[ions]_t$			$[SO_4^{2-}]_t$	$[NO_3^-]_t$	$[H_2PO_4^-]_t$	C_b	$[Cl^-]_w$
CAW		$\Sigma[ions]_w$		$[SO_4^{2-}]_w$	$[NO_3^-]_w$	$[H_2PO_4^-]_w$	$[HCO_3^-]_w$	$[Cl^-]_w$
AAF			$\Sigma[ions]_f$	$[SO_4^{2-}]_f$	$[NO_3^-]_f$	$[H_2PO_4^-]_f$	0	0
Ca^{2+}	$[Ca^{2+}]_t$	$[Ca^{2+}]_w$	$[Ca^{2+}]_f$	0	$[Ca(NO_3)_2]$	0	0	0
Mg^{2+}	$[Mg^{2+}]_t$	$[Mg^{2+}]_w$	$[Mg^{2+}]_f$	$[MgSO_4]$	$[Mg(NO_3)_2]$	0	0	0
K^+	$[K^+]_t$	$[K^+]_w$	$[K^+]_f$	$[K_2SO_4]$	$[KNO_3]$	$[KH_2PO_4]$	0	0
NH_4^+	$[NH_4^+]_t$	$[NH_4^+]_w$	$[NH_4^+]_f$	0	$[NH_4NO_3]$	0	0	0
Na^+	$[Na^+]_w$	$[Na^+]_w$	0	0	0	0	0	0
H^+	0	0	$[H^+]_f$	0	$[HNO_3]$	$[H_3PO_4]$	0	0

Notation

The column labeled by CCS and the row labeled by CAS indicate the concentrations (meq L^{-1}) of cations and anions in the target nutrient solution, respectively, which are calculated using the equations of Table 5.

The column labeled by CCW and the row labeled by CAW indicate the concentrations (meq L^{-1}) of cations and anions in the tap water, respectively.

The column labeled by CAF and the row labeled by AAF indicate dosages (meq L^{-1}) of cations and anions to be added through fertilizers, respectively, which should be calculated as described in section 6, step 5.

The crossing points between cations (rows) and anions (columns) indicate dosages (meq L^{-1}) of fertilizers to be added to tap water in order to achieve the target EC, pH and macronutrient ratios in the nutrient solution, which are calculated using the equations of Table 6.

Table 8. Calculation of the dosages (meq L^{-1}) of macronutrient fertilizers that should be added to tap water of a particular ionic composition to prepare a nutrient solution for roses whose composition corresponds to given* target values of pH, EC, and macronutrient ratios

Anions/ cations	CCS	CCW	CAF	SO_4^{2-}	NO_3^-	$H_2PO_4^-$	HCO_3^-	Cl^-
CAS	17.19			2.44	11.46	1.21	0.79	1.30
CAW		6.57		1.27	0.20	0	3.80	1.30
AAF			13.64	1.17	11.26	1.21	0	0
Ca^{2+}	6.84	3.30	3.54	0	3.54	0	0	0
Mg^{2+}	2.63	1.74	0.89	0.89	0	0	0	0
K^+	4.73	0.10	4.63	0.28	3.14	1.21	0	0
NH_4^+	1.56	0	1.56	0	1.56	0	0	0
Na^+	1.43	1.43	0	0	0	0	0	0
H^+	0	0	3.01	0	3.01	0	0	0

*The calculations are based on the following data: i) target EC =1.90 dS m^{-1}; ii) target pH = 5.7; iii) target concentration ratios (meq/meq): K : Ca : Mg = 4.5 : 6.5 : 2.5, N : K = 2.75, $[NH_4^+]/([NH_4^+]+[NO_3^-]) = 0.12$, $[H_2PO_4^-]/([SO_4^{2-}]+[NO_3^-]+[H_2PO_4^-]) = 0.08$; iv) ionic composition of tap water (meq L^{-1}) : Ca = 3.30, Mg=1.74, K=0.10, NH_4^+ = 0, Na = 1.43, SO_4^{2-} = 1.27, NO_3^- = 0.20, $H_2PO_4^-$ = 0, HCO_3^- = 3.80, Cl^- = 1.30; v) tap water pH = 7.90.

$W_{Ca(NO_3)_2}$ the equivalent weight corresponding to Ca (108.05) should be introduced as $E_{Ca(NO_3)_2}$. Moreover, to take into account the amount of NH_4NO_3 that is included in the commercially used calcium nitrate, $[NH_4NO_3]$ should be replaced by $[NH_4NO_3] - 0.1[Ca(NO_3)_2]$ when using (12) to calculate the weight of ammonium nitrate ($W_{NH_4NO_3}$).

7. The amounts of micronutrient fertilizers which are required to achieve the target trace element concentrations are calculated using the formula

$$W_j = (10^3 n_j)^{-1} [G]_j\, M_j\, V_j\, A_j, \quad j = Fe, Mn, Zn, Cu, B \text{ and } Mo \qquad (13)$$

where W_j is the amount (g) of the fertilizer containing the j (Fe, Mn, Zn, Cu, B and Mo) micronutrient which is needed to prepare a certain volume (V_j in m³) of stock solution containing the j micronutrient; n_j is the number of gr-atoms of the j micronutrient in one mol of the related fertilizer; $[G]_j$ is the difference between the target concentration (µmol L^{-1}) of the j micronutrient in the nutrient solution and that found in the tap water; M_j is the molecular weight of the fertilizer containing the j micronutrient; and A_j is the dilution ratio (concentration factor) of the stock solution containing the fertilizer of the j micronutrient. However, the concentration of Fe in the tap water is not taken into account when applying (13) to calculate the weight of iron fertilizer because after addition of the stock solutions most of this iron precipitates, mainly in the form of iron phosphate (De Rijck and Schrevens, 1998c).

The results of the water analysis should be taken into account when specifying the target values of EC, macronutrient concentration ratios, and micronutrient concentrations. In a few exceptional cases, some target values may be not feasible due to the composition of the tap water. In these cases, the target values should conform to the available water quality. For instance, if the water contains 36.5 mg L^{-1} Mg, which is equal to 3 meq L^{-1} Mg, it is not possible to obtain a nutrient solution having a target EC of 1.8 dS m^{-1} and a K : Ca : Mg equal to 2.5 : 3 : 1 (on equivalent basis). In this case, one should either increase the target EC or decrease the target (K+Ca) : Mg ratio, or install a reverse osmosis system to change the mineral composition of the tap water.

The method proposed above for calculating the composition of nutrient solutions and the amounts of the required fertilizers is based on the assumption that no Na$^+$, Cl$^-$, HCO$_3^-$ and Si is added to the nutrient solution through fertilizers, no H_2SO_4 or HCl is used to adjust the pH, and only ammonium nitrate is used to supply NH_4^+. Moreover, it is assumed that no calcium sulphate, calcium phosphate and magnesium phosphate are used as fertilizers, because of their insufficient solubility to the water. However, the above calculation scheme is capable of further extension to make provision for Na$^+$, Cl$^-$ and Si addition also, for use of H_2SO_4 or HCl to adjust pH, as well as for the supply of ammonium and phosphorus in the form of $(NH_4)_2SO_4$ or $NH_4H_2PO_4$.

7. Perspectives

Future developments in hydroponics are mainly focused on a more extensive automation of the nutrient solution management, particularly in closed systems in which the excess nutrient solution is recycled, as well as on a complete standardization of the substrate analysis in order to obtain more reliable results and to facilitate their interpretation.

The invention of more sophisticated automation equipment is very important especially for the development of more efficient techniques of nutrient solution recycling. The currently applied techniques to automatically replenish the drain solution are based on real-time measurements of only one variable, EC. A widely used technique to replenish and reuse the drain solution involves mixing the latter with water at a ratio which is automatically adjusted by means of on-line measurements of the EC of the mixture in order to achieve a preset EC (E_m). Thus, the quantities of stock solutions which are injected to the mixture of drain solution and water in order to obtain the target EC (E_t) are constant in all watering applications. According to another method, the EC (E_r) and the volume (V_r) of the drain solution to be reused are measured in real-time prior to each watering application. Subsequently, the amount of stock solutions which are required to obtain the target EC (E_t) are automatically calculated as functions of E_r and V_r and injected to the mixture of drain solution and water by means of a computer controlled system (Savvas and Manos, 1999).

The nutrient and salt concentrations in the drain solution are dependent on current plant uptake which fluctuates widely during the cropping period, as has already been pointed out in section 5.4. However, no reliable equipment to monitor the individual nutrient concentrations in the drain solution is currently available (Albery et al. 1985; Gieling et al. 1988; Morard, 1996; Savvas and Manos, 1999). This is the main obstacle to further improvements in recycling techniques, since such facilities would enable automated calculation and injection of nutrients and water to the drain solution at amounts resulting in the reconstitution of the target nutrient composition. Hence, a prerequisite for further developments in the automation of closed hydroponics is the invention of reliable equipment enabling real-time measurements of all or at least some nutrients in the drain solution.

Soil-borne diseases may occur not only in soil grown crops but also in hydroponics, if root infecting pathogens are transmitted to the rooting medium. Especially in closed hydroponic systems, the infection of only a few or even one plant may result in the dispersal of the plant pathogen over the whole crop via the recycled nutrient solution. Thus, the development of reliable techniques to provide efficient disinfection of the drainage prior to reuse is a challenge for the hydroponic technology. A review of methods employed in the disinfection of the drain solution in closed hydroponic systems has been given by Runia (1995). According to recent reports, heat treatment, ultra violet radiation and slow sand filtration seem to be the most promising methods of nutrient solution disinfection in closed hydroponics (Runia et al. 1996; Runia and Amsing, 1999; Van Os et al. 1996; Van Os, 1999) but further improvements of the currently applied technologies are required.

When the plants are grown in recirculating nutrient solution or even on rockwool, the nutritional status in the root environment can be readily determined by sampling nutrient solution directly from the root zone. However, when the plants are grown on substrates such as perlite, pumice, zeolite, peat, choir dust, etc., a direct analysis of the nutrient solution which is present in the root environment is not possible. When such porous materials are employed, substrate analysis is an important tool in managing the nutrition of the crop. In these cases, the nutrient status in the root zone is determined by means of a water extract which is obtained from a substrate sample (Sonneveld, 1999). However, to obtain reliable results and to interpret them, the analytical methods should be standardized. Moreover, skilful experimental work is needed in order to establish guidelines for each particular substrate related to both sampling and interpretation of the analytical data.

References

Adams, P. 1980. Nutrient uptake by cucumbers from recirculating solution. *Acta Horticulturae*, **98**: 119-126.

Adams, P. 1988. Some responses of tomatoes grown in NFT to sodium chloride. *In:* Proceedings, 7th International Congress on Soilless Culture. ISOSC, Wageningen, The Netherlands, pp. 59-71.

Adams, P. 1991. Effects of increasing the salinity of the nutrient solution with major nutrients or sodium chloride on the yield, quality and composition of tomatoes grown in rockwool. *Journal of Horticultural Science* **66**: 201-207.

Adams, P. and D.M. Massey. 1984. Nutrient uptake by tomatoes from recirculating solutions. *In:* Proceedings, 6th International Congress on Soilless Culture. ISOSC, Wageningen, The Netherlands, pp. 71-79.

Adams, P. and M.M. Grimmett. 1986. Some responses of tomatoes to the concentration of potassium in recirculating nutrient solutions. *Acta Horticulturae* **178**: 29-35.

Adams, P. and L.C. Ho. 1989. Effects of constant and fluctuating salinity on the yield, quality and calcium status of tomatoes. *Journal of Horticultural Science* **64**: 725-732.

Adams, P. and L.C. Ho. 1992. The susceptibility of modern tomato cultivars to blossom-end rot in relation to salinity. *Journal of Horticultural Science* **67**: 827-839.

Adams, P., Graves, C.J. and G.W. Winsor. 1989. Some responses of cucumber, grown in beds of peat, to micronutrients and pH. *Journal of Horticultural Science* **64**: 293-299.

Albery, W.J., Haggett, B.G.D. and R. Svanberg. 1985. The development of sensors for hydroponics. *Biosensors* **1**: 369-397.

Benoit, F. and N. Ceustermans. 1995. Horticultural aspects of ecological soilless growing methods. *Acta Horticulturae* **396**: 11-23.

Ben-Zioni, A.Y., Vaadia, Y. and S.H. Lips. 1971. Nitrate uptake by roots as regulated by nitrate reduction products of the shoots. *Physiologia Plantarum* **24**: 288-290.

Blaabjerg, J. 1983. Physical and chemical compositions of the inactive growing medium Grodan and its fields of application and extension. *Acta Horticulturae* **133**: 53-57.

Boertje. G.A. 1986. The effect of the nutrient concentration in the propagation of tomatoes and cucumbers on rockwool. *Acta Horticulturae* **178**: 59-65.

Caro, M., Cruz, V., Cuartero, J., Estañ M.T. and M.C. Bolarin. 1991. Salinity tolerance of normal-fruited and cherry tomato cultivars. *Plant and Soil* **136**: 249-255.

Challinor, P.F., Fuller, M.P. and J. Parkinson. 1996. Growth and development of sweet pepper plants on unloaded and nutrient loaded clinoptilolite zeolite. In: Proceedings, 9th International Congress on Soilless Culture. ISOSC, Wageningen, The Netherlands, pp. 105-122.

Chance, W.O. (III), Somda, Z.C. and H.A. Mills. 1999. Effect of nitrogen form during the flowering period on zucchini squash growth and nutrient element uptake. Journal of Plant Nutrition 22: 597-607.

Cooper, A.J. 1975. Crop production in recirculating nutrient solution. Scientia Horticulturae 3: 251-258.

Cooper, A.J. 1979. The ABC of NFT. Grower Books, London.

Daum, D. and M.K. Schenk. 1998. Influence of nutrient solution pH on N_2O and N_2 emissions from a soilless culture system. Plant and Soil 203: 279-287.

David, P.P., Nelson, P.V. and D.C. Sanders. 1994. A humic acid improves growth of tomato seedling in solution culture. Journal of Plant Nutrition 17: 173-184.

De Kreij, C. 1995. Latest insights into water and nutrient control in soilless cultivation. Acta Horticulturae 408: 47-61.

De Kreij, C. 1999. Production, blossom-end rot, and cation uptake of sweet pepper as affected by sodium, cation ratio, and EC of the nutrient solution. Gartenbauwissenschaft 64: 158-164.

De Kreij, C. and B. Van der Hoeven. 1996. Effect of humic substances, pH and its control on growth of chrysanthemum in aeroponics. In: Proceedings, 9th International Congress on Soilless Culture. ISOSC, Wageningen, The Netherlands, pp. 207-230.

De Kreij, C. and P.C. Van Os. 1988. Production and quality of gerbera in rockwool as affected by electrical conductivity of the nutrient solution. In: Proceedings, 7th International Congress on Soilless Culture. ISOSC, Wageningen, The Netherlands, pp. 255-264.

De Kreij, C., Janse, J., Van Goor, B.J. and J.D.J Van Doesburg. 1992. The incidence of calcium oxalate crystals in fruit walls of tomato (Lycopersicon esculentum Mill.) as affected by humidity, phosphate and calcium supply. Journal of Horticultural Science 67: 45-50.

De Kreij, C., Voogt, W., Van Den Bos, A.L. and R. Baas. 1997. Voedingsoplossingen gesloten teeltsystemen (Nutrient solutions for closed cultivation systems). Brochures 1-16. Research Station for Floriculture and Glasshouse Vegetables (PBG), Naaldwijk, The Netherlands.

De Rijck, G. and E. Schrevens. 1997a. Elemental bioavailability in nutrient solutions in relation to dissociation reactions. Journal of Plant Nutrition 20: 901-910.

De Rijck, G. and E. Schrevens. 1997b. pH influenced by the elemental composition of nutrient solutions. Journal of Plant Nutrition 20: 911-923.

De Rijck, G. and E. Schrevens. 1998a. Elemental bioavailability in nutrient solutions in relation to complexation reactions. Journal of Plant Nutrition 21: 849-859.

De Rijck, G. and E. Schrevens. 1998b. Elemental bioavailability in nutrient solutions in relation to precipitation reactions. Journal of Plant Nutrition 21: 2103-2113.

De Rijck, G. and E. Schrevens. 1998c. Composition of the mineral composition of twelve standard nutrient solutions. Journal of Plant Nutrition 21: 2115-2125.

De Rijck, G. and E. Schrevens, 1999. Guidelines to optimize the macrocation and macroanion composition of nutrient solutions using mixture theory. Journal of Agricultural Engineering Research 72: 355-362.

Ehret, D.L. and L.C. Ho. 1986a. The effects of salinity on dry matter partitioning and fruit growth in tomatoes grown in nutrient film culture. Journal of Horticultural Science 61: 361-367.

Ehret, D.L. and L.C. Ho. 1986b. Translocation of calcium in relation to tomato fruit growth. Annals of Botany 58: 679-688.

Elia, A., Conversa, G., Serio, F. and P. Santamaria. 1996. Response of eggplant to NH_4 : NO_3 ratio. In: Proceedings, 9th International Congress on Soilless Culture. ISOSC, Wageningen, The Netherlands, pp. 167-180.

Errebhi, M. and G.E. Wilcox. 1990. Tomato growth and nutrient uptake pattern as influenced by nitrogen form ratio. *Journal of Plant Nutrition* **13**: 1031-1034.
Feigin, A. 1990. Interactive effects of salinity and ammonium/nitrate ratio on growth and chemical composition of melon plants. *Journal of Plant Nutrition* **13**: 1257-1269.
Feigin, A., Zwibel, M., Rylski, I., Zamir, N. and N. Levav. 1979. The effect of ammonium/nitrate ratio in the nutrient solution on tomato yield and quality. *Acta Horticulturae* **98**: 149-160.
Feigin, A., Ginzburg, C., Ackerman, A. and S. Gilead. 1984. Response of roses growing in a volcanic rock substrate to different NH_4/NO_3 ratios in the nutrient solution. In: Proceedings, 6th International Congress on Soilless Culture. ISOSC, Wageningen, The Netherlands, pp. 207-213.
Gaugh, C. and G.E. Hobson. 1990. A comparison of the productivity, quality, shelf-life characteristics and consumer reaction to the crop from cherry tomato plants grown at different levels of salinity. *Journal of Horticultural Science* **65**: 431-439.
Gericke, W.F. 1937. Hydroponics-crop production in liquid culture media. *Science* **85**: 177-178.
Gericke, W.F. 1938. Crop production without soil. *Nature* **141**: 536-540.
Gieling, T.H., Van Os, E.A. and A. De Jager. 1988. The application of chemo-sensors and bio-sensors for soilless cultures. *Acta Horticulturae* **230**: 357-361.
Gislerod, H.R. and P. Adams. 1983. Diurnal variations in the oxygen content and acid requirement of recirculating nutrient solutions and in the uptake of water and potassium by cucumber and tomato plants. *Scientia Horticulturae* **21**: 311-321.
Gizas, G., Savvas, D. and I. Mitsios. 1999. Availability of macrocations in perlite and pumice as influenced by the application of nutrient solutions having different cation concentration ratios. International Symposium on Growing Media and Hydroponics. August 31- September 6, 1999. Kassandra, Halkidiki, Greece. *Acta Horticulturae (in press).*
Gosselin, A., Charbonneau, J., Vezina, L.-P. and M.-J. Trudel. 1988. Restrictions of vegetative growth of tomato plants imposed by altering the nitrogen concentration and the electrical conductivity of the nutrient solution. *Acta Horticulturae* **222**: 71-78.
Grattan, S.R. and C.M. Grieve. 1999. Salinity-mineral relations in horticultural crops. *Scientia Horticulturae* **78**: 127-157.
Graves, C.J. 1983. The nutrient film technique. *Horticultural Reviews* **5**: 1-44.
Hall, D.A. 1983. The influence of nitrogen concentration and salinity of recirculating solutions on the early-season vigour and productivity of glasshouse tomatoes. *Journal of Horticultural Science* **58**: 411-415.
Hanan, J.J. 1998. *Greenhouses: Advanced Technology for Protected Cultivation.* CRC Press, Boca Raton, Florida, USA.
Helal, H.M. and K. Mengel. 1981. Interaction between light intensity and NaCl salinity and their effects on growth, CO_2 assimilation, and photosynthate conversion in young broad beans. *Plant Physiology* **67**: 999-1002.
Hendrix, J.E. 1967. The effect of pH on the uptake and accumulation of phosphate and sulphate ions by bean plants. *American Journal of Botany* **54**: 560-564.
Heuer, B. 1991. Growth, photosynthesis and protein content in cucumber plants as affected by supplied nitrogen form. *Journal of Plant Nutrition* **14**: 363-373.
Hewitt, E.J. 1966. Sand and water culture methods used in the study of plant nutrition. Technical communication No. 22 (revised). Commonwealth Bureau of Horticulture and Plantation Crops, East Malling, Maidstone, Kent, UK.
Hoagland, D.R. and D.I. Arnon. 1950. The water-culture method for growing plants without soil. *Calif. Agric. Exp. St., Circ.* **347** (Revised by D.I. Arnon).
Islam, A.K.M.S., Edwards, D.G. and C.J. Asher. 1980. pH optima for crop growth. Results of a flowing solution culture experiment with six species. *Plant and Soil* **54**: 339-357.
Jacobson, L. 1951. Maintenance of iron supply in nutrient solutions by a single addition of ferric potassium ethylenediamine tetra-acetate. *Plant Physiology* **26**: 411.
Johnson, R.W., Dixon, M.A. and D.R. Lee. 1992. Water relations of the tomato fruit during growth. *Plant Cell and Environment* **15**: 947-953.

Jones (Jr), J.B. 1982. Hydroponics: its history and use in plant nutrition studies. *Journal of Plant Nutrition* **5**: 1003-1030.
Knop, W. 1859. Ein Vegetationsversuch. *Die Landwirtschaftlichen Versuchs-Stationen* **1**: 181-202.
Kirkby, E.A. and K. Mengel. 1967. Ionic balance in different tissues of the tomato plant in relation to nitrate, urea or ammonium nutrition. *Plant Physiology* **42**: 6-14.
Letey, J., Jarrell, W.M. and N. Valoras. 1982. Nitrogen and water uptake patterns and growth of plants at various minimum solution nitrate concentrations. *Journal of Plant Nutrition* **5**: 73-89.
Liang, Y.C. 1999. Effects of silicon on enzyme activity and sodium, potassium and calcium concentration in barley under salt stress. *Plant and Soil* **209**: 217-224.
Maas, E.V. and G.J. Hoffman. 1977. Crop salt tolerance—current assessment. *Journal of the Irrigation and Drainage Division, ASCE* **103** (IR2) 115-134.
Maaswinkel, R.H.M. and G.W.H. Welles. 1986. Factors influencing glassiness in lettuce. *Netherlands Journal of Agricultural Science* **34**: 51-65.
Marschner, H. 1995. *Mineral Nutrition of Higher Plants*. 2nd Edition, Academic Press, London, UK.
Massey, D.M. and G.W. Winsor. 1980a. Some responses of tomatoes to nitrogen in recirculating solutions. *Acta Horticulture*, **98**: 127-133.
Massey, D.M. and G.W. Winsor. 1980b. Some responses of tomato plants to phosphorus concentration in nutrient film culture. *In:* Proceedings, 5th International Congress on Soilless Culture. ISOSC, Wageningen, The Netherlands, pp. 205-213.
Massey, D.M., Hayward, A.C. and G.W. Winsor. 1984. Some responses of tomatoes to salinity in nutrient-film culture. Annual Report of the Glasshouse Crops Research Institute for 1983: 60-62.
Mavrogianopoulos, G.N., Spanakis, J. and P. Tsikalas. 1999. Effect of carbon dioxide enrichment and salinity on photosynthesis and yield in melon. *Scientia Horticulturae* **79**: 51-63.
Meiri, A. and A. Polljakoff-Mayber. 1970. Effects of various salinity regimes on growth, leaf expansion and transpiration rate of bean plants. *Soil Science* **109**: 26-34.
Mengel, K. and E.A. Kirkby. 1987. *Principles of Plant Nutrition*. International Potash Institute, Bern, Switzerland.
Menzies, J., Bowen, P. and D. Ehret. 1992. Foliar applications of potassium silicate reduce severity of powdery mildew on cucumber, muskmelon, and zucchini squash. *Journal of the American Society for Horticultural Science*, **117**: 902-905.
Miyake, Y. and E. Takahashi. 1978. Silicon deficiency of tomato plant. *Soil Science and Plant Nutrition* **24**: 175-189.
Miyake, Y. and E. Takahashi. 1983. Effect of silicon on the growth of solution-cultured cucumber plants. *Soil Science and Plant Nutrition* **29**: 71-83.
Mizrahi, Y. and D. Pasternak. 1985. Effects of salinity on quality of various agricultural crops. *Plant and Soil* **89**: 301-307.
Morard, P. 1996. Possible use of ion selective electrodes for nutrient solutions in recirculated systems. *In:* Proceedings, 9th International Congress on Soilless Culture. ISOSC, Wageningen, The Netherlands, 1996, pp. 291-298.
Morard, P. and B. Benavides. 1990. Relative accumulation of macronutrient ions in different parts of cucumber (*Cucumis sativus*). *Scientia Horticulturae* **44**: 17-30.
Moustafa, A.T. and J.V. Morgan. 1983. Influence of solution concentration on growth, flower quality and nutrient uptake in spray chrysanthemum. *Acta Horticulturae* **133**: 13-24.
Nelson, P.V. 1998. *Greenhouse Operation and Management* (5th Edition). Prentice-Hall Inc., New Jersey, USA.
Nichols, M.A., Fisher, K.J., Morgan, L.S. and A. Simon. 1994. Osmotic stress, yield and quality of hydroponic tomatoes. *Acta Horticulturae* **361**: 302-310.
Nukaya, A., Voogt, W. and C. Sonneveld. 1991. Effects of NO_3, SO_4, and Cl ratios on tomatoes grown in recirculating system. *Acta Horticulturae* **294**: 297-304.

Nukaya, A., Goto, K., Jang, H., Kano, A. and K. Ohkawa. 1995. Effect of K/Ca ratio in the nutrient solution on incidence of blossom-end rot and gold specks of tomato fruit grown in rockwool. *Acta Horticulturae* **396**: 123-130.

Ottosson, L. 1977. Vegetable production on mineral wool. *Acta Horticulturae* **58**: 147-152.

Papadopoulos, A.P. and S. Pararajasingham. 1998. Effects of controlling pH with hydrochloric acid on the growth, yield and fruit quality of greenhouse tomato grown by nutrient film technique. *Hort Technology* **8**: 193-198.

Papadopoulos, A.P., Pararajasingham, S. and X. Hao. 1999. Fertilizer substitutions in hydroponically grown greenhouse tomatoes. *Hort Technology* **9**: 59-65.

Pardossi, A., Tognoni, F. and G. Bertero. 1987. The influence of nutrient solution concentration on growth, mineral uptake and yield of tomato plants grown in NFT. *Advances in Horticultural Science* **1**: 55-60.

Pardossi, A., Bagnoli, G., Malorgio, F., Campiotti, C.A. and F. Tognoni. 1999. NaCl effects on celery (*Apium graveolens* L.) grown in NFT. *Scientia Horticulturae* **81**: 229-242.

Petersen, K.K., Willumsen, J. and K. Kaack. 1998. Composition and taste of tomatoes as affected by increased salinity and different salinity sources. *Journal of Horticultural Science* **73**: 205-215.

Raviv, M., Krasnovsky, A., Medina, S. and R. Reuveni. 1998. Assessment of various control strategies for recirculation of greenhouse effluents under semi-arid conditions. *Journal of Horticultural Science* **73**: 485-491.

Resh, H.M. 1997. *Hydroponic Food Production* (5th Edition). Woodbridge Press Publishing Company, Santa Barbara, California.

Runia, W.T. 1995. A review of possibilities for disinfection of recirculation water from soilless cultures. *Acta Horticulturae* **382**: 221-229.

Runia, W.T. and J.J. Amsing. 1999. Disinfection of recirculation water from closed cultivation systems by heat treatment. *In:* International Symposium on Growing Media and Hydroponics. August 31 - September 6, 1999. Kassandra, Halkidiki, Greece. Abstracts, p. 25.

Runia, W.T., Michielsen, J.M.G.P., Kuik, A.J. and E.A. Van Os. 1996. Elimination of root infecting pathogens in recirculation water by slow sand filtration. *In:* Proceedings, 9th International Congress on Soilless Culture. ISOSC, Wageningen, The Netherlands, 1996, pp. 395-407.

Sachs, J. 1859. über den Einfluss der chemischen und physikalischen Beschaffenheit des Bodens auf die Transpiration der Pflanzen. *Die Landwirtschaftlichen Versuchs-Stationen* **1**: 203-240.

Sachs, J. 1861. Vegetationsversuche mit Ausschluss des Bodens, über die Nährstoffe und sonstigen Ernährungsbedingungen von Mais, Bohnen und anderen Pflanzen. *Die Landwirtschaftlichen Versuchs-Stationen* **3**: 30-44.

Santamaria, P., Elia, A. Gonnella, M. and F. Serio. 1996. Ways of reducing nitrate content in hydroponically grown leafy vegetables. *In:* Proceedings, 9th International Congress on Soilless Culture. ISOSC, Wageningen, The Netherlands, pp. 437-451.

Savvas, D. 1992. Vegetatives und generatives Wachstum bei Auberginen (*Solanum melongena* L.) in Hydrokultur in Abhängigkeit von der elektrischen Leitfähigkeit der Nährlösung. Ph. D. Thesis, University of Bonn, Germany.

Savvas, D. and F. Lenz. 1994a. Influence of salinity on the incidence of the physiological disorder 'internal fruit rot' in hydroponically-grown eggplants. *Angewandte Botanik* **68**: 32-35.

Savvas, D. and F. Lenz. 1994b. Einfluss einer NaCl-Salzbelastung auf das vegetative und generative Wachstum von Aubergine (*Solanum melongena* L.) in Hydrokultur. *Gartenbauwissenschaft* **59**: 172-177.

Savvas, D. and F. Lenz. 1995. Nährstoffaufnahme von Aubergine (*Solanum melongena* L.) in Hydrokultur. *Gartenbauwissenschaft* **60**: 29-33.

Savvas, D. and G. Manos. 1999. Automated composition control of nutrient solution in closed soilless culture systems. *Journal of Agricultural Engineering Research* **73**: 29-33.

Savvas, D. and K. Adamidis. 1999. Automated management of nutrient solutions based on target electrical conductivity, pH, and nutrient concentration ratios. *Journal of Plant Nutrition* **22**: 1415-1432.
Savvas, D. and F. Lenz. 2000a. Effects of NaCl or nutrient-induced salinity on growth, yield, and composition of eggplants grown in rockwool. *Scientia Horticulturae* **84**: 37-47.
Savvas, D. and F. Lenz. 2000b. Response of eggplants grown in recirculating nutrient solution to salinity imposed prior to the start of harvesting. *Journal of Horticultural Science & Biotechnology* **75**: 262-267.
Schippers, P.A. 1980. Composition changes in the nutrient solution during the growth of plants in recirculating nutrient culture. *Acta Horticulturae* **98**: 103-117.
Schrevens, E. and J. Cornell, 1993. Design and analysis of mixture systems: Applications in hydroponic, plant nutrition research. *Plant and Soil* **154**: 45-52.
Schwarz, M. 1995. *Soilless Culture Management.* Advanced Series in Agricultural Sciences, Vol. 24. Springer-Verlag, Berlin, Heidelberg.
Seemann, J. and C. Critchley. 1985. Effects of salt stress on the growth, ion content, stomatal behaviour and photosynthetic capacity of a salt sensitive species, *Phaseolus vulgaris* L. Planta **164**: 151-162.
Shannon, M.C., McCreight, J.D. and J.H. Draper. 1983. Screening tests for salt tolerance in lettuce. *Journal of the American Society for Horticultural Science* **108**: 225-230.
Shannon, M.C. and C.M Grieve. 1999. Tolerance of vegetable crops to salinity. *Scientia Horticulturae* **78**: 5-38.
Shinohara, Y. and Y. Suzuki. 1988. Quality improvement of hydroponically grown leaf vegetables. *Acta Horticulturae* **230**: 279-286.
Siminis, C., Loulakis, M., Kefakis, M., Manios T. and V. Manios, 1998. Humic substances from compost affects nutrient accumulation and fruit yield in tomato. *Acta Horticulturae* **469**: 353-358.
Sonneveld, C. 1981. Items for application of macro-elements in soilless culture. *Acta Horticulturae* **126**: 187-195.
Sonneveld, C. 1982. A method for calculating the composition of nutrient solutions for soilless cultures. Translated edition. Informatiereeks, No 57. Glasshouse Crops Research Station Naaldwijk, The Netherlands.
Sonneveld, C. 1987. Magnesium deficiency in rockwool-grown tomatoes as affected by climatic conditions and plant nutrition. *Journal of Plant Nutrition* **10**: 1591-1604.
Sonneveld, C. 1989. Rockwool as a substrate in protected cultivation. *Chronica Horticulturae* **29**: 33-36.
Sonneveld, C. 1995. Fertigation in the greenhouse industry. *In:* Proceedings of the Dahlia Greidinger International Symposium on Fertigation. Technion - Israel Institute of Technology, Haifa, Israel, 121-140.
Sonneveld, C. 1999. Chemical analysis in substrates and hydroponics—Use and interpretation. International Symposium on Growing Media and Hydroponics. August 31 - September 6, 1999. Kassandra, Halkidiki, Greece. Abstracts, p. 31.
Sonneveld, C. and S.S. De Bes. 1984. Micronutrient uptake of glasshouse cucumbers grown on rockwool. *Communications in Soil Science and Plant Analysis* **15**: 519-535.
Sonneveld, C. and C. De Kreij. 1999. Response of cucumber (*Cucumis sativus* L.) to an unequal distribution of salts in the root environment. *Plant and Soil* **209**: 47-56.
Sonneveld, C. and N. Straver. 1994. Nutrient solutions for vegetables and flowers grown in water or substrates. 10th Edition. Serie: Voedingsoplossingen Glastuinbouw, No 8, PBG Naaldwijk - PBG Aalsmeer, The Netherlands, 45 pp.
Sonneveld, C. and A.M.M. Van der Burg. 1991. Sodium chloride salinity in fruit vegetable crops in soilless culture. *Netherlands Journal of Agricultural Science* **39**: 115-122.
Sonneveld, C. and S.J. Voogt. 1980. The application of manganese in nutrient solutions for tomatoes grown in a recirculating system. *Acta Horticulturae* **98**: 171-178.
Sonneveld, C. and W. Voogt. 1985. Growth and cation absorption of some fruit-vegetable crops grown on rockwool as affected by different cation ratios in the nutrient solution. *Journal of Plant Nutrition* **8**: 585-602.

Sonneveld, C. and W. Voogt. 1990. Response of tomatoes (*Lycopersicon esculentum*) to an unequal distribution of nutrients in the root environment. *Plant and Soil* **124**: 251-256.

Sonneveld, C. and W. Voogt. 1993. The concentration of nutrients for growing *Anthurium andreanum* in substrate. *Acta Horticulturae* **342**: 61-67.

Sonneveld, C. and G.W.H. Welles. 1984. Growing vegetables in substrates in the Netherlands. *In:* Proceedings, 6th International Congress on Soilless Culture. ISOSC, Wageningen, The Netherlands, pp. 613-631.

Sonneveld, C. and G.W.H. Welles. 1988. Yield and quality of rockwool-grown tomatoes as affected by variations in EC-value and climatic conditions. *Plant and Soil* **111**: 37-42.

Sonneveld, C., Baas, R., Nijssen, H.M.C. and J. de Hoog. 1999. Salt tolerance of flower crops grown in soilless culture. *Journal of Plant Nutrition* **22**: 1033-1048.

Steiner, A.A. 1961. A universal method for preparing nutrient solutions of a certain desired composition. *Plant and Soil* **15**: 134-154.

Steiner, A.A. 1966. The influence of the chemical composition of a nutrient solution on the production of tomato plants. *Plant and Soil* **24**: 454-466.

Steiner, A.A. 1976. Nomenclature with hydroponics. *In:* Proceedings, 4th International Congress on Soilless Culture. IWOSC, Wageningen, The Netherlands, pp. 19-20.

Steiner, A.A. 1980. The selective capacity of plants for ions and its importance for the composition and treatment of the nutrient solution. *Acta Horticulturae* **98**: 87-97.

Steiner, A.A. 1984. The universal nutrient solution. *In:* Proceedings, 6th International Congress on Soilless Culture. ISOSC, Wageningen, The Netherlands, pp. 633-649.

Stensvand, A. and H.R. Gislerod. 1992. The effect of the NO_3/NH_4 ratio of the nutrient solution on growth and mineral uptake in *Chrysanthemum* X *morifolium*, *Passiflora caerulea*, and *Cordyline fruticosa*. *Gartenbauwissenschaft* **57**: 193-198.

Tapia, M.L. and R. Dabed. 1984. Nutrient uptake by sweet pepper grown in quartz. *In:* Proceedings, 6th International Congress on Soilless Culture: ISOSC, Wageningen, The Netherlands, pp. 683-696.

US Salinity Laboratory Staff. 1954. Diagnosis and improvement of saline and alkali soils. USDA, Agric. Handbook No. 60.

Van der Boon, J.J.W. 1988. Effect of EC, and Cl and NH_4 concentration of nutrient solutions on nitrate accumulation in lettuce. *Acta Horticulturae* **222**: 35-42.

Van Goor, B.J., De Jager, A. and W. Voogt. 1988. Nutrient uptake by some horticultural crops during the growing period. *In:* Proceedings, 7th International Congress on Soilless Culture. ISOSC, Wageningen, The Netherlands, pp. 163-176.

Van Ieperen, W. 1996. Effects of different day and night salinity levels on vegetative growth, yield and quality of tomato. *Journal of Horticultural Science* **71**: 99-111.

Van Os, E.A. 1999. Design of sustainable hydroponic systems in relation to environmentally friendly disinfection methods. International Symposium on Growing Media and Hydroponics. August 31 - September 6, 1999. Kassandra, Halkidiki, Greece. Abstracts, p. 23.

Van Os, E.A., Bruins, M.A., Van Buuren, J. and D.J. Van der Veer. 1996. Physical and chemical measurements in slow sand filters to disinfect recirculating nutrient solutions. *In:* Proceedings, 9th International Congress on Soilless Culture. ISOSC, Wageningen, The Netherlands, 1996, pp. 313-327.

Verdonck, O., De Vleeschauwer, D. and M. De Boodt. 1981. The influence of the substrate to plant growth. *Acta Horticulturae* **126**: 251-258.

Verdonck, O., Penninck, R., De Groot, M. and P.A.M. Heymans. 1984. Argex an expanded clay used as a substrate in hydroponics. *In:* Proceedings, 6th International Congress on Soilless Culture. ISOSC, Wageningen, The Netherlands, pp. 705-711.

Verwer, F.L.J.A.W. 1976. Growing horticultural crops in rockwool and nutrient film. *In:* Proceedings, 4th International Congress on Soilless Culture. IWOSC, Wageningen, The Netherlands: pp. 107-119.

Verwer, F.L.J.A.W. 1978. Research and results with horticultural crops in rockwool and nutrient film. *Acta Horticulturae* **82**: 141-147.

Verwer, F.L. and J.J.C. Welleman. 1980. The possibilities of Grodan rockwool in horticulture. *In:* Proceedings, 5th International Congress on Soilless Culture. ISOSC, Wageningen, The Netherlands, pp. 263-277.

Voogt, W. 1988a. K/Ca ratios with butterhead lettuce grown in recirculating water. *In:* Proceedings, 7th International Congress on Soilless Culture. ISOSC, Wageningen, The Netherlands, pp. 469-482.

Voogt, W. 1988b. The growth of beefsteak tomato as affected by K/Ca ratios in the nutrient solution. *Acta Horticulturae* **222**: 155-165.

Wendt, T. 1982. Nitratgehalte von Gemüse in NFT. Gemüse 15: 154-156.

Willumsen, J. 1980. pH of the flowing nutrient solution. *Acta Horticulturae* 98: 191-199.

Willumsen, J., Petersen, K.K. and K. Kaack. 1996. Yield and blossom-end rot of tomato as affected by salinity and cation activity ratios in the root zone. *Journal of Horticultural Science* **71**: 81-98.

Wreesmann, C. 1996. Chelated micro-nutrients for soilless culture. *In:* Proceedings, 9th International Congress on Soilless Culture. ISOSC, Wageningen, The Netherlands, pp. 559-572.

Color plate 3.1. Eggplant fruits affected by the calcium related physiological disorder 'internal fruit rot' (IFR).

4

Managing Crop Load

Sally A. Bound

Tasmanian Institute of Agricultural Research, 13 St John's Avenue, New Town, Tasmania 7008, Australia

1. Introduction

The science of growing fruit trees may be relatively well understood, but the science of cropping has encountered many difficulties. Although it can be difficult to achieve adequate fruit set on some fruit crops in some growing regions, in most fruit production areas there is a tendency for many perennial fruit trees to exhibit irregularity in cropping, alternating between the production of heavy crops with light crops (biennial bearing). This irregularity of flowering and cropping is a major economic constraint of many horticulturally important perennial species.

Crop regulation (thinning) is the removal of excess or unwanted flowers and/or fruit from the tree and is one of the most important orchard practices in ensuring regular cropping and good fruit quality. Effective crop regulation eliminates biennial bearing, increases fruit size and improves fruit colour and quality. Provided tree and spur vigour are adequate, fruit size at harvest is directly related to the timing and degree of thinning.

Thinning can be done either by hand, which is very time consuming and expensive, mechanically or with chemicals Unlike hand thinning, with chemicals the bulk of the thinning can be accomplished at blossom time and during the early post-bloom period.

Chemical thinning uses either hormonal-type growth regulators or caustic (desiccating) materials to reduce the amount of flowers and/or fruit and to overcome biennial bearing. This may require the use of several different chemicals and rates during the bloom and post-bloom period. The degree of thinning achieved and the effect on return bloom the following spring depends on a number of factors, including species/cultivar, tree health, tree age, rootstock, blossom density, previous crop, proximity to pollinisers, weather, chemicals used, and application method and time.

Growers now have a number of thinning materials available. Usually two or more chemicals are used within a thinning program, particularly for hard-to-thin cultivars. Surfactants and oils are often added to chemical thinners to increase thinning effectiveness. Advances in spray technology have also assisted in the improvement in predictability of crop regulation techniques.

2. History of Crop Regulation

The need to thin fruit from bearing trees has been recognised since fruit trees have been cultivated, with hand thinning of fruit an accepted practice to improve fruit size and quality. However, it was not until Russel and Pickering (1919) showed that the alternate bearing habit of apples could be controlled by thinning at the bloom stage instead of 6-8 weeks after bloom that the first real attempts were made to control cropping. This led to blossoms being thinned with small scissors during the 1920s. By 1925, Bagenal et al. recognized that the drop of immature fruit from a healthy tree was increased by chemical sprays applied for pest control. They observed that lime sulphur (calcium polysulfide) induced excessive drop of young apples.

The quest for chemicals to reduce fruit set began in earnest in the 1930s with the testing of the common spray materials of that era as pre-bloom sprays. During the depression years there was a low demand for fruit and chemical agents capable of entirely preventing fruit set were sought. In 1934, work in the USA by Auchter and Roberts, found that tar oil distillates were effective in killing flower buds of some apple cultivars but caused injury to vegetative tissues. Soon after, the emphasis shifted away from total to partial crop removal, leaving enough fruit to produce a full crop and ensure adequate return bloom the following season.

By 1939, dinitro-ortho-cyclohexylphenol was found to be promising as a thinning agent (Magness et al. 1940) leading to the exploration of dinitro compounds in the 1940s. Phenol forms caused russeting and mis-shapen fruit, however both acid and sodium dinitro-ortho-cresylate (DNOC) produced satisfactory results as blossom thinners on apples and stonefruits and to a lesser extent on pears. At the same time the hormone-type chemicals naphthalene acetic acid (NAA) and naphthalene acetamide (NAAm) were found to reduce fruit set when applied at blossom time (Gardener et al. 1939). Further work with both DNOC and NAA showed that chemical thinning was far more effective than hand thinning, resulting in larger fruit at harvest and increased return bloom.

Although high rates of NAA (50 ppm) were able to reduce apple fruit set by up to 77%, severe injury in the form of epinasty, scorching and leaf drop resulted (Burkholder and McCown, 1941). Similar concentrations of NAAm caused no visible foliage injury, but it was not as effective as NAA at reducing fruit set. Studies were conducted to obtain both the most efficient timing and concentration and by the mid 1940s it had been shown that NAA was effective as a post bloom thinner at 2-3 weeks after bloom (Davidson et al. 1945). Work on olives demonstrated that NAA was effective at reducing fruit set when applied at full bloom (FB), or completely preventing fruit set at higher concentrations (Hartmann, 1952).

Research in Australia in 1961 (Miller, 1985) demonstrated that the addition of the wetting agent Tween 20 (polyoxyethylene sorbitan monolaurate) to sprays of NAA resulted in effective thinning of apples at much lower concentrations of NAA—thus greatly reducing the phytotoxicity of this chemical.

Further work with NAA in Australia has shown that the most effective time of application for this chemical under Australian conditions is from full bloom up to seven days after full bloom (dAFB) (Bound et al. 1991b). Apart from being less effective, applications any later than 10 dAFB tended to cause unacceptably high levels of pygmy fruit. These researchers also highlighted the negative interactions of NAA with plant growth regulators containing gibberellins.

The effectiveness of 2,4,5-trichlorophenoxyacetic acid (2,4,5-T) on thinning of stonefruits, including apricots, peaches and French prunes was demonstrated by Crane (1953), Harris et al. (1953), and Harris and Hansen (1955). Harris and Smith (1957) reported successful thinning of pecan nut crops with the sodium salt of 1,2-dihydro-pyridazine-3,6-dione (maleic hydrazide) and 2,4,5-T.

During the late 1950s and 1960s several new post-bloom thinners were introduced (Batjer and Westwood, 1960; Westwood, 1965), including the insecticide carbaryl (1-naphthyl-N-methylcarbamate) which reduced fruit set when applied from 15 - 30 days after bloom. The response to carbaryl was consistent across several apple varieties but was ineffective on pears and stonefruits. Other post-bloom thinners developed during this period were 6-methyl 2,3-quin-oxalinedithiol cyclic carbonate (MQCC) which was effective on apple; 2-chloroethyl phosphonic acid (ethephon)—effective on apple and stonefruit; and 3-chlorophenoxy-alpha-propionamide (3-CPA) which was examined for use on peaches, apricots and plums. The fungicide thiram (bis (dimethyl thio-carbomoyl) disulphide) was also shown to have a thinning effect on some apple varieties (Veinbrandts, 1967).

Work by Bradley and Crane (1960) demonstrated that gibberellin A_3 (GA_3) inhibited floral bud development in prunus species, resulting in the production of fewer flowers the following spring and thus a reduced cropping potential. However, at this time, attempts to use gibberellins in commercial thinning were not successful.

The concept of serial spraying (more than one thinning spray per season) was introduced in Australia in 1963. Apple varieties which were difficult to thin responded well to a blossom application of DNOC followed by NAA 10 days later. On easy to thin varieties carbaryl was more consistent than NAA.

In 1968, work in the USA found that the best method of determining time of application of thinning sprays was calendar days after full bloom. Full bloom could be determined easily and was found to be fairly consistent for large blocks of trees. Fruit size was tested as a measure of developmental stage but it was found that there was too much size variation between and within trees.

The development of ethephon continued after 1968 but it was found that fruitgrowth was retarded by the high concentrations considered necessary to thin fruit, thus size at harvest was not increased in spite of reduced fruit numbers. Consequently ethephon was only considered useful for complete fruit removal. However, more extensive work in Australia showed that while lower concentrations were effective over the period from full bloom up to six weeks AFB there were two highly sensitive periods—one at full bloom and one five weeks AFB (Veinbrandts and Hutchinson, 1976). Ethephon was also shown to directly stimulate flower bud initiation, thus strongly promoting return bloom. It has been shown to be effective in a wide range of crops, including apple (Veinbrandt and Hutchinson, 1976), plum (Martin et al. 1975), Valencia orange (Hutton, 1992), and coffee (Browning and Cannell, 1970). Ethephon replaced DNOC as a blossom thinner in Australia in the late 1970s.

By the early 1980s work in both North Carolina, USA and in Australia demonstrated that spray volume was a major factor in the response of thinners applied with air-blast sprayers. Fruit thinning was improved further by increasing spray volumes rather than increasing chemical concentration.

Considerable advances with spray application technology have been made in the 1990s, with a move away from the traditional high volume air-blast sprayers fitted with hydraulic nozzles to either controlled droplet application (CDA) technology or the fitting of low pressure, low volume nozzles to conventional airblast sprayers. These developments have allowed effective application of thinning chemicals at considerably reduced water volumes.

The search for new thinners, both blossom and post-bloom chemicals is ongoing. Increased environmental and public health concerns have led to the banning or phasing out of many chemicals and has encouraged the development and introduction of non-persistent chemicals with low toxicity levels. It was shown as early as 1984 that 6-benzyladenine (BA) could thin fruit. Work carried out in both the USA and Australia during the following decade has led to the development of BA as a post bloom thinner in Australia and in the USA with other countries working towards registration.

Many desiccating chemicals have been trialed as blossom thinners during the 1980s and 1990s. Most have been unsuccessful because of lack of thinning at low concentrations and phytotoxic side effects at higher concentrations. However, a few have been shown to be capable of thinning without excessive phytotoxicity. Sulfcarbamide (1-aminomethanamide dihydrogen tetraoxosulphate) is now used in the USA although it is ineffective under Australian conditions (S.A. Bound and K.M. Jones, unpublished data). Ammonium thiosulphate (ATS) is being developed for both pome and stone fruits in Australia, USA and Europe, while chemicals such as endothal (dipotassium 7-oxoabicyclo (2,2,1) heptane-2,3,-dicarboxylate) and pelargonic acid are also showing promise.

3. Flowering and Fruit Set

Fruit set is the rapid growth of the ovary that usually follows pollination and fertilization. This is normally accompanied by other changes, such as wilting of petals and stamens and the abscission of flowers that fail to set. Fruit set is influenced by a number of factors, including strength of bloom, pollination, weather conditions and competition between blooms.

Many fruit species are, by nature, biennial bearing. If left to themselves they tend to produce a heavy crop one year (on-year) followed by a light or no crop the next year (off-year). Biennial bearing in apple, pear, stonefruits and filbert is due to lack of floral initiation in the on-year (Fig. 1) which leads to no or poor return bloom the following year. Pistachio alternate bearing is due to a different mechanism—dropping of flower buds after they have initiated (Crane, 1971), while in walnut and pecan female flowers are initiated just prior to anthesis rather than in the previous year (Westwood, 1978).

Once started, this alternate cropping cycle is largely self-perpetuating as crop size has a major influence on the potential crop for the following year. However, by careful tree management this biennial bearing cycle can be broken to allow consistent cropping from year to year. Flower-bud formation, fruit set and shoot growth are closely interrelated and

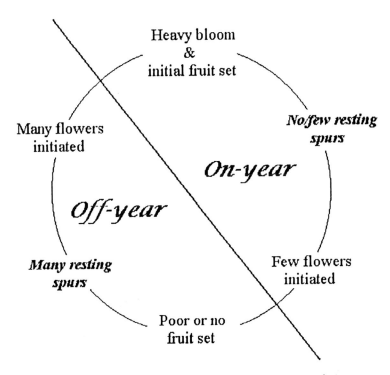

Fig. 1. Typical biennial bearing cycle as exhibited in pome fruits.

any attempt to affect one is likely to affect the others. There is an antagonistic relationship between vegetative shoot growth and fruiting. In vigorous trees re-orientation of branches from the vertical to the horizontal has the effect of reducing vegetative shoot growth and enhancing flower production and fruit set.

To prevent biennial bearing the delicate balance between fruit set, flower bud formation and shoot growth must be controlled. While a number of cultural techniques such as pruning, tying down of branches, fruit thinning by hand, use of dwarfing rootstocks, and fertilizer regimes are used, the use of plant growth regulators have proved to be invaluable.

4. Crop Regulation

4.1. Purpose of Thinning

As most fruit trees will set more fruit than required, flower or fruitlet thinning is undertaken to reduce the crop load and to stimulate floral initiation for next year's crop. This achieves a regular bearing pattern, increased fruit size, improved fruit colour and quality, reduced limb breakage and minimises hand thinning. Added advantages of a well structured, timely thinning program are often an increase in fruit firmness (and thus keeping quality) and fruit sugar content.

Understanding the implications of crop loading is the basis of thinning. Crop load can be expressed as either number of fruit per cm^2 trunk cross-sectional area (TCSA) or per 100 blossom clusters. In citrus a counting frame is used to determine fruiting density.

It is not well appreciated that the number of fruit per tree and mean fruit weight are inversely related. Trees with larger fruit generally have fewer fruit than trees with smaller fruit. In addition, in crops which are prone to biennial bearing less than half the spurs on the tree should be carrying fruit to ensure adequate return bloom.

4.2. Pruning

Pruning should be considered an integral part of any thinning program. Well sized, high quality fruit are associated with open, healthy trees with large leaves on both spurs and shoots. Tree canopies need to be opened up by removing old, weaker wood.

The amount of structural wood in a tree creates an internal limiting factor on the potential productivity of the tree as it competes with the fruit for carbohydrates. Because a small tree has a much smaller structural wood mass it can convert a higher percentage of its photosynthetic production into fruit rather than wood. This difference in tree structure helps to account for the better yield efficiency usually observed in smaller trees.

The healthier and more vigorous the flower buds the better the fruit will be. In pome fruits, the best fruit occurs on younger spurs (2-4 year old), hence the tree needs to be kept young by appropriate pruning so that fruiting occurs on young wood.

Adequate light penetration is also important for the production of good quality fruit, particularly where colour is required.

4.3. Time of Thinning

If excess fruit is removed before fruit cell division is complete and total fruit cell numbers have been determined there is the likelihood that cell numbers can be increased. When thinning is delayed until several weeks after flowering substantial nutrient reserves are wasted producing fruit which are later removed. It is more efficient to remove excess fruit early so these reserves go to fruit destined to remain on the tree. The longer the delay in thinning the less effective it is in increasing fruit size (Fig. 2).

4.4. Level of Thinning

An upper limit of fruit size exists for each variety. This limit is influenced by orchard management and seasonal conditions. Once the crop load of a tree has been reduced to a level that will allow the remaining fruit to achieve this size, further thinning will only serve to reduce yield. The aim is to produce the greatest number of fruit of the desired size and quality without impacting on return bloom.

4.5. Methods of Thinning

There are three methods by which thinning can be accomplished: hand, mechanical, or chemical.

4.5.1. Hand Thinning

Hand thinning is simply pulling or breaking off flowers or fruit with fingers, or cutting with small scissors or secateurs. Early hand thinning at blossom time is difficult to achieve accurately as it is impossible to ascertain which flowers will set. Hand thinning is very time consuming and

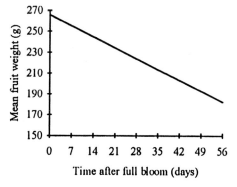

Fig. 2. The effect of time of thinning on fruit size of 'Fuji' apple (data from Jones et al. 1992).

expensive. Hand thinning is often carried out in countries where labour costs are low, however in practical terms it is difficult to complete hand thinning before the end of the cell division period or before flower initiation is complete. This means that limits have already been set on fruit size and quality as cell numbers within the fruit have already been determined, and there is a high likelihood that trees will be pushed into a biennial bearing cycle.

Hand thinning still has a place in the modern orchard but it should be viewed as a normal 'follow up' operation after the chemical thinning program to remove damaged or small fruit, or to break up bunches. It can also be used on young trees and on economically valuable or newer cultivars where interactions with thinning chemicals is not properly understood.

Fruit size should be the basis for hand thinning. Historically, fruitlets were thinned to a pre-determined spacing; however, small fruit will never catch up in size with larger fruit; hence size thinning should be practiced. Size thinning is the selective removal of small, weak fruit regardless of spacing.

4.5.2. Mechanical Thinning

Mechanical thinning may be conducted in a number of ways. A direct blast of high pressure water can effectively remove flowers and/or fruitlets. A stiff bristled brush or rake can be used to 'sweep' off some of the small fruitlets. Other physical methods to remove flowers or fruitlets include flailing the tree with rope or switches, use of poles, or shaking of the tree or limbs. These procedures tend to create an uneven distribution of fruit, cause further damage to fruit remaining on the tree, or preferentially remove larger fruit, leaving the smaller fruit.

4.5.3. Chemical Thinning

Chemical thinning uses caustic materials or hormonal-type growth regulators to reduce the amount of flowers and/or fruit. Numerous factors can affect the degree of thinning and the effect on return bloom the following spring. These include species/cultivar, tree health, tree age, rootstock, vigour, blossom density, proximity to pollinisers, weather, chemicals used and the application method.

The advantages of chemical thinning over hand or mechanical thinning are: reduced labour costs, better fruit size and quality, more even fruit distribution within the tree and better return bloom on biennial cultivars. These advantages should be weighed against the possible disadvantages which include: risk of frost after early sprays, risks of over- or under-thinning, and fruit or foliage injury. Chemical thinning is also very weather dependent and often varies greatly without apparent reason. Further complications can occur when chemical thinning agents interact negatively with other agrochemicals. Bound et al. (1991b) and Jones et al. (1991) have described the consequences of some of these interactions.

There are now a number of chemical thinning agents available worldwide, however not all chemicals are effective on all fruit crops and there are variations in response under differing climatic conditions. In addition, within each fruit species there is considerable variation between cultivars/selections in sensitivity to chemical thinning agents. Surfactants and oils are often added to chemical thinners to increase thinning effectiveness.

Unlike hand thinning, with chemicals the majority of the thinning can be accomplished at blossom time and during the early post-bloom period.

Ideally a thinning program should consist of a blossom thinner followed by a post-bloom thinner. If the chemical thinners have been effective, then all that should be required is a subsequent light hand thin to remove damaged fruit or break up any remaining bunches.

4.6. Chemical Type and Mode of Action

Chemicals used for thinning can be classified according to application timing:
1. bloom or blossom thinners which are applied at flowering, and
2. post-bloom thinners which are applied after flowering is complete—usually anywhere from 10 to 40 days after flowering.

Another way of classifying thinning chemicals is by their mode of action. Thinners can work in one of two ways:
1. As *growth regulators*, altering complex physiological processes in the tree, or reducing the ability of fruitlets to compete for resources by stimulating abortion of seeds or through other poorly understood mechanisms. These chemicals can be either blossom or post-bloom thinners.
2. As *blossom desiccants*, also referred to as *blossom burners*. This group of chemicals acts by desiccating the stigma, thus preventing pollination.

Correct application timing of desiccating chemicals is critical as they need to be applied during blossom after sufficient flowers have set fruit. The aim is to allow the early opening flowers to pollinate and set fruit as they tend to produce the largest fruit, then apply the desiccant to remove the later flowers—this may mean more than one application of the desiccant in crops, cultivars or regions with an extended flowering period. The mode of action of desiccants makes them less dependent on weather conditions for their effectiveness than hormonal-type blossom thinners such as NAA or ethephon. However under conditions of high humidity or when rewetting occurs soon after application they can be re-activated, in some cases causing severe burning, damaging buds, fruit and leaves.

4.7. Chemical Thinning Agents

While there are some chemicals which are used world-wide for thinning, there are differences between countries and growing regions on recommended application timing and concentrations. Some of these differences are due to fundamental differences in climate and culture in the various growing regions. Others, however, are the result of the differing degrees of uptake of new research and technology. The more commonly used chemical thinning agents are summarised in Table 1. While concentration is important, the time of application of chemical thinning agents has far greater impact on the efficiency of thinning.

Table 1. Chemicals commonly used for thinning

Generic name	Trade name	Chemical name	Crop
DNOC*	Elgetol	4,6-dinitro-ortho-cresol	apple, pear, peach, plum, apricot
NAA		1-naphthaleneaceticacid	apple, pear, olive, citrus
NAD	Amid-thin	1-naphthaleneacetamide	apple, pear
carbaryl	Sevin[1,2] Thinsec[3]	1-naphthyl (N)-methyl carbamate	apple
Thiram	TMTD	bis (dimethyl thio-carbomoyl) disulphide	apple
MQCC	Morestan	6 methyl 2,3-quinoxalinedithiol cyclic carbonate	apple
ethephon	Ethrel/CEPA	2-chloroethyl phosphonic acid	apple, citrus, peach, plum, apricot
3-CPA	Fruitone CPA	3-chlorophenoxy-alpha-propionamide	peach, plum, apricot
BA benzyladenine	CyLex[1]/Accel[2]/ Paturyl[3]	N-(phenyl)-1H-purine 6-amine	apple, pear
ATS	Culminate[1]	ammonium thiosulphate	apple, pear, Nashi, peach, plum, apricot
sulfcarbamide	Wilthin[2]	1-aminomethanamide di-hydrogen tetraoxosulphate	apple, peach, nectarine
Oxamyl	Vydate[2]	2-oxoethanoimido-thioic acid methyl ester	apple
pelargonic acid	Thinex[2]	nonanoic acid	apple
endothall	ThinRite[2]	7, oxabicyclo (2,2,1) heptane—2-3 dicarboxylic acid	apple
	Armothin	fatty amine polymer	stonefruit
GA_3	Release	gibberellic acid	stonefruit

[1]Australia; [2]USA; [3]Europe,
* not available since 1989

5. Factors Affecting Thinning Response

Thinning decisions are difficult as many factors affect the end result. These include tree history, pollination, pruning, choice of thinning method, spray volume, and climatic effects (Fig. 3). While each factor can individually influence the outcome of thinning programs, more commonly there are interactions between several factors.

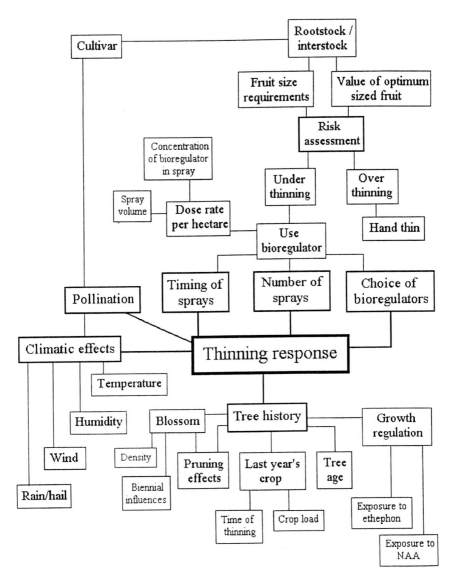

Fig. 3. Interacting factors involved in the crop regulation process.

6. Impact of Chemical Thinning on Fruit Quality

The earlier thinning is performed, regardless of the method of thinning, the larger the fruit size at harvest. Thinning before cell division is complete normally results in higher cell numbers in the fruit thus leading to firmer fruit.

While there are major benefits in using chemical thinning agents there can also be drawbacks, particularly if insufficient attention is paid to optimum application conditions. NAA can result in the formation of pygmy fruit in apple (Bound et al. 1991b). At higher concentrations or later application timings both NAA and ethephon are likely to depress fruit size, counteracting the benefits gained by early thinning. Ethephon also has a tendency to flatten fruit. This fruit flattening effect can be counteracted by the use of formulations of BA plus GA_{4+7}, however this is an added cost to the grower.

Both NAA and carbaryl can reduce seed numbers in pome fruit or cause abortion of fertilized seeds. This problem of reduction in seed numbers increases proportionately the later after full bloom that the NAA is applied, or with carbaryl the nearer to full bloom (Jones et al. 1998). As seed number affects fruit quality (Bramlage et al. 1990) and fruit size this is a most undesirable side effect.

If applied under cool temperatures or high humidity carbaryl may cause skin russeting, thus downgrading fruit. NAA may also cause russet under humid conditions. Application of any chemical after a prolonged cool wet period is likely to result in an increase in fruit russet.

BA has the added advantage over other thinners of increasing fruit size independently of the thinning effect. It does this by increasing the number of cells in the fruit, leading to firmer fruit. Fruit soluble solids may also be increased by BA.

In some circumstances sulfcarbamide can cause fruit dimpling which resembles insect damage (Warner, 1999).

7. Crop Regulation in Various Fruit Crops

7.1. Apple (*Malus domestica* Borkh.)

Most apple cultivars require some degree of thinning. In order to crop trees consistently every year over 90% of the flowers/potential fruitlets need to be removed from the tree within six weeks of flowering. If a tree with good blossom (on-year) is allowed to thin naturally, about 50-60% of the fruitlets will abscise during the period from flowering to about 60 dAFB. However, this is not sufficient for thinning purposes and most abscission occurs too late to obtain fruit size benefits and prevent biennial bearing.

There has been considerable research undertaken on chemical thinning of apples, with the result that there is a wide range of chemical thinning agents available for use on apples. Recommendations for time of application of chemicals varies between countries, and even growing regions within countries.

In Australia it is recommended NAA be applied from full bloom up to 7 days AFB, however in most other areas NAA is applied much later as a post-bloom spray. Despite the fact that NAA can thin most cultivars between full bloom and 21 dAFB, the earlier it is applied the better the response in fruit size. Late applications also tend to encourage the formation of pygmy fruit (Bound et al. 1991b). NAA has a tendency to interact with other plant growth regulators containing GA_{4+7}. Schwallier (1996) states that NAA is not compatible with *Promalin*® (a 50 : 50 formulation of BA + GA_{4+7} used to improve fruit shape) or with *Accel*® (a formulation of BA containing 0.2% GA_{4+7}) on red Delicious and should not be applied on the same tree in the same year. However, Bound et al. (1997) have successfully integrated *Cytolin*® (registered as *Promalin*® in the USA) into thinning programs by reducing the dosage rate of NAA. NAA is also compatible with *CyLex*®—a formulation of BA used in Australia which, unlike *Accel*®, does not contain GA_{4+7} (Bound, unpublished data). In the USA, NAA is commonly used in a tank mix with carbaryl (Schwallier, 1996).

Ethephon can be a vigorous thinner; it can completely remove weak spurs or deplete fruit positioned low on the tree more easily than from the more vigorous higher spurs. The effective time frame for ethephon use is from the balloon blossom stage to 7 dAFB (Jones et al. 1998). Ethephon can also be used to remove the blossom from the late flowering spurs when they reach balloon blossom. Although ethephon also thins effectively at around 40 dAFB it does not improve fruit size and has the added disadvantage that it causes yellowing of the fruit at harvest.

Carbaryl is the main thinner used in many growing regions, including Europe and New Zealand. The use of carbaryl in Britain as early as petal fall (Knight and Spencer, 1982) has been advocated to achieve optimum fruit size; however because the flowering period is extended in Australia, use of carbaryl is strongly discouraged in this country to avoid bee deaths—the recommended application timing is from 14 to 60 dAFB (Jones et al. 1998). Australian growers often apply carbaryl as a tank mix with thiram (Jones et al. 1998) for added thinning effect.

The synthetic cytokinin 6-benzyladenine (BA) has recently been developed as a post-bloom thinner for apple. Recommendations for its use in Australia are to apply ethephon or NAA at full bloom followed by 150 ppm BA (*CyLex*®) from 10 to 23 dAFB, depending on cultivar. In Washington it is recommended that BA (*Accel*®) be used in combination with carbaryl to give an additive thinning response. BA works best when applied under warm conditions and early in a warming trend.

Since DNOC was withdrawn in 1989 (Williams, 1994) many desiccating agents have been assessed, but very few have shown potential as thinners. One of the most promising is ammonium thiosulphate (ATS) (Bound, 1995). Recommendations for the use of ATS in Australia are to apply two or three times during the bloom period, depending on the duration of flowering. Work with endothallic acid in several countries has demonstrated superiority to other desiccants such as sulfcarbamide in both effectiveness and safety from damage at effective levels (Williams et al. 1995).

7.2. European Pear (*Pyrus communis*)

In many countries the thinning of pear flowers or fruitlets is uncommon and, compared with apples, there has been relatively little attention given to developing thinning programs.

The need for thinning pears varies by region and cultivar. According to Williams and Edgerton (1981) in California 'Bartlett' pears are seldom thinned, however in Washington and Oregon they require extensive thinning to attain acceptable market size. Lombard (1982) suggests that only 3-5% of the flowers need to set for an adequate pear crop. As with apple, the earlier the thinning is done the greater the effect on improving the size of the remaining fruit. According to Westwood (1978) pears initiate flowers later than apples, but as long as thinning is done by 60 days after full bloom it can aid in return bloom. Williams and Edgerton (1981) state that hand thinning of 'Bartlett' pears after 70 days from full bloom does not result in an appreciable increase in the size of the remaining fruit. The same environmental and physiological factors affecting the response of apple fruit to chemical thinning agents also affect pears.

In Washington State, USA. Tukey and Williams (1972) and Williams (1974) found NAA would thin pears at 15 dAFB. In Romania, Modoran et al. (1979) found that carbaryl was a better thinner of 'Williams' Bon Chretien' than NAA, while Link (1976) also found that carbaryl would thin pears. Australian growers have had little success with NAA or carbaryl on pears (Bound et al. 1991a). Current recommendations in the USA for thinning of 'Bartlett' pears are to apply either NAD or NAA at 10 to 15 ppm from 15 to 21 dAFB.

Ethephon has also been shown to be an effective thinner of some pear cultivars in some regions. Selimi and Gibbs (1976, 1978) used ethephon to thin 'Williams' Bon Chretien' in the Goulburn Valley in Australia, while Bound et al. (1991a) successfully used ethephon to thin 'Winter Cole' in the cooler Tasmanian climate. Knight (1982) also found that ethephon, used in conjunction with gibberellic acid, reduced fruit number of 'Conference' pears in the UK. However, in Norway, Meland and Gjerde (1996) saw no thinning effect on several pear cultivars, while in Egypt, Stino et al. (1980) increased fruit set on 'Le Conte' pear trees with ethephon.

The use of blossom desiccants for thinning of pears was first discussed by Weaver (1972) who recommended DNOC. Meland and Gjerde (1996) found that lime sulphur produced some thinning on two out of four cultivars tested, however there was little impact on return bloom.

In Australia the effect of the blossom desiccants ATS and endothal have been examined on the cultivars 'Winter Cole' and 'Packham's Triumph'. Bound et al. (1999) and Bound and Mitchell (2000a) have recommended that ATS thins most effectively following two applications during the flowering period.

Bound et al. (1999) and Bound and Mitchell (2000b) have also examined the effect of the post-bloom thinner benzyladenine (BA) on 'Packham's Triumph'. They found that 100 to 150 ppm BA applied from 10 to 35 dAFB reduced the crop load significantly.

7.3. Asian Pear (*Pyrus pyrifolia*)

All Asian pear varieties require heavy thinning to obtain good fruit size, to ensure annual cropping, and to avoid limb breakage. Thinning is generally done by hand as the development of safe effective chemical thinners for Asian pears has been slow. According to Beutel (1989), in the USA some growers blossom thin by removing all but 2-3 flowers per cluster, but most growers wait until after fruit set before reducing the crop load to 1 fruit per spur and spacing fruit 4-6 inches apart. However in Australia the thinning process involves spur pruning followed by bud thinning, flower removal and finally post-bloom hand thinning. Mitchell and Bound (1999) have examined the effects of the growth regulators ethephon and benzyladenine and the desiccant ammonium thiosulphate (ATS) on 'Nashi' pears in Australia. This work has resulted in the successful development of a chemical thinning program using ATS as a blossom thinner (Bound and Mitchell, 2000c).

7.4. Citrus (*Citrus* spp.)

Citrus cultivars commonly alternate between heavy and light crops. The cycle is often initiated by climatic events such as heat waves, or by orchard management practices such as deep tillage. Trees with this biennial cycle produce poorer fruit (Lenz and Cary, 1969), and production, handling and marketing of this fruit are less efficient (Gallasch, 1974).

Application of crop regulation techniques in citrus orchards helps reduce the incidence of biennial bearing, and improve fruit size and packouts. Citrus growers commonly measure crop load with the aid of a counting frame (a cubic wire frame measuring 0.5 metre along each side which is placed so the outer foliage of the tree coincides with the outer edge of the counting frame). All fruitlets inside the frame are counted with several counts performed per tree at different points, this is then averaged to obtain an average figure per tree. The aim is have the following fruit densities: 'navels', 'valencias' and 'tangelos', 5-6 fruit per cube; 'mandarins', 10-15 fruit per cube.

There are a number of different ways to regulate crop load in citrus orchards.

7.4.1. Pruning

In addition to increasing light and spray penetration, pruning helps to encourage a lighter crop of green blossom and thus helps to reduce crop load. Pruning should be the first step towards crop regulation.

7.4.2. Hand Thinning

Hand thinning is practiced extensively in Australia to control crop load and fruit on mandarins and tangelos. Hand thinning normally commences once the natural fruit drop has stopped, removing all deformed, marked or smaller than average sized fruit (Anonymous, 1998). The same author also raises the important point of considering only what is on the tree, and not looking at how much fruit is on the ground.

7.4.3. Chemical Thinning

Recommendations for timing of chemical thinning agents for citrus crops is that stage of development when natural fruit drop is almost complete. Ethephon has been shown to reduce crop load across a range of citrus fruits (Chapman, 1984; Koller et al. 1986; Hutton, 1992; Gallasch, 1998). A light spray just wetting leaves and fruit, of 250 to 300 ppm ethephon is recommended in Australia for citrus thinning (Gallasch,1998). Heavy applications are likely to result in excessive leaf drop.

Although NAA is not recommended in California for thinning citrus crops because of the high degree of variability in thinning (Coggins, 1997), in Florida, NAA is recommended for fruit thinning of tangerines, murcotts and tangelos (Davies et al. 1998). However, the caution is given that applications made to trees of low vigour and/or under stress may result in severe overthinning.

7.5. Olive (*Olea europea* L.)

Olives have a strong tendency toward alternate bearing. Yield records from most olive producing countries show a marked fluctuation from year to year. According to Hartmann (1952) hand thinning failed to gain widespread acceptance as a regular orchard operation as it is an overwhelming task, particularly on a large acreage. The main benefits of thinning olives are: increased fruit size, increased flesh-pit ratio, higher oil content, earlier fruit maturity, and a reduction in the tendency towards biennial bearing.

Work by Hartmann (1952) in California showed that a reduction in the set of olive fruits can be accomplished by spraying NAA at concentrations of 40 to 50 ppm at full bloom, while concentrations of 75 to 100 ppm result in complete prevention of fruit set. However Hartmann suggests that spray thinning at the blossom stage is not recommended since it is impossible to predict the need for thinning at this time. His work also found that concentrations of 100 to 125 ppm NAA in a 1.5% summer oil emulsion gave satisfactory thinning if applied 14 to 25 dAFB when the fruit was 3 to 5 mm in diameter. Weaver (1972) reports similar results by Lavee and Spiegel in Israel in the late 1950s.

While NAD is also effective for thinning of olive, it is not as effective as NAA (Lavee and Spiegel-Roy, 1967). To achieve equal activity, concentrations of NAD 1.5 to 2 times those of NAA are required.

7.6. Stone Fruits

Peach (*Prunus persica* L.), plum (*Prunus domestica, Prunus salicina*) and apricot (*Prunus armeniaca* L.) cultivars require thinning to produce fruit of optimal size and quality. Use of chemical thinning agents has definite advantages over both hand thinning and the various mechanical methods which have been developed. The major advantage of chemical thinning is the earliness and speed of thinning in removing the excess fruit, thus allowing the remaining fruit to develop without serious competition. Most of the work with chemical thinning agents suitable for stone fruits has been conducted on peaches, some of which applies to nectarines. Some work has been carried out on plums and very little on apricots. As expected, there is variation in response to chemicals by various cultivars. Baroni et al. (1986) showed this variable response for peaches, Suranyi (1986) for apricots, and Belmars and Keulemans (1987) for plums.

While numerous chemicals have been screened for peach fruit thinning activity, including hormonal-type chemicals, desiccants and herbicides there has been little success in developing commercially viable chemical thinning compound for use on stonefruit (Southwick et al. 1996).

Most stonefruits bloom very early in the spring and this tends to be a major deterrent to blossom thinning as inclement weather can lead to erratic fruit set and the risk of frosts is high.

According to Sefick (1975) peach thinning is inseparable from profitable peach growing. Thinning increases fruit size, improves fruit colour and quality, reduces limb breakage, promotes general tree vigour, and increases the effectiveness of the pest control program. Relying on hand thinning alone is not satisfactory as thinning cannot be accomplished throughout the entire orchard in time for optimum results. Nickell (1982) suggests that set on 10 to 15% of the flowers usually results in a full crop.

GA_3 has been reported as having good potential for chemically thinning peaches (Southwick et al. 1995) and apricots (Southwick et al. 1997) by decreasing flower density the following spring. The blossom desiccant ATS is now used in Australia for reducing fruit set in plums and low chill peaches. ATS and ArmorThin can also successfully thin apricots (Bound, 1995; L. Mitchell, personal communication).

Ethephon has been used successfully as a blossom thinner for plum (Martin et al. 1975), however its effect can be erratic.

7.7. Kiwifruit (*Actinidia deliciosa* Planch.)

In kiwifruit, yield has a significant influence on biennial bearing and fruit size. Both pruning and thinning are used for controlling crop load, with time of thinning having a major impact on fruit size. Lahav et al. (1989) demonstrated that vines thinned at the budswell stage carry larger fruit than vines thinned after fruit set. They also found that the more fruits on the vine the smaller the individual fruits.

7.8. Miscellaneous

Overbearing in the early stages of growth is a problem in young coffee (*Coffea arabica* L.) as it can impair later reproductive capacity. Ethephon has been found to produce large reductions in the percent of fruit retained (Browning and Cannell, 1970).

Japanese persimmon fruit (*Diospyros kaki*) can be thinned with NAA applied within 30 days after full bloom (Nickell, 1982).

8. Conclusions

Achieving an optimum crop load is a major challenge facing perennial fruit growers every year. Achievement of this goal delivers optimum yields of well sized, high quality fruit annually. Crop regulation is a complex issue as it is influenced by an array of factors including cultivar, tree history, tree growth habit, pollination, climate and method of thinning. Unfortunately, thinning strategies cannot be transposed from one cultivar to another as each cultivar responds differently to chemical thinning agents. The development of reliable chemical thinning strategies is therefore an ongoing process as new cultivars and chemicals become available.

While thinning is a proven technique for reducing biennial bearing and improving fruit size and quality, it will not compensate for poor management of other orchard operations. Where an orchard is subjected to water stress or nutrition is lacking, thinning is unlikely to substantially increase fruit size or quality—the benefits of thinning are greatest where good management of all aspects of fruit production are employed.

Although thinning with bioregulators is now regularly practiced in many perennial fruits, particularly apple, predictability of results has been a major factor in determining confidence in the practice. Unpredictability gives rise to risk and lack of confidence, and a decline in use of bioregulators to thin fruit. While most orchardists are concerned about overthinning, the more common underthinning is possibly more counter productive as it means extensive hand thinning is required leading to high costs, poor fruit quality and the potential for biennial bearing.

Thinning recommendations need to be based on rigorous scientific findings and must be repeatable. Often research has involved only one or two aspects of the problem, leaving other important factors unexplored. As a consequence, application of findings has led to inconsistencies and difficulties in application, and a subsequent mistrust of recommendations. This has resulted in many countries still relying heavily on hand thinning, unlike the USA and Australia where aggressive thinning policies have been adopted by many orchardists.

References

Anonymous. 1998. To thin or not to thin. *Australian Citrus news* **74**: 7.

Auchter, E.C. and J.W. Roberts. 1934. Experiments in the spraying of apples for the prevention of fruit set. *Proceedings of the American Society of Horticultural Science* **30**: 22-25.

Bagenal, N.B., Goodwin, W., Salmon, E.S. and W.M. Ware. 1925. Spraying experiments against apple scab. *Journal of the Ministry of Agriculture (Great Britain)* **32**: 137-150.

Baroni, G., Tonutti, P. and A. Ramina. 1986. Integrated chemical thinning of peach cultivars. *Rivista della Ortoflorofrutticoltura Italiana* **70**: 215-225.

Batjer, L.P. and M.N. Westwood. 1960. 1-Naphthyl N-methylcarbamate, a new chemical for thinning apples. *Proceedings of the American Society of Horticultural Science* **75**: 1-4.

Belmars, K. and J. Keulmans. 1987. Timely fruit thinning of plums. Often an essential culture measure. *Boer en de Tuinder* **93**: 23-24.

Beutel, J.A. 1989. Asian Pears. Family Farm Series, Small Farm Centre, University of California.

Bound, S.A. 1995. Assessment of the desiccant FPX0259 as a blossom thinner for pome and stone fruit. Report of 1994/95 trial work submitted to Ferro Corporation (Australia) Pty Ltd.

Bound, S.A. and L. Mitchell, 2000a. The effect of blossom desiccants on crop load of Packhams Triumph pear. Presented at the VIII International Symposium on Pear, Ferrara-Bologna, Haly, Sept 4-9, 2000.

Bound, S.A. and L. Mitchell. 2000b. A new post-bloom thining agent for Packhams Triumph Pear. Presented at the VIII International Symposium on Pear, Ferraia-Bologna, Haly, Sept. 4-9, 2000.

Bound, S.A. and L. Mitchell. 2000c. Chemical thinning in Nashi. Presented at the VIII International Symposium on Pear, Ferrara-Bologna, Haly, Sept. 4-9, 2000.

Bound, S.A., Jones, K.M. and T.B. Koen. 1991a. Ethephon concentration and timing effects on thinning Winter Cole pears. *Australian Journal of Experimental Agriculture* **31**: 133-136.

Bound, S.A., Jones, K.M., Koen, T.B., Oakford, M.J., Barrett, M.H. and N.E. Stone. 1991b. The interaction of Cytolin and NAA on cropping red Delicious. *Journal of Horticultural Science* **66**: 559-567.

Bound, S.A., Jones, K.M. and M.J. Oakford. 1997. Integrating Cytolin into a chemical thinning program for red 'Delicious' apple. *Australian Journal of Experimental Agriculture* **37**: 113-118.

Bound, S.A., Mitchell, L. and S. Daniels. 1999. Thinning options for Packham pear. Poster presented at Snack Fruit 99, the combined National Fruit Industry Conference, Canberra, 19-23 July 1999.

Bradley, M.V. and J. Crane. 1960. Gibberellin-induced inhibition of bud development in some species of *Prunus*. *Science* **131**: 825-826.

Bramlage, W.J., Weis, S.A. and D.W. Green. 1990. Observations on the relationships among seed number, fruit calcium, and senescent breakdown in apples. *HortScience* **25**: 351-353.

Browning, G. and M.G.R. Cannell. 1970. Use of 2-chloroethane phosphonic acid to promote the abscission and ripening of fruit of *Coffea arabica* L. *Journal Horticultural Science* **45**: 223-232.

Burkholder, C.L. and M. McCown. 1941. Effect of scoring and of a naphthyl acetic acid and amide spray upon fruit set and of the spray upon preharvest fruit drop. *Proceedings of the American Society of Horticultural Science* **38**: 117-120.

Chapman, J.C. 1984. Ethephon as a fruit thinning agent for 'Murcott' mandarins. *Scientia Horticulturae* **24**: 135-141.

Coggins, C.W. 1997. Citrus plant growth regulators—general information. University of California IPM Pest Management Guidelines: Citrus, UC DANR Publication 3339.

Crane, J.C. 1953. Further responses of the apricot to 2,4,5-trichlorophenoxyacetic acid application. *Proceedings of the American Society of Horticultural Science* **61**: 163-174.

Crane, J.C. 1971. The unusual mechanism of alternate bearing in the Pistachio. *HortScience* **6**: 489-490.

Davidson, J.H., Hammer, O.H., Reimer, C.A. and W.C. Dutton. 1945. Thinning apples with the sodium salt of naphthyl acetic acid. Michigan Agricultural Experiment Station Quarterly Bulletin No. 27: 352-356.

Davies, F.S., Ismail, M.A., Stover, E.W., Tucker, D.P.H. and T.A. Wheaton. 1998. Florida Citrus pest management guide: plant growth regulators. Institute of Food and Agricultural Sciences, University of Florida, Fact Sheet HS-108.

Gallasch, P.T. 1974. Regulating Valencia orange crops with CEPA (2-chloroethyl phosphonic acid)—preliminary studies. *Australian Journal of Experimental Agriculture and Animal Husbandry* **14**: 835-838.

Gallasch, P.T. 1998. Thinning citrus fruitlets to increase fruit size. SARDI Citrus Information, South Australian Research and Development Institute. Retrieved by Netscape Navigator Version 3.01 on 30/8/99, URL: http://www.sardi.sa.gov.au/

Gardner, F.E., Marth, C. and L.P. Batjer. 1939. Spraying with plant growth substances for control of the preharvest drop of apples. *Proceedings of the American Society of Horticultural Science* **37**: 415-428.

Harris, R.W. and C.J. Hansen. 1955. The effect of growth regulators on the drop of French prune. *Proceedings of the American Society of Horticultural Science* **66**: 79-83.

Harris, R.W. and C.L. Smith. 1957. Chemical thinning of Pecan Nut crops. *Proceedings of the American Society of Horticultural Science* **70**: 204-208.

Harris, R.W., Crane, J.C., Hansen, C.J. and R.M. Brooks. 1953. 2,4,5-T sprays on stone fruit. *California Agriculture* **7**: 8-9.

Hartmann, H.T. 1952. Spray thinning of olives with naphthalenacetic acid. *Proceedings of the American Society of Horticultural Science* **59**: 187-195.

Hutton, R.J. 1992. Improving fruit size and packout of late Valencia oranges with ethephon fruit-thinning sprays. *Australian Journal of Experimental Agriculture* **32**: 753-758.

Jones, K.M., Koen, T.B., Bound, S.A., Oakford, M.J., Pettenon, S. and T. Rudge. 1991. Effects of paclobutrazol and carbaryl on the yield of Hi-Early red 'Delicious' apples. *Journal of Horticultural Science* **66**: 159-163.

Jones, K.M., Bound, S.A., Koen, T.B. and M.J. Oakford. 1992. Effect of timing of hand thinning on the cropping potential of red Fuji apple trees. *Australian Journal of Experimental Agriculture* **32**: 417-420.

Jones, K.M., Bound, S.A. and P. Miller. 1998. *Crop Regulation of Pome Fruit in Australia.* Tasmanian Institute of Agricultural Research, University of Tasmania.

Knight, J.N. 1982. Regulation of cropping and fruit quality of Conference pear by the use of gibberellic acid and thinning. II. The effect of ethephon as a flower thinner when used in conjunction with gibberellic acid application. *Journal of Horticultural Science* **57**: 61-67.

Knight, J.N. and J.E. Spencer. 1982. Chemical fruit thinning. Report of the East Malling Research Station for 1981, 29-30.

Koller, O.C., Marodin, G.A.B., Manica, I., Schwarz, S.F., deBarros, I.B.I. and C.I.N. Barradas. 1986. Response of mandarin (*Citrus deliciosa* Tenore) cv. Montenegrina to chemical and manual fruit thinning. *Proceedings of the Interamerican Society for Tropical Horticulture* **30**: 45-57.

Lahav, E., Korkin, A. and G. Adar. 1989. Thinning stage influences fruit size and yield of Kiwifruit. *HortScience* **24**: 438-440.

Lavee, S. and P. Spiegel-Roy. 1967. Effect of time of application of two growth substances on the thinning of olive fruit. *Proceedings of the American Society of Horticultural Science* **91**: 180-186.

Lenz, F. And P.R. Cary. 1969. Relationships between the vegetative and reproductive growth in Washington Navel orange as affected by nutrition. *In:* Proceedings of the First International Citrus Symposium, Vol 3, pp 1615.

Link, H. 1976. Possibilities of using growth substances in fruit growing. *Mitteilungen fur den Obstbau* **20(3)**: 57-65.

Lombard, P.B. 1982. Pear pollination and fruit set. *In:* The Pear—Cultivars to Marketing, T. Van der Zwet and N.F. Childers (Eds), Horticultural Publications, Florida, 91-103.

Magness, J.R., Batjer, L.P. and C.P. Harley. 1940. Spraying apples for blossom removal. *Proceedings of the American Society for Horticultural Science* **37**: 141-146.

Martin, G.C., Fitch, L.B., Sibbett, G.S., Carnill, G.L. and D.E. Ramos. 1975. Thinning French prune (*Prunus domestica* L.) with (2-chloroethyl) phosphonic acid. *Journal of the American Society for Horticultural Science* **100**: 90-93.

Meland, M. and B. Gjerde. 1996. Thinning apples and pears in a nordic climate. I. The effect of NAA, ethephon and lime sulphur on fruit set, yield and return bloom of four pear cultivars. *Norwegian Journal of Agricultural Sciences* **10(4)**: 437-451.

Miller, P. 1985. Apple thinning in Australia. Horticultural Research Institute, Knoxfield, Victorian Department of Agriculture and Rural Affairs.

Mitchell, L. and S.A. Bound. 1999. Chemical thinning of Nashi. Poster presented at Snack Fruit 99, the combined National Fruit Industry Conference, Canberra, 19-23 July 1999.
Modoran, I., Pasc, I., Prica, D., Vladeanu, D. and D. Modoran. 1979. Research on regulating the yield of apple and pear trees to ensure regular bearing. *Lucarile Stiintifice ale Institutulini de Cercetari pentru Pomicultura, Pitesti* **6**: 79-89.
Nickell, L.G. 1982. *Plant Growth Regulators: Agricultural Uses.* Springer-Verlag, Berlin, Heidelberg, New York.
Russel, H.A. and S. Pickering. 1919. Science and Fruit Growing. Macmillan, New York.
Schwallier, P.G. 1996. Apple thinning guide. Great Lakes Publishing Company, Sparta, USA.
Sefick, H.J. 1975. Hand-thinning of peaches. *In:* The Peach, N.F. Childers (Ed.), Horticultural Publications, New Jersey, USA, 270-276.
Selimi, A. and J.F. Gibbs. 1976. Temperate tree fruits. Tatura Horticultural Research Station, Annual Research Report 1975-76, p. 13.
Selimi, A. and J.F. Gibbs. 1978. Pears. Irrigation Research Institute, Tatura. Biennial report for 1976-77 and 1977-78, p. 23.
Southwick, S.M., Weis, K.G., Yeager, J.T. and H. Zhou. 1995. Controlling cropping in 'Loadel' cling peach using gibberellin: Effects on flower density, fruit distribution, fruit firmness, fruit thinning and yield. *Journal American Society for Horticultural Science* **120**: 1087-1095.
Southwick, S.M., Weis, K.G. and J.T. Yeager. 1996. Chemical thinning of stone fruits in California. *Good Fruit Grower* **47(7)**: 34-35, 43, 52.
Southwick, S.M., Yeager, J.T. and K.G. Weis. 1997. Use of gibberellins on 'Patterson' apricot (*Prunus armeniaca*) to reduce hand thinning and improve fruit size and firmness: Effects over three seasons. *Journal of Horticultural Science* **72**: 645-652.
Stino, G.R., Khattab, M.M. and N.M. Gharabawy. 1980. Effect of ethephon on thinning Le Conte pear trees. Research Bulletin, Faculty of Agriculture, Ain Shams University. No. 1319.
Suranyi, D. 1986. The role of ecological factors affecting the thinning of apricot. *Acta Horticulturae* **192**: 73-80.
Tukey, R.B. and M.W. Williams. 1972. Chemical thinning of Bartlett pears. Tree Fruit Production Series, Cooperative Extension Service, Washington State University, No. EM3516.
Veinbrandts, N. 1967. Thiram thins Granny Smith apples. *The Fruit World & Market Grower* **68(11)**: 52.
Veinbrandts, N. and J.F. Hutchinson. 1976. Studies in the use of 2-chloro-ethylphosphonic acid (ethephon) as a thinning agent for Jonathan apples. *Australian Journal of Experimental Agriculture and Animal Husbandry* **16**: 937-942.
Warner, G. 1999. New apple thinners won't be here this season. *Good Fruit Grower* **50(7)**: 6-7.
Weaver, R.J. 1972. *Plant Growth Substances in Agriculture.* W.H. Freeman and Company, San Francisco, 335-352.
Westwood, M.N. 1965. A cyclic carbonate and three new carbamates as chemical thinners for apple. *Proceedings of the American Society of Horticultural Science* **86**: 37-40.
Westwood, M.N. 1978. Temperate-Zone Pomology. W.H. Freeman and Company, San Francisco, 199-219.
Williams, M.W. 1974. Chemical thinning and guide for size-thinning Bartlett pears. *In:* Proceedings of the 69th Annual Meeting of the Washington State Horticultural Association (USDA: Wenatchee, Washington D.C.).
Williams, M.W. 1994. Factors influencing chemical thinning and update on new chemical thinning agents. *Compact Fruit Tree* **27**: 115-122.
Williams, M.W. and L.J. Edgerton. 1981. Fruit thinning of apples and pears with chemicals. United States Department of Agriculture, Agriculture Information Bulletin Number 289.
Williams, M.W., Bound, S.A. Hughes, J. and S. Tustin. 1995. Endothall: A blossom thinner for apples. *HortTechnology* **5**: 257-259.

5

Low-alcohol Grape and Fruit Wine

Gary J. Pickering
Associate Professor, Cool Climate Oenology-Viticultuse Institute, Brock University, St. Cathariner, ON L25 3A1, Canada.

Abbreviations

CO_2	carbon dioxide
DLRAW	dealcoholized, low, and reduced-alcohol wine
GOX	glucose oxidase
RO	reverse-osmosis
SCC	spinning cone column
SO_2	sulphur dioxide
v/v	volume per volume

1. Introduction

Grape wines with a reduced-alcohol (ethanol) content have been commercially available for over two decades, and as a generic group of products, has posed a number of technical and marketing challenges. While there have been significant developments over the last few years in the production methods employed and consequent improvements in wine quality, these are generally not appreciated by consumers and other sectors of the wine industry.

Apples, pears, plums, cherries, berries and currants are the most commonly used fruits in fruit-wine production, and as they typically contain considerably less fermentable sugar than grapes, their wines are intrinsically lower in alcohol content. However, with many fruit wine styles, sugar addition (chaptalisation) to the juice or fermenting must is practised in order to yield wine of a higher alcoholic strength. In contrast, a number of 'low' and particularly dealcoholized fruit wines are available commercially.

This chapter primarily reviews the technologies used to produce dealcoholized, low, and reduced-alcohol wine, although consideration is also given to the key quality, sensory, economic and marketing issues associated with wine having a reduced-alcohol content. Wines with a reduced-alcohol content are conventionally classified as either dealcoholized or no alcohol (< 0.5% v/v), low-alcohol (0.5% to 1.2% v/v), or reduced-alcohol (1.2% to 5.5%-6.5% v/v). These alcohol value ranges can vary between countries. For convenience, all three categories (dealcoholized, low, and reduced-alcohol wine) are referred to as DLRAW.

2. Production Methods

Techniques for producing DLRAW have been available since the early 1900s. Although commercial production now relies almost exclusively on membrane and modified distillation-based systems, a number of alternative approaches have been used in the past or put forward in research and patent literature. These methods are summarized in Table 1 and expanded on below. While most of the technologies reviewed have been advanced for use with grape wine or juice, many are equally applicable to the production of low and dealcoholized fruit wine. Consideration of the effects of production methods on the quality and sensory aspects of the finished wine has been left, where possible, to later in the chapter.

Table 1. Methods for producing dealcoholized, low and reduced-alcohol wine

Principal	*Method*
Reduction of fermentable sugar concentration in fruit or juice	*Use of unripe fruit*
	Juice dilution
	Freeze concentration and fractionation
	Enzymes
Removal of alcohol from wine	*Thermal*: distillation under vacuum or atmospheric pressure; evaporation; freeze concentration and fractionation
	Membrane: reverse osmosis; dialysis
	Adsorption: resins; silica gel
	Extraction: organic solvents; supercritical CO_2
Other	*Arrest fermentation early*
	Dilution of wine
	Low-alcohol producing yeast
	Combinations of above methods

2.1. Thermal and Distillation Processes

Distillation using either evaporators or distillation columns is the most common thermal-based method for removing alcohol from wine. Until recently, the process of dealcoholization required heating and the evaporation of 50 to 70% of the wine to reduce the alcohol content to below 0.5% v/v. The original pressure boiling pan and distilling vessel were replaced by vacuum distillation apparatus enabling removal of ethanol at much lower temperatures. Single or multiple step evaporators can be used. There have been numerous variations and modifications of the distillation/evaporation principle, most of which are patented (Thumm, 1975; Deglon, 1975; Boucher, 1983, 1985, 1988; Schobinger et al. 1986b;

Trothe, 1990). Many of these modifications incorporate one or more non-thermal methods, improved aroma recovery techniques, shorter processing times, lower temperatures, and include the addition of blended grape juice or concentrate to the alcohol-reduced wine. This latter step is primarily for adjustment of sensory properties.

2.1.1. Spinning Cone Column

The spinning cone column (SCC) is a modern, multi-stage strip column first developed in the USA in the 1930s and modified more recently in Australia. It enjoys increasing popularity amongst DLRAW producers. The SCC is a gas-liquid contacting device consisting of a vertical countercurrent flow system that contains a succession of alternate rotating and stationary metal cones whose upper surfaces are wetted with a thin film of liquid (wine or juice). The liquid flows down the upper surfaces of the stationary cones under the influence of gravity and moves down the upper surfaces of the rotating cones in a thin film by the action of an applied centrifugal force. Vapour flows up the column, traversing the spaces between the successive fixed and rotating cones (Fig. 1). A commercial SCC used for producing DLRAW is shown in Fig. 2.

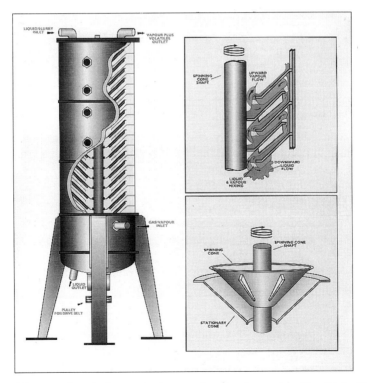

Fig. 1. Diagram of a spinning cone column and mode of operation. *(Reprinted with permission from Flavourtech Pty Ltd).*

Fig. 2. A commercial spinning cone column used for the production of dealcoholized, low, and reduced-alcohol wine. *(Reprinted with permission from Flavourtech Pty Ltd)*.

One processing option using the system is given below:

Juice extracted and fermented conventionally ➲ Wine aromas removed using SCC ➲ Alcohol removed using SCC ➲ Wine aromas added back to dealcoholized wine ➲ Blending with full-strength wine, juice, or juice concentrate ➲ Filtration ➲ Bottling

Advantages of the SCC system include high efficiency, low liquid residence times, low entrainment, minimal thermal damage, ability to handle highly viscous juice, and good energy efficiency (Sykes et al. 1992; Gray, 1993; Pyle, 1994). Whether dealcoholized, low, or reduced-alcohol wine is desired, the same basic process is applied, with the volume processed through the column and blending options governing final alcohol content. Interestingly, the SCC process is also used to finely adjust alcohol levels in full-strength premium wines. A further advantage of distillation and evaporation techniques generally is that extracts, minerals, and other non-volatile components in the original wine are preserved.

The chief technical disadvantage of SCC is that some heating of the wine is still required for the dealcoholization step (carried out at around 38° C).

2.1.2. Freeze Concentration

Another thermal method, although infrequently used, is freeze concentration. Water in wine can be removed by freezing and the alcohol in the residual liquid can be removed by vacuum distillation. Also the wine can be cooled until crystals are formed which are separated and later thawed. Low-alcohol wine that results can be adjusted to any alcohol content with the separated alcohol fraction. The process is relatively delicate and expensive (Schobinger et al. 1986a; Villettaz, 1987).

2.2. Membrane Processes

Semipermeable membranes by which alcohol can be separated from fermented beverages have been available since the 1970s. Reverse-osmosis (RO) and dialysis are two processes that make use of such membranes and both are used commercially. They have the advantage of working at low temperatures of approximately 5-10° C where there is minimal negative influence on taste (Schobinger et al. 1986a).

2.2.1. Reverse-Osmosis

Overall reverse-osmosis is probably the most widely used technique at present for reducing the alcohol content in wine. It involves pressure filtration of the wine through a fine porous membrane that is permeable to alcohol and water, but not too many of the dissolved extract components. However, some aroma compounds (e.g. esters and aldehydes), organic acids and potassium can diffuse through with the alcohol (Schobinger et al. 1986a, Villettaz, 1986). Figures 3 and 4 summarize the process and equipment used.

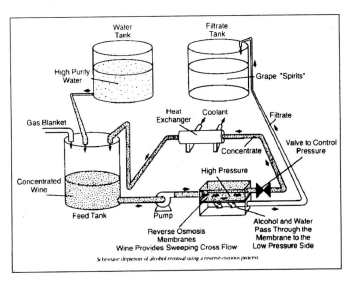

Fig. 3. Processing scheme for alcohol removal using a reverse-osmosis process. (From Meier, 1992. Reprinted with permission from Ryan Publications Pty Ltd).

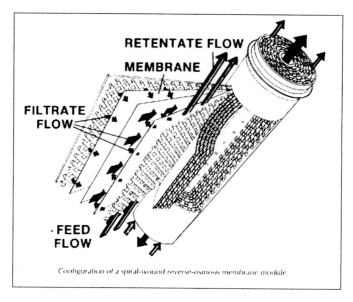

Fig. 4. Schematic of a spiral-wound reverse-osmosis membrane module showing flow of feed, filtrate, and retentate. *(From* Meier 1992. *Reprinted with permission from Ryan Publications Pty Ltd).*

With the use of a proper support system and sufficient pressure, RO can reduce the alcohol content of wine to almost any degree desired. Additional advantages include the reductive environment that can be maintained during processing and good energy efficiencies. However, as water is removed along with ethanol, it must be added back to the concentrated wine or added to the wine before use of RO. This creates legal problems in countries where the addition of water to wine is prohibited. Bui et al. (1986) circumvented the problem of water addition by using a double RO process that produces low-alcohol and alcohol-enriched wines simultaneously. Other processing variations based on RO also have been advanced (Cuénat et al. 1985; Weiss, 1987; Chinaud et al. 1991).

2.2.2. Dialysis

In contrast to RO which uses hydrostatic pressure as the driving force, dialysis uses differences in concentrations for substance transport. With dialysis, water is used to provide the concentration gradient, allowing net movement of ethanol and low molecular weight compounds out of the wine and into the water. The advantages of this process include functioning without pressure, no increases in concentration nor dilution required, no cooling of the system, and the loss of CO_2 is less (Schobinger et al. 1986a). Wucherpfennig et al. (1986) describe a method in which the wine is dialysed against wine dealcoholized by vacuum distillation rather than against water; as the concentration gradient exists only for alcohol, they claim little change is observed in the concentration of other components.

2.3. Extraction Processes

Wine can be extracted directly by organic solvents such as pentane and hexane, or the alcohol and aroma-containing condensate resulting from wine evaporation can also be extracted. In both cases, the aroma compounds are largely in solution. Potential disadvantages of direct extraction of wine include thermal damage, and particularly the presence of solvent residuals in the extract. The process is not used commercially. In the case of liquid-liquid extraction, the solvent must be food appropriate, and CO_2 is most commonly used.

A variation known as high pressure extraction uses both extraction and distillation principles. Wine is extracted using liquid CO_2 which, under specific pressure and temperature conditions, has similar properties to solvents. Through subsequent and differential temperature and pressure adjustments, the extracted wine components precipitate and ethanol and aroma compounds can be separated. The aroma component can be returned later to the extracted wine. The process reportedly results in relatively good quality products, but can be expensive (Schobinger et al. 1986a).

2.4. Removal of Juice Sugar

Harvesting fruit at an early stage of development and subsequent vinification results in wine of reduced alcohol content; however 'unripe' aromas and unacceptably high acid levels in the finished wine result in a product of inferior quality.

Removal of a portion of the fermentable sugar from the juiced fruit allowed to reach full ripeness and subsequent anaerobic fermentation is another method for reducing the alcohol content of the finished wine. A number of variations have been developed using this approach, with recent work focusing on the use of exogenous enzyme systems. A discussion of this follows.

2.4.1. Glucose Oxidase

Glucose oxidase (EC 1.1.3.4) (GOX) is an aerobic dehydrogenase that catalyses the oxidation of glucose to gluconolactone in the presence of molecular oxygen. In a subsequent step, gluconolactone is hydrolysed non-enzymatically to gluconic acid (Fig. 5).

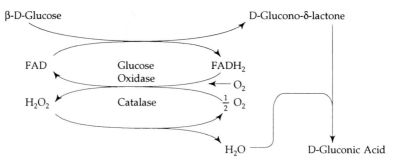

Fig. 5. Reaction mechanism of the glucose oxidase/catalase system.

The use of this enzyme in the production of reduced-alcohol grape wines was introduced by Villettaz (Villettaz, 1986, 1987) and Heresztyn (1987). As the fermentable sugar fraction of grape juice is approximately 50% glucose and 50% fructose (some variation exists due to variety, season, and ripeness), fermenting the glucose-depleted juice after treatment with GOX produces a wine with approximately half of the potential alcohol content. Application with fruit juices is also possible, although the relatively high fructose : glucose ratio in many fruits may necessitate the use of glucose isomerase to convert (isomerise) the desired amount of fructose into glucose.

Pickering et al. (1999a) reported on trials designed to optimize the glucose conversion by GOX in grape juice and showed that the low pH of the juice is the dominant factor in limiting glucose conversion. A processing scheme is given in Fig. 6.

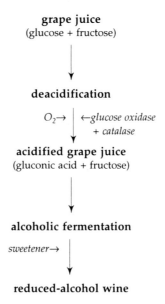

Fig. 6. Simplified flow diagram of reduced-alcohol white wine production using glucose oxidase.

Chemical evaluation of white wine produced using this approach showed that up to 40% reduction in potential alcohol yield can be obtained, high levels of gluconic acid are retained in the finished wine, and relatively little change in other non-volatile compounds occurs. In addition, a relatively higher concentration of esters and fatty acids was reported. The wines have a deeper colour, and show increased SO_2-binding power compared to control wines (Pickering et al. 1999b, c, e).

With respect to sensory characteristics, these GOX-treated wines are acidic due to the high level of gluconic acid, with sweetening (unfermented juice, juice concentrate, sugar) recommended to address this imbalance. Most aroma, aroma-by-mouth, and mouthfeel characteristics, however, appear to be relatively unaffected (Pickering et al. 1999d). Glucose oxidase is not on the schedule of permitted wine additives in many countries, although various other enzymes sourced from *Aspergillus niger* are, and it is widely used throughout the food industry. The technology is also limited at present to reduced-alcohol white wine, although application to red wine production may be possible with modified processing.

2.4.2. Other Approaches

A method has been patented which involves separating fruit juice into a high sugar and low sugar fraction by freezing, to form a slush (Lang and Casimir, 1986). The slush is then filtered using a custom built extractor. Volatile components from the juice of the high sugar fraction are stripped using a spinning cone fractional distillation column and added to the low sugar fraction, which is then fermented.

More common, although with limited commercial application at present, are biological approaches to sugar reduction. A process has been described for the preparation of low sugar or sugar-free fruit juices based on continuous or semi-continuous culture with yeast, with the conditions resulting in metabolism of sugar to CO_2 and water rather than to ethanol (Kappeli, 1989). A procedure has been advocated involving fermentation of grape must using a 'special yeast' under controlled aeration conditions, with (according to the patent application) typical fermentation aroma compounds formed and fruit-derived aroma compounds liberated. When the desired alcohol concentration is reached, aeration is terminated and the yeast separated from the fermented must by microfiltration (Grossmann et al. 1991).

Following a broad screening of yeast strains, Kolb et al. (1993) found that *Pichia stipitis* was particularly well suited to juice sugar removal. Their claims include the elimination of more than 50% of juice sugar within 20 hours, no requirement for added nutritive or other substances, and minimum adverse effects on the sensory and functional qualities of the juice.

Smith (1995) studied the effects of temperature and aeration on the reduction of sugar content and production of alcohol by selected yeast strains in Müller-Thurgau grape juice. She also combined inoculation with selected yeast and short-term controlled aeration of the juice with an anaerobic fermentation using *Saccharomyces cerevisiae* to produce reduced-alcohol wine. With *Pichia stipitis* or *Candida tropicalis* as the aerobes, production of wines with 25-30% less alcohol and an acceptable taste were reported.

2.5. Blends and Dilution

Where permissible, dilution with water is the simplest means of reducing alcohol concentration, with flavour enhancement able to offset flavour dilution. Alcohol dilution can also be achieved through blending full strength, reduced-alcohol, or partially fermented wine with fruit juice(s), as in the commercial production of wine coolers. Some developments in this area include the formation of a reduced-alcohol wine product from blending grape must with kiwifruit juice (*Actinidia chinensis* Planch), and low-alcohol wine coolers from blends of red table wine and blood orange juice (Anelli et al. 1986; Maccarone et al. 1993). The correction of sensory imbalances in DLRAW through the use of blending is discussed in more detail later.

2.6. Other Methods

Alcohol can be adsorbed onto porous resins such as styrol/divinylbenzol-copolymers or alternatively silica gels can be used. These processes are more suitable for laboratory rather than large-scale production.

The *early arrest of fermentation* will produce wine of reduced-alcohol content, and can be used successfully with both grape and fruit musts. There can be some structure to these wines, although the method is obviously restrictive in terms of the styles that can be produced; it is best used when the product is the low-alcohol version of a wine style that is traditionally sweet. As the residual sugar content is high, the wine has to be stabilized, which is usually achieved by pasteurization and SO_2 addition.

The use of *lower efficiency fermentation yeasts* is a possibility for the future, and research currently under-way is investigating reducing the amount of ethanol produced by *Saccharomyces cerevisiae* by diverting sugar metabolism into glycerol production through genetic manipulation. A related project in progress is the screening for yeast strain with reduced alcohol production (Henschke, 1995).

Finally, production methods may incorporate a combination of the techniques outlined above to achieve the alcohol reduction desired and to improve the quality.

3. Quality Considerations

3.1. Fruit Quality

It is a long established wine-making paradigm that high-quality wine cannot be produced from anything less than high-quality fruit. In the early days of DLRAW production, the fruit was seldom of this standard. The grapes used were often considered of inadequate quality for companies' main-stream full-strength wines, and were relegated to 'lesser' products such as DLRAW. Often this fruit was from heavily cropped vines or non-premium grape varieties, which also limited the quality potential of the final wines.

Fortunately this situation is improving now, with most of the top producers paying greater attention to the quality of source fruit. Many now also include in their range a selection of single-variety wines produced from the classic *vinifera* cultivars; a relatively recent development. Even with adequate fruit quality, it is still difficult to produce high quality products if the base wine (in the case of dealcoholisation) is of low quality. Gross winemaking faults or low-aroma intensity often will be carried through to the final wine.

3.2. Sensory Properties

Reduced sensory quality compared with full-strength wine has frequently been associated with DLRAW, and in particular, problems with flavour imbalance and a lack of 'body' (Schobinger and Dürr, 1983; Schobinger, 1986; Howley and Young, 1992). In addition, below approximately 6% v/v, alcohol-reduced wines are often perceived as wine-like beverages rather than wine, with the taste of completely dealcoholized wines described as lying between a conventional wine and grape juice (Schobinger, 1986; Schobinger et al. 1986a). These altered sensory properties can occur as a result of the processing required to produce DLRAW or as a direct consequence of the reduced ethanol content.

Aroma compounds are lost in the dealcoholizing process, particularly with some evaporation-based techniques. The literature suggests that extraction and dialysis methods may have the greatest retention of aroma compounds (Duerr and Cuénat, 1988; Schobinger et al. 1986a; Wucherpfennig et al. 1986). Acids and salts also can be partly removed in the dealcoholizing processes, further affecting sensory properties. Thermal based methods in particular have been criticised for imparting undesirable 'cooked' and other flavours, while a loss of flavour intensity and 'wine characters' has been reported with reverse-osmosis based production (Neubert, 1976; Schobinger et al. 1986a). Advances in evaporative and membrane-based technologies over the last decade have gone some way to addressing these limitations.

However, regardless of which method is employed, aroma loss and modification of other flavour components is reported with increasing alcohol removal (Génat et al. 1985; Schobinger, 1986; Schobinger et al. 1986a; Lynch, 1988; Noble, 1995).

Alcohol, as well as possessing taste properties (sweetness and bitterness) and thermal effects, may also play an important role as a taste and aroma enhancer. With increasing dealcoholization, acidity, bitterness and astringency are heightened, often to the point of imbalance, as the softening and harmonizing effect of alcohol is increasingly lost. In addition, there is a long-held belief that there is a loss of 'body' in DLRAW associated with their reduced-alcohol content. The data of Pickering et al. (1998) confirmed lower perceived viscosity and weight in wines containing 0-10% alcohol v/v, while suggesting that this may not hold for wines above 10-12%. Further research is required into the role of alcohol in wine mouthfeel, particularly given the weight of anecdotal evidence equating increasing alcohol concentration with increasing palate weight and 'body'. Production options for maximizing mouthfeel in DLRAW are discussed in the next section.

The sensory properties of DLRAW therefore, will vary depending on the alcohol concentration, and in turn this will help dictate what flavour adjustments are required to maximize their sensory quality and acceptability.

3.2.1. Flavour Adjustment

Although various claims, especially in patent applications, ascribe improved sensory characteristics to particular production methods, an absence of formal sensory evaluation is the rule rather than the exception. However there have been a few key developments that have undoubtedly lead to improved quality. The reduction in processing temperatures now possible under vacuum (and other modifications) has minimized the cooked notes associated with many distillation/evaporation methods. Isolation and return of aroma compounds to the wine is now practised, and is almost invariably necessary to improve the aroma quality and wine character. In one of the few comparative studies in the literature, Carnacini et al. (1989) compared the elimination of alcohol and recovery of aroma compounds from wine by permoseparation on a RO membrane, dialysis, and extraction with subcritical and supercritical CO_2, and concluded that extraction with supercritical CO_2 is the most promising process.

Blending, using full-strength wine, pomace wash extracts, juice or juice concentrate is an important processing option in regaining some aroma and mouthfeel balance in finished wines (Schobinger and Dürr, 1983; Génat et al. 1985; Schobinger, 1986; Fetter and Schoeller, 1989; Petershans, 1989). An important principle when blending with grape juice or concentrate is to achieve the necessary balance while minimizing the grape juice character. The use of these adjuncts raises the residual sugar level of the finished wine, generally limiting production to 'non-dry' styles. Other important parameters to consider when optimizing the blend are the phenolic and CO_2 content, sugar-free extract, sugar/acid ratio, and the juice aroma. For grape wines, relatively high levels of sugar and acid are necessary for a satisfactory sensory impression, although deacidification using malolactic fermentation or calcium carbonate also can be carried out in the case of exceptionally high juice or wine acidity. Recommended ratios and ranges for many of these composition variables have been documented (Schobinger and Dürr, 1983; Schobinger, 1986; Schobinger et al. 1986a).

4. Marketing and Economics

4.1. Economic Considerations

4.1.1. Costs

The cost of production (e.g. capital, equipment, and operating) is obviously a critical factor governing the viability and profit margins of DLRAW, and is variable depending on the process used. Due to commercial sensitivities it is difficult to obtain reliable data on relative production costs. Significant capital investment in equipment is required for the latest membrane and 'spinning-cone' methods, with the latter being significantly more expensive. These two processes are now the most prevalent in commercial DLRAW production.

Once equipped, however, the production costs can be remarkably inexpensive. Using traditional evaporative techniques, the cost to produce reduced-alcohol wine can be as low as one cent per litre (Scudamore-Smith and Moran, 1997), and operating costs for RO-based production may average between one and three cents ($US) per litre (Meier, 1992). This contrasts with other processes, such as the use of GOX to reduce the juice sugar concentration prior to fermentation, where the cost of the enzyme preparation may add up to 40 cents ($US) per litre, although initial set-up costs are likely to be minimal (Pickering, 1997).

4.1.2. Savings

There also exists some cost recovery for producers using methods based on alcohol removal from fully fermented wine; the alcohol fraction may be retrieved and used for other products (such as brandy) or on-sold. Outsourcing the dealcoholizing step to companies with the requisite equipment is practised and makes financial sense for small to medium size wineries with, for instance, a DLRAW brand as a support for their full-strength wines.

A catalyst helping to stimulate commercial interest in DLRAW over the last decade has been the reduced sales tax and duty afforded these products in many countries. Interestingly, however, a number of (adverse) disparities in the tax structure for DLRAW compared with reduced-alcohol beer can be found (e.g. the federal sales tax on wine in

Australia). The tax rates and tariffs applicable to DLRAWs are usually based on their respective alcohol content, and can vary considerable between countries. Thus, when developing DLRAWs, the interaction between wine alcohol concentration and market destination creates important tax and duty considerations.

However, regardless of the economics of production and any advantage from reduced taxes and duty, there exists considerable marketing challenges with DLRAW.

4.2. Marketing Issues

4.2.1. The Problem

Although DLRAWs have been on the international market since at least the 1970s, early consumer demand was generally low. During the 1980s increased information on and media publicity about health and well-being, along with increased pressure from drink-drive campaigns, led many consumers to reassess their alcohol intake. During this period other alcohol-reduced beverages, particularly beers, lagers, and wine coolers, were successfully developed and marketed (Howley and Young, 1992). The 1990s saw a decrease in wine consumption on the part of the more frequent wine consumers, and particularly amongst the traditional wine drinking countries, with concern over alcohol identified as an important reason for this (d'Hauteville, 1994).

These trends have been largely responsible for the increased interest from producers and researchers over the last 15 years in lower alcohol alternatives to full-strength wines. However, despite what might appear to be a very favourable environment for increased promotion of DLRAW, only a small number of producers can claim sustained success in the market. Reliable data on market share for DLRAW products is sparse, however, for the UK it is likely to be in the order of 2-6%. Anecdotal evidence suggests that the North American market is showing slow growth, while the Australian and New Zealand markets remain static and low. Principal markets for the largest Australian producer have been the UK, Canada, and Scandinavia, with an annual growth rate of up to 5 per cent (pers. comm., Alfredo Calle, Southcorp Wines Pty Ltd).

Overall, DLRAWs as a generic group of wine products have performed well below market expectations and potential. Negative taste association (and its connections to attitudes), labelling, insufficient promotion, and inadequate positioning have been advanced as major reasons for this (Howley and Young, 1992; d'Hauteville, 1994).

4.2.2. Attitudes

An attitude of 'snobbishness' towards DLRAW is prevalent amongst many wine drinkers, and indeed many in the wine industry itself. It is an attitude that extends to fruit wines. These negative attitudes are aided and abetted by many wine writers and key industry figures, who serve important roles as opinion leaders. It is well established in current literature that opinion leaders in their communications affect the behaviour of other consumers. This chain-like process has probably contributed to the reduced market performance of DLRAW.

Differences in acceptance of the concept of low-alcohol wine vary between countries. In a recent survey of three EC countries, German respondents had the highest rate of rejection, followed by the French, with British respondents showing the highest level of acceptance of low-alcohol wines (d'Hauteville, 1994). Even in the British group, however, the acceptance rate was low—only one quarter of the sample accepted the concept of low-alcohol wine. Anecdotal evidence suggests that consumer acceptance of DLRAW is also low in Australia and New Zealand.

The critical question, therefore, is what is responsible for this low level of acceptance? Based on survey data and development of an attitudinal model, the relative importance of a number of determinant factors affecting acceptability of low-alcohol wine was examined by d'Hauteville (1994). These factors were perception of product attributes, involvement in and consumption habits for wine, the consumption situation, consumption habits for light products, and personal innovativeness. The study concluded that taste quality, or rather the perception that 'lack of alcohol will harm the taste of the wine' was the most important attribute affecting consumer acceptance.

4.2.3. Labelling

Legislation governing the labelling of DLRAWs and restrictive definitions of wine are seen as impediments to market development. For instance, in the EC, DLRAW products cannot be labelled as 'low-alcohol wine'. When produced from fully fermented musts and subsequently dealcoholized to < 1.2 % alcohol, they may be labelled as 'dealcoholized wine'. However, when produced from partially fermented juice to an alcohol content of < 5.5%, they must be labelled 'partially fermented must'. The use of these and similar terms on labels may adversely affect purchasing behaviour by creating the perception of a lower quality product or confusing potential purchasers. Producers who wish to label their products as 'wine' must opt for the more expensive process of full fermentation. There is a need for research into the effects of labelling and general packaging on consumer perception of and purchasing behaviour towards DLRAW.

Figure 7 shows a limited selection of DLRAWs produced in Australia and North America.

Fig. 7. A small selection of dealcoholized, low and reduced-alcohol wines from Australia and North America. *(Reprinted with permission from Ariel Vineyards and Southcrop Wines Pty Ltd).*

4.3. Promotion and Positioning

Promotional investment and activity for DLRAW in Australasia generally have been minimal, and parallel sales. However, in the Northern Hemisphere, particularly the UK, the relatively small number of brands that have received substantial promotional and advertising support enjoy clear market dominance (Howley and Young, 1992).

Over the last decade, approaches to promoting DLRAW have included advancing them on the drink-drive platform, associating them as a healthy, low-calorie alternative or as a sociably responsible choice. Recent approaches to promotion, particularly in North America, have focused on the health benefits of DLRAW. A number of marketing campaigns have differentiated their products from full-strength wines by highlighting the recently established beneficial effects of wine consumption (e.g. red wine lowers risk of coronary heart disease), while alluding to the negative effects of alcohol on health (e.g. increased risk of cancer and liver disease). That is, wines with all of the positive and none of the negative health effects. Increased attention also has been paid to niche market segments such as pregnant women, breast-feeding mothers, and consumers unable to take alcohol for medical reasons.

The d'Hauteville study (d'Hauteville, 1994) suggests that while there is some acceptance of low-alcohol wines from both wine drinkers and the light product market segments, they are best positioned as a 'wine' rather than a 'diet' alternative in order to optimize market performance.

Many producers see image as a limiting factor, with consumers still associating DLRAW as being closer to grape juice than wine despite advances in technology that have meant most are more wine than juice-like now and of higher general quality. Commentators also have stressed the need for greater branding and image construction, especially in markets such as the UK where there are over 40 DLRAW products now available.

5. The Future

Predictions for DLRAW vary considerably amongst industry commentators. Many believe there is little future for low and reduced-alcohol grape wine, seeing market potential only for non-alcoholic wine (< 0.5%), particularly amongst the ageing population. Others, particularly in North and South America, believe there will be increased market interest in wines of 5-7% alcohol. Further research has been advocated into the development of DLRAWs that are more compatible with meals, particularly in those countries (e.g. France) where wine and food consumption are strongly associated.

The significant advances in the quality of DLRAW over recent years have not been effectively communicated to consumers. There is a consen-

sus amongst those involved in the industry on what needs to occur for these wines to meet their market potential, specifically:
- increased credibility and acceptance through advocacy by industry and opinion makers
- increased awareness and familiarity of the products amongst consumers, given that penetration is low in most markets
- sustained promotion and advertising campaigns.

6. Conclusions

Although dealcoholized, low and reduced-alcohol wines are consistent with modern trends towards healthy lifestyles and reduced calorie intake, they have not achieved their full market potential. While the sensory quality in the early years was in many instances unsatisfactory, significant improvements in production methods have occurred with membrane and 'spinning-cone' based technologies now the most commonly used. These techniques have undoubtedly lead to improvements in quality, particularly from lower processing temperatures, improved aroma retention and recovery and a better understanding of the role of adjuncts in addressing flavour imbalances.

Despite these technical improvements, there still exists consumer resistance towards DLRAW in many market sectors. This may be largely attributable to ongoing limitations in sensory quality and to promotional issues, including limited advertising budgets and a low level of awareness of recent improvements in quality.

7. Acknowledgements

I wish to thank Alfredo Calle (Southcorp Wines Pty Ltd), Craig Rosser (Ariel Vineyards), and Chris Gray (Flavourtech Pty Ltd) for their technical contributions. Sincere thanks to Dr Art Thomas (Eastern Institute of Technology, Hawke's Bay) for proofing and improving the manuscript.

References

Anelli, G., Massantini, R. and M. Contini. 1986. Kiwine: nuova bevanda a basso tenore alcolico a base di mosto d'uva. Industrie delle bevande, giugno: 180-184.
Low-alcohol drinks. Which? December: 576-577.
Low alcohol: A must for the future? *Decanter, December:* 85-86.
Alcohol-free wine market has a bright future. *The Grocer, October,* **22**: 166.
Technologien zur Herstellung von alkoholarmen und alkoholfreien Fruchtweinen. *Fluessiges-Obst* **50 (12):** 666-669.
Preparation of wine having a low calorie content and a reduced alcohol content. United States Patent 4 405 652, 20/9, *In:* U. Schobinger, P. Dürr, and R. Waldvogel, 1986a. Die Entalkoholisierung von Wein und Fruchtweinen (eine neue Möglichkeit zur Herstellung von zuckerreduzierten Fruchtsaftgetränken). *Schweiz. Zeitschrift für Obst- und Weinbau* **122 (4):** 98-110.
Boucher, A.R. 1985. Verfahren und Vorrichtung zur Herstellung eines alkoholfreien Weingetrünks. Deutsche Offenlegungsschrift DE 3 429 777 A1, 28/2 *In:* U. Schobinger, P. Dürr, and R. Waldvogel, 1986a. Die Entalkoholisierung von Wein und Fruchtweinen

(eine neue Möglichkeit zur Herstellung von zuckerreduzierten Fruchtsaftgetränken). *Schweiz. Zeitschrift für Obst- und Weinbau* **122** (4): 98-110.
Boucher, A.R. 1988. Preparation of alcohol free wine. United States Patent US 4 775 538 (FSTA vol. 21 (1989) No. 2, 2 V 43).
Bui, K., Dick, R., Moulin, G. and P. Galzy. 1986. A reverse osmosis for the production of low alcohol content wine. *Am. J. Enol. Vitic.* **37(4)**: 297-300.
Carnacini, A., Marignetti, N., Antonelli, A., Natali, N. and S. Migazzi. 1989. Alcohol removal and aroma recovery from wine by permoseparation, dialysis, and extraction with CO_2. Industrie delle Bevande 18(102): 257-264. (Subfile: FSTA 00394216 90-07-h0053).
Chinaud, N., Broussaus, P. and G. Ferrari. 1991. Application of reverse osmosis to dealcoholization of wines. Connaissance de la Vigne et du Vin 25(4): 245-250. (Subfile: FSTA vol. 24(1992)-9H117).
Herstellungsmoglichkeiten entalkoholisierter Weine. *Weinwirtshaft-Technik* **125(5)**: 50-59.
Cuénat, Ph., Kobel, D., Crettenand, J. and J.-M. Girard. 1985. Sur les potentialités de l'osmose inverse pour la désalcoolisation partielle ou totale des vins—Exemples d'application. *Revue suisse Vitic. Arboric. Hortic.* **17(6)**: 367-371.
Déglon, H. 1975. Procédé pour la séparation des alcools et essences volatiles des autres produits dans les mélanges obtenus par fermentations. Schweizer patent 564 603, 15.6
In: U. Schobinger, P. Dürr, and R. Waldvogel, 1986a. Die Entalkoholisierung von Wein und Fruchtweinen (eine neue Möglichkeit zur Herstellung von zuckerreduzierten Fruchtsaftgetränken). *Schweiz. Zeitschrift für Obst- und Weinbau* **122** (4): 98-110.
Duerr, P. and P. Cuéna., 1988. Production of dealcoholized wine. *In:* Proceedings of the Second International Symposium for Cool Climate Viticulture and Oenology, Auckland, New Zealand, 11-15 January 1988. R.E. Smart, R.J. Thornton, S.B. Rodriguez and J.E. Young (Eds), NZ Society for Viticulture and Oenology, PO Box 90-276, Auckland Mail Centre, NZ. pp. 363-364.
Fetter, K. and S. Schoeller. 1989. Entalkoholisierte Weine. Geschmackliche Analyse. Weinwirtschaft-Technik, 125(1): 21-23. (subfile: FSTA 00385610 90-01-h0041).
The effect of ethanol, catechin concentration, and pH on sourness and bitterness of wine. *Am. J. Enol. Vitic.* **45**, 1, 6: 10.
Gray, C. 1993. History of the spinning cone column. Juice technology workshop, Cornell University, New York State, Special Report no. 67, 31-37.
Grossmann, M., Kruse, R. and W. Heintz. 1991. Verfahren zur herstellung alkoholarmer bis alkoholfreier getränke. *German Federal Republic Patent Application* DE 39 39 064 (FSTA vol. 24 (1992) No. 4, 4 H 154).
d'Hauteville, F. 1994. Consumer acceptance of low alcohol wines. *International Journal of Wine Marketing* 6(1): 35-48.
Marketing changes, legal questions and the production of dealcoholized wines. *In:* Proceedings 9th International Oenological Symposium, Cascais, Portugal, 24-26 May 1990. International Association for Modern Winery Technology and Management, zum Abtweingarten 15, D-79241 Ihringen, Germany, pp. 154-165.
Henschke, P.A. 1995. Microbiology Group Report. *In:* The 1995 Australian Wine Research Institute Annual Report, PO Box 197, Glen Osmond, South Australia, pp. 18-19.
Heresztyn, T. 1987. Conversion of glucose to gluconic acid by glucose oxidase enzyme in Muscat Gordo juice. *The Australian Grapegrower and Winemaker*, April, pp. 25-27.
Howley, M. and N. Young. 1992. Low alcohol wines: The consumers choice? *International Journal of Wine Marketing* 4(3): 45-46.
Process for manufacture of sugar-free or low-sugar fruit juice, and fruit juice prepared by this process. Swiss Patent No. CH 668-887 AS.
How low-alcohol wines are seen by the public. Off Licence News, August 1, 18.
Kolb, E., Wiesenberger, A., Stahl, U. and K. Harwart. 1993. Biological processes for the partial reduction of sugar in fruit juices. *In:* Proceedings of the International Federation of Fruit Juice Producers Symposium, Budapest, 4-7 May 1993, pp. 7-20.

Lang, T.R. and D.J. Casimir. 1986. Low alcohol wine. European Patent Application EP 0 177 282 A2.
The mouthfeel of beer—a review. *J. Inst. Brew.* **99**: 31-37.
Lynch, L. 1988. Low-alcohol: A must for the future? *Decanter, December* 1988, 85: 86.
Maccarone, E., Nicolosi Asmundo, C., Cataldi Lupo, M.C., Campisi, S. and B. Fallico. 1993. Oranwine. A new wine cooler with blood orange juice. *Italian J. Food Sci.* **5**(4): 397-408 (FSTA, 1996).
Process for alcohol elimination from wine and beer by solid CO_2. *Industri delle Bevande* **21**(121): 369-374. (Subfile: FSTA 00443962 93-01-h0005.)
Taste interaction of ethyl alcohol with sweet, salty, sour and bitter compounds. *J. Sci. Fd Agric.* 21, December: 653-655.
Meier, P.M. 1992. The reverse osmosis process for wine dealcoholization. The Australian Grapegrower and Winemaker 348, December: 9-10.
Neubert, S. 1976. *Über die Anwendung der Umkehrosmose in der Getränkeindustrie*, Dissertation der Justus-Liebig-Universität Giessen. In: U. Schobinger, and P. Duerr, 1986. Diatetische und sensorische Aspekte alkoholfreier Weine. *Alimenta* **22**: 33-36.
Noble, A.C. 1995. Application of time-intensity procedures for the evaluation of taste and mouthfeel (1994 Honorary Research Lecture). *Am. J. Enol. Vitic.*(1): 128-133.
Petershans, H. 1989. Weinherstellungsverfahren. German Federal Republic Patent Application, no: DE 37 22 535 A1. (subfile: FSTA 00376547 89-06-v0062).
Pickering, G.J. 1997. The production of reduced-alcohol wine using glucose oxidase. PhD thesis. Lincoln University, New Zealand.
Recent research on the use of glucose oxidase for the production of reduced alcohol wines. In: Proceedings of the 11th International Oenological Symposium, 3-5 June 1996, Sopron, Budapest, Hungary; International Association for Winery Technology & Management, zum Kaiserstuhl 6, D-79206, Breisach, Germany, pp. 153-179.
Pickering, G.J., Heatherbell, D.A., Barnes, M.F. and L.P. Vanhanen. 1998a. The Effect of Ethanol Concentration on the Temporal Perception of Viscosity and Density in White Wine. *Am. J. Enol. Vitic.* **49**(3): 306-318.
The production of reduced-alcohol wine using glucose oxidase-treated juice—effects on wine composition, stability, SO_2-binding and sensory properties. In: Proceedings of the XXIII World Congress on the Vine and Wine, II—Oenologie; 22-27 June 1998, Lisbon, Portugal; OIV, 18 rue d'Aguesseau, 75008 Paris, France, II: 161-169.
Pickering, G.J., Heatherbell, D.A. and M.F. Barnes. 1999a. Optimizing glucose conversion in the production of reduced alcohol wines from glucose oxidase treated musts. *Food Research International* **31**(10): 685-692.
Pickering, G.J., Heatherbell, D.A. and M.F. Barnes. 1999b. The production of reduced-alcohol wine using glucose oxidase-treated juice I—Composition. *Am. J. Enol. Vitic.* **50**(3), 291-298.
Pickering, G.J., Heatherbell, D.A. and M.F. Barnes. 1999c. The production of reduced-alcohol wine using glucose oxidase-treated juice II—SO_2-binding and stability. *Am. J. Enol. Vitic.* **50**(3), 299-306.
Pickering, G.J., Heatherbell, D.A. and M.F. Barnes. 1999d. The production of reduced-alcohol wine using glucose oxidase-treated juice III—Sensory. *Am. J. Enol. Vitic.* **50**(3), 307-316.
Pickering, G.J., Heatherbell, D.A. and M.F. Barnes. 1999e. GC-MS analysis of reduced-alcohol Müller-Thurgau wine produced using glucose oxidase-treated juice. Food Research International (submitted).
Pyle, L. 1994. Processed foods with natural flavour: the use of novel recovery technology. Nutrition and Food Science, No. 1, 12-14 (FSTA subfile: 00467941 94-04-e0028).
How low can you go? *Decanter, March* 1989: 75-78.
Alkohlfreier Wein - Hestellungsverfahren und Sensorische Aspekte. *Mitt. Gebiete Lebensm. Hyg.* **77**: 23-28.
Schobinger, U. and P. Dürr 1983. Diatetische und sensorische Aspekte alkoholfreier Weine. *Alimenta*, **22**(2): 33-36.

Schobinger, U., Waldvogel, R. and P. Durr. 1986b. Verfahren zur herstellung von alkoholfreiem wein oder fruchtwein. Schweizer Patent CH 654 023 A5 *In:* P. Duerr and P. Cuénat, 1988. *Production of Dealcoholized Wine*. Proceedings of the Second International Symposium for Cool Climate Viticulture and Oenology, Auckland, New Zealand, 11-15 January 1988. R.E. Smart, R.J. Thornton, S.B. Rodriguez, and J.E. Young (Eds), NZ Society for Viticulture and Oenology, PO Box 90-276, Auckland Mail Centre, NZ. pp. 363-364.

Schobinger, U., Waldvogel, R., and P. Durr. 1982. Verfahren zur herstellung von alkoholfreiem wein oder fruchtwein. Internationale Patentanmeldung WO 82/02723, 19/8. *In:* U. Schobinger, P. Dürr, and R. Waldvogel, 1986a. Die Entalkoholisierung von Wein und Fruchtweinen (eine neue Möglichkeit zur Herstellung von zuckerreduzierten Fruchtsaftgetränken). *Schweiz. Zeitschrift für Obst- und Weinbau* **122**(4): 98-110.

Schobinger, U., Durr, P. and R. Waldvogel. 1986a. Die Entalkoholisierung von Wein und Fruchtweinen (eine neue Möglichkeit zur Herstellung von zuckerreduzierten Fruchtsaftgetranken). *Schweiz. Zeitschrift für Obst- und Weinbau*, **122**(4): 98-110.

Scudamore-Smith, P. and J. Moran, 1997. A growing market for reduced alcohol wines. *Australian and New Zealand Wine Industry Journal* **12**(2): 165-166.

Verfahren zum vermindern des alkoholgehaltes alkoholischer getranke. European patent application EP 0 397 642 A1 (FSTA vol. 23 (1991) No. 6 6 V 18).

Herstellungs-möglichkeiten entalkoholisierter weine. Weinwirtschaft Technik No. 5 19 Mai: 50-59.

Impact of anti-alcohol campaigns on the wine industry. *In:* Proceedings 9th International Oenological Symposium, Cascais, Portugal 24-26 May 1990, International Association for Modern Winery Technology and Management, zum Abtweingarten 15, D-79241 Ihringen, Germany, pp. 126-137.

Smith, P.M. 1995. Biological processes for the reduction of alcohol in wines. Dissertation (M. Appl. Sci.), Lincoln University, New Zealand.

Alkoholarme Weine and Sekte - theoretische und technologische Aspekte ihrer Herstellung. *Lebensmittel industrie* **37**(4): 162-164.

Sykes, S.J., Casimir, D.J. and R.G.H. Prince. 1992. Recent advances in spinning cone column technology. *Food Australia*, **44**(10): 462-464.

Thumm, H.J. 1975. Low alcohol wine. Austral. Patent 66 366, 24.4 *In:* U. Schobinger, P. Dürr, and R. Waldvogel. 1986a. Die Entalkoholisierung von Wein und Fruchtweinen (eine neue Möglichkeit zur Herstellung von zuckerreduzierten Fruchtsaftgetränken). *Schweiz. Zeitschrift für Obst- und Weinbau* **122**(4): 98-110.

Trothe, R. 1990. Verfahren zur entalkoholisierung oder alkoholreduzierung von weinen und fruchtweinen. German Democratic Republic Patent DD 283 153 (FSTA vol. 23 (1991) No. 4, 4 V 63).

Experimental production of light wines. *Enotecnico* **22**(3): 287-304 (subfile: FSTA 00343439 87-09-h0129).

Villettaz, J-C. 1986. Method for production of a low alcoholic wine and agent for performance of the method. European patent No: EP 0 194 043 A1.

Villettaz, J-C. 1987. A new method for the production of low alcohol wines and better balanced wines. *In:* Proceedings of Sixth Australian Wine Industry Technical Conference, 14-17 July 1986, Adelaide, T. Lee (Ed.). Australian Industrial Publishers, Adelaide, pp. 125-128.

Determination and removal of gluconic acid in reduced alcohol wine and high acid grape juice. M. Appl. Sc. Thesis, Lincoln University, New Zealand.

Weiss, M. 1987. Verfahren zum herabsetzen des alkoholgehalts alkoholhaltiger getränke, insbesondere wein und schaumwein. German Federal Republic Patent DE 34 13 085 C2 (FSTA vol. 20 (1988) No. 1, 1 V 161).

Wucherpfennig, Von K., Millies K.D. and M. Christmann. (1986). Hestellung entalkohololisierter weine unter besonderer berücksichtigung des dialyseverfahrens. *Die Weinwirtschaft Technik*. 122 (9): 346-354 (FSTA vol. 19 (1987) No. 7, 7 H 128).

6

Paprika Spice Production

Andreas Klieber

Senior Lecturer, Postharvest Horticulture, The University of Adelaide Waite Campus, Department of Horticulture, Viticulture and Oenology, PMB 1, Glen Osmond SA 5064, Australia

Abbreviations

°C	degree Celsius
μL	microliter
ASTA	American Spice Trade Association
BT	*Bacillus thuringiensis*
cfu	colony forming units
CIE	Commission Internationale de l'Eclairage (International Commission on Illumination)
ELISA	enzyme linked immunoabsorbent assay
h°	hue angle in CIE Lab system
HPLC	high-performance liquid chromatography
L	lightness value in CIE Lab system
NIST	National Institute of Standards and Technology, USA
ppb	parts per billion (μg/kg)
ppm	parts per million (mg/kg)
RH	relative humidity
SHU	Scoville Heat Units
SICU	Standard International Colour Units
spp	species

1. Introduction

To achieve the best possible yield and quality of paprika spice, the entire production system from site selection, to production and postharvest handling must be considered. This review of the paprika spice system draws on recent research publications as well as a research report from the Rural Industries Research and Development Corporation, Australia (Klieber, 2000). An extensive review of fresh capsicum or bell pepper, production has been conducted by Wien (1997) who summarized production from seed to vegetative growth to fruit development, especially with regard to glasshouse production. Some aspects are relevant, but paprika for spice production is field grown and different considerations for production, harvesting and processing apply.

1.1 Historic and Economic Significance

Capsicum species belong to the Solanaceae genus that also includes tomatoes and potatoes among others (Bosland et al. 1996). The centre of their origin and diversity lies in Bolivia and Peru (Purseglove et al. 1981; Bosland et al. 1996). After the discovery of the Americas capsicum spice was distributed to Europe from 1494 (Szucs, 1975) and was believed to be 'pepper', which botanically belongs to the *Piper* family. Now *Capsicum* species are grown all around the world (Somos, 1984) and are essential to many cuisines.

A number of categories of *Capsicum* species can be distinguished according to genetics, intended use and fruit characteristics such as colour, pungency (hotness) and shape. Botanically, these fruit are berries, but consumers often consider them as vegetables especially if they are consumed in salads. Sweet, non-pungent fruit are generally called capsicum or bell pepper, but some other types such as pimiento are also not hot (Bosland et al. 1996). The term 'paprika' is used differently in different regions of the world. In Europe, and particularly in Hungary, it may refer to *Capsicum* species in general and therefore include pungent and non-pungent types. In North America and Australia, paprika generally refers to non-pungent *Capsicum* spice powder that may contain only very low levels of hot compounds (Bosland et al. 1996). For this review, paprika refers to cultivars that produce non-pungent red spice powder, even though some considerations that affect pungency of other types will be highlighted. Pungent fruit are also differentiated from non-pungent paprika in that they are generally smaller and lighter in red colour (Purseglove et al. 1981). The colour of paprika spice is always red, and therefore fruit need to be fully ripened from their green, unripe state. The main use of paprika spice is to add colour and flavour to cooked dishes, but it can also be added to sauces or salads or as a colourful sprinkle on top of a dish. Another use of paprika fruit is for the extraction of their red pigments as food and cosmetic colourants.

Capsicum annuum is commonly grown for paprika spice production, as it produces large fruit that can be easily harvested, handled and processed. It is also considered to be one of the most easily grown species. Other species that may be grown for fresh consumption of smaller, pungent fruit include *C. chinense*, e.g. Habanero, *C. frutescens*, e.g. Tabasco, *C. pubescense*, e.g. Manzano, and *C. baccatum*, e.g. Aji (Bosland et al. 1996).

Overall worldwide *Capsicum* production grew from 11 million tonnes in 1990 to 16 million tonnes in 1997, with 9.5 million tonnes produced in Asia alone (FAO, 1999). It is difficult to estimate the amount of capsicum for spice production, due to differences in definitions and general data for all *Capsicum* types being collected. However, 30,000-35,000 tonnes of paprika spice were traded internationally in the early 1980s (Smith, 1982), mostly in Europe and North America, with an annual increase in demand of 5-10%. Extrapolating this data would mean an international trade of more than 100,000 tonnes and a value greater than US$ 300 million in 1999. The overall production is likely to be much higher though, as domestic consumption is not included in international trade figures. Spain, Hungary and other Eastern European countries were the main producers of paprika spice (Smith, 1982), contrary to chillies that are predominantly produced in Asia.

2. Paprika Quality

The quality of paprika spice is judged using a range of attributes. Colour hue and intensity are the most obvious attributes, but sweetness and flavour are also important. Pungency may be assessed to ensure that it is at low levels or at a desired higher level for 'Hungarian' style pungent paprika. In addition limits of impurity, levels of microbial contamination with fungi, yeasts, salmonella and coliforms, particle size, moisture content and other parameters may be specified by spice traders. For example, international trade often requires a mold count of less than 300 colony forming units (cfu) g^{-1} (Deak et al. 1986). Standard techniques for microbial counts, sugar level, moisture content and other parameters can be employed; therefore colour, pungency and mycotoxins will be mainly considered here.

Paprika fruit are red coloured due to red-pigmented carotenoids, mainly capsanthin, capsorubin, zeaxanthin and cryptoxanthin (Lease and Lease, 1956). In intact fruit, carotenoids are very stable, however, they are easily destroyed by autoxidation from exposure to heat, light and oxygen when paprika is dried and ground into spice (Bunnell and Bauerfeind, 1962; Britton and Hornero-Mendez, 1997). Over time the spice loses quality as it becomes more orange and less intensely coloured, and with excessive heat exposure, the powder can become brown. An additional benefit of the carotenoids is that they have provitamin A activity (Somos, 1984; Minguez-Mosquera and Hornero-Mendez, 1993). Vitamin A is essential in the human diet to prevent blindness and daily requirements can be met by 3-4 g of paprika powder (Somos, 1984). Recent research is also pointing to anti-cancer properties of carotenoids (Hertog et al. 1997).

The colour quality of paprika is determined as either extractable red colour or surface colour. Extractable colour is the standard method employed in the spice industry and is commonly measured by the official method of the American Spice Trade Association (ASTA) (Woodbury, 1990). The method relies on a simple extraction of the carotenoids from the dried powder with acetone and measuring the absorbance of the extract at 460 nm compared to an ASTA glass reference standard (NIST SRM 2030a Glass Filter, Gaithersburg Maryland). The lower acceptable limit for trade of paprika is generally 180 ASTA units, but higher levels are very desirable as it results in a better quality and allows for the loss of red colouration that occurs during storage, as discussed in section 3.4.3. Alternatively standard international colour units (SICU) can be used, with 100 ASTA units equating to 4000 SICU (Bosland, 1993).

Surface colour is not generally measured, but it provides an indication of the visual appearance of the paprika spice. The CIE system is most often used and the two values that are most useful are the lightness (L) value and the hue angle (h°). The hue angle will be closer to 0° if the paprika is more red and 90° if it is more yellow (Wall, 1994); therefore a hue angle close to 0° is desired. A lower L value indicates a darker colour that may indicate a more intense colour; however, if paprika shows quality defects, from excessive drying temperatures or fungal development, it will be brown and dark and will therefore have an artificially low L value. Therefore, the industry generally relies on extractable colour and observation of visual defects of the spice.

Pungency of *Capsicum* species is due to compounds called capsaicinoids (Collins and Bosland, 1994). Chemically they are amino amides of vanillylamine and C_9-C_{11} branched fatty acids (Quinones-Seglie et al. 1989) with about 90% of fruit capsaicinoids being capsaicin and dihydrocapsaicin. Other capsaicinoids are nordihydrocapsaicin, norcapsaicin, homocapsaicin, nornorcapsaicin and homodihydrocapsaicin. Production of capsaicinoids in *Capsicum* fruit is controlled by genetics, competition between fruit and the environment and mainly occurs in the placenta (Fujiwake et al. 1982; Iwai et al. 1977) and the seed (Purseglove et al. 1981). Capsaicinoids are perceived as hot as they open calcium channels in nerve cells of the skin and mucous membranes; this leads to nerve signals that follow similar neural paths as for thermal heat (Clapham, 1997). If these nerve cells are continuously stimulated with capsaicinoids, calcium mediated cell death occurs which may explain the gradual desensitization in long-term consumers of hot *Capsicum* spice. To produce non-pungent paprika, selection of a cultivar that contains no or very low levels of capsaicinoids is necessary.

Commercially heat levels are measured as Scoville Heat Units (SHU), named after the inventor of an organoleptic test that determines pungency by dilution series (Collins and Bosland, 1994). A representative ground sample is used and capsaicinoids measured using high-performance liquid chromatography (HPLC) (American Spice Trade Association, 1985), but recently near infrared spectrophotometry has also been examined (Asian Vegetable Research and Development Centre, 1993). Paprika generally has no or only traces of pungency, while hot chilli spice often has levels of around 10,000-60,000 SHU.

Paprika is susceptible to mycotoxin contamination (Scott and Kennedy, 1973; Seenappa, et al. 1980; Udagawa, 1982; Putzka, 1994). Aflatoxins are the most potent carcinogens of these and are most commonly found in paprika. Other health effects include acute toxicity and impaired mental development. The most potent aflatoxins are B_1, B_2, G_1 and G_2 all of which have been found in paprika. Aflatoxins are only produced by some strains of *Aspergillus flavus* and *Aspergillus parasiticus* if they encounter appropriate environments. Of 44 soil isolates of *A. flavus* 16 produced aflatoxins (Saito et al. 1976), and 28% of *A. flavus* strains isolated from spices (Udagawa, 1982) and 25% of *A. flavus* strains isolated from paprika (Ath-Har et al. 1988) produced aflatoxins. Environmental conditions can inhibit the development of aflatoxins of these strains. During production and storage of paprika fruit a relative humidity (RH) below 70% (Seenappa et al. 1980) and temperatures above 45° C will prevent aflatoxin formation, whereas the spice needs to be dried to a moisture content below 14% and/or stored below 20° C (Anonymous 1994). As a preventative measure paprika fruit that become infected with mould in the field or after harvest need to be excluded through sorting. During drying, temperatures should be above 45° C and it should occur quickly to reduce the moisture content to safe levels. The spice is very hydroscopic after grinding and it needs to be packaged quickly in a low humidity environment within a moisture barrier. Also sound sanitation practices need to be implemented to minimize the level of fungal contamination of the environment. However, once aflatoxin contamination has occurred it is not possible to destroy them by processing or cooking, as they are chemically very stable (Macdonald and Castle, 1996).

Other mycotoxins that may be found are ochratoxin A (Patel et al. 1996), produced by *Aspergillus ochraceus* and *Penicillium viridicatum*, fumonisins (Patel et al. 1996), produced by *Fusarium* spp., patulin (Frank, 1977), produced by *Penicillium expansum*, *Penicillium urticae* and *Byssochlamys nivea*, and the common fungus *Alternaria alternata*, which is often found in paprika, may also be a source of mycotoxins (Wall and Biles, 1994).

Of particular public health concern is the widespread, and often high, aflatoxin contamination found in paprika and chilli products. In an American study aflatoxins levels above legal limits were found in 17/24 chilli samples (Scott and Kennedy, 1973), in a Japanese study in 2/12 chilli and 5/12 paprika samples (Tabata et al. 1993) in Germany in 55% of chilli/paprika samples (Putzka, 1994) and in over 80% of samples in Australia (Ulieber, 2000). The maximum legal limit of aflatoxins in human food is 20 ppb in the US (Food and Drug Administration, 1999), 5 ppb (5μg/kg) in Australia (Australian and New Zealand Food Authority, 1996) and 2 ppb in Germany (Putzka, 1994). Some reported levels in contaminated samples were 525 ppb in Ethiopia (Fufa and Urga, 1996), up to 48 ppb in India (Jaffar et al. 1994), 51 ppb in the UK (Macdonald and Castle, 1996) 234 ppb in Italy (Finoli and Ferrari, 1994) and 89 ppb in Australia (Ulieber, 2000).

The traditional way of quantifying aflatoxins and other mycotoxins is to use HPLC methods, but today enzyme-linked immunoabsorbent assays (ELISA) and affinity columns are routinely used, as they are more sensitive (Adachi et al. 1991) and need no sophisticated equipment.

3. Paprika Spice Production System

Paprika spice production can be generally summarized as a sequential system. Initially, a market for the product needs to be determined, then the appropriate seed material and production site need to be selected. Production factors such as fertilizer, irrigation, pest and disease control need to managed, and the harvest maturity must be determined for successful harvesting of the crop. This is followed by drying and grinding of the paprika fruit and the powder is stored until it is consumed or incorporated into a food product.

Generally individual parts of this system have been researched, mostly relating to general field production practices (Indira et al. 1985; Meena-Nair et al. 1990; Grattidge, 1993), glasshouse production (van Uffelen and Elgersma, 1990) or breeding programs (Tewari, 1987; Fuentes and Mora, 1988; Wright and Walker, 1991; Hibberd et al. 1992).

3.1. Seed and Site Selection

As mentioned previously, paprika cultivars belong to C. *annuum* as this species has a large fruit size and is relatively easy to cultivate. Paprika cultivars must contain no, or low levels of hotness, unless a spicy 'Hungarian' paprika is desired. Also a cultivar with a high red colour intensity needs to be chosen. Seed companies are a first contact to source seed that may be suitable for production under local conditions. In addition paprika cultivar evaluations are being conducted in several locations around the world. In the absence of reliable data for a given area, it is recommended to trial new cultivars on a small scale first, as crop management inputs, yield and quality can be different for a cultivar when it encounters different climatic and soil conditions.

Paprika yield and quality are greatly influenced by the chosen site. According to its origin it is a warm season crop that performs poorly below 15-16°C and is easily killed by light frost (Bosland et al. 1994; Matta and Cotter, 1994). At low temperatures seed emergence is also slowed down and is completely prevented below 15° C (Lorenz and Maynard, 1980). To obtain uniform plant stands and good yields, rapid emergence is essential (Matta and Cotter, 1994). The other factor that determines yield is fruit set, which is prevented below 16° C and above 32° C. In addition, pungency of hot cultivars is reduced by low growth temperatures (Cotter, 1980; Klieber, 2000).

A well drained, moisture-holding loam or sandy loam that includes some organic matter is the preferred soil type for paprika (Bosland et al. 1994; Matta and Cotter, 1994). A neutral to slightly alkaline pH of 7.0-8.5 is recommended. To prevent root diseases from inadequate drainage the site should be levelled to a slope of 0.01-0.03%. However, an adequate water supply needs to be available, either through natural rainfall or through irrigation, as water stress reduces yields dramatically through fruit and flower abscission (Matta and Cotter, 1994). Plants are especially susceptible during flowering and yield may also be reduced due to inhibition of pollination (Haigh et al. 1996). To prevent disease, paprika needs to be rotated in the field with other crops on a 3-4 year rotation (Bosland et al. 1994).

Yields are also reduced by excessive wind exposure, as bacterial diseases increase and plant size and leaf area are reduced; good windbreaks are therefore advisable (Hodges et al. 1996).

3.2. Managing Production Factors

3.2.1. Crop Establishment

Paprika plants are best established in the field in mid-spring, as they are frost sensitive and as ripening is slowed by low soil temperatures (Bosland et al. 1994). Fruit are consequently harvested from mid- to late-summer. Repeated harvests from the same plant are only feasible if there is no frost risk later in the season, but quality may also be reduced as fruit growth and ripening is slowed at lower air temperatures.

Paprika is established in the field by transplanting seedlings or by direct seeding (Carter, 1994). Direct seeding generally uses dry seed that are directly placed into the soil. This can result in slow and non-uniform germination if the soil temperature is too low. A number of techniques have been tested to overcome this, with variable success. Seeds can be hydrated to 10-25% moisture content or and primed for 3-7 days with polyethylene glycol, re-dried to their original moisture level and then planted (Carter, 1994).

A sowing depth of 1.9-2.5 cm is recommended (Bosland et al. 1994; Carter, 1994) and soil crusting from wet-dry cycles needs to be avoided during paprika seedling emergence, as they are relatively weak. Crusting may be avoided by using sprinkler irrigation (Carter, 1994). For single rows using flat beds with 90-100cm spacing, a seeding rate of 2.2-4.4kg/ha is generally used. Once plants are 10-15 cm tall they are thinned to a 30 cm within row spacing (Matta and Cotter, 1994). Thinning costs can be avoided by maintaining a high planting density. Plant populations of 20 plants/m^2 may be needed for maximum yield (Kahn, 1992) and within row spacings of less than 10 cm for row spacings of 0.9 m have been advocated (Kahn et al. 1997). A good yield can also be achieved for 5 cm within row spacings in double rows that are separated by 65 cm (Klieber, 2000). The size and yield of each plant is reduced by not thinning, but due to the increased density the overall yield is high (Fig. 1). Plants grown under high density also are less branched and lend themselves to machine harvesting (Levy et al. 1989; Klieber, 2000).

For transplants, seeds are germinated and grown for about 5-6 weeks under protection to 15-20 cm height. They are then generally transplanted into the field with a spacing of 30 cm within rows and 90-100 cm between rows (Carter, 1994). However, higher densities of 18 cm within row spacing in double rows that are planted 65 cm apart can also result in good yields (Klieber, 2000). During transplanting, the roots need to be maintained intact, and root establishment can be aided using phosphorous solutions. The structure of transplants is different to direct seeded plants as they are usually shorter, more massive and have more nodes

Fig. 1. High density planting for machine harvesting.

and branches (Bosland et al. 1994; Cooksey et al. 1994). As transplants are more likely to shed leaves and stems during mechanical harvest, direct seeding is the preferred method of stand establishment for machine harvesting. However, if early manual crop harvest is desired, transplanting, row covers and plastic mulches can be used in conjunction (Matta and Cotter, 1994), but outcomes may be variable (Bosland et al. 1994) and earliness may not be important for spice production.

On deciding the establishment method, higher planting and establishment costs of transplanting need to be compared to the increased costs of seeds and cultivation, if the stand is thinned, of direct seeding.

3.2.2. Nutrition and Water Requirements

As for all crops, fertilizer application is determined according to soil analysis. For paprika yield and quality, the two most important nutrients are nitrogen and phosphorous. Applying nitrogen before planting can help seedling vigour, provided soil nitrogen otherwise is below 20 ppm nitrate (Bosland et al. 1994). Generally nitrogen (N) is applied at a rate of 2.2-5.7 kg nitrogen/ha if it is banded 7-10 cm below the seed at the time of seeding, or it can be broadcast and disced into the soil at a rate of 22-33 kg N/ha (Bosland et al. 1994). If fertilizer is banded, 34 kg/ha of P_2O_5 is banded at the same time if soil levels are low, as this results in higher yields and a more red fruit (Matta and Cotter, 1994). Further N sidedressings of 22-34 kg N/ha may be applied when the first flower buds appear and when the first fruit set, but excess N produces low-yielding, large plants, increases disease levels and fruit ripening is delayed (Bosland et al. 1994). In addition phosphorous (P) sidedressings and foliar micronutrient sprays may be necessary depending on soil conditions. Tissue analysis is the best indicator of mineral deficiencies and toxicity (Goldberg, 1995). Of the micronutrients iron (Fe), Manganese (Mn) and Zinc (Zn), only Fe was found to be related to paprika quality as increased Fe increased the red colour content. Calcium (Ca) deficiency can lead to blossom-end rot of fruit and magnesium (Mg) deficiency will lead to yellowing between leaf veins (Grattidge, 1993).

As mentioned above, paprika need a good water supply. Water deficit stress causes flowers and small fruits to abscise and also increases the risk of fruit blossom end rot due to insufficient calcium uptake. Water deficit stress is especially detrimental during flowering, fruit set and as fruit develop. This is due to fruit and flower abscission in response to water stress, which reduces yields dramatically (Fig. 2). Conversely, overwatering and water logging result in serious root diseases (Dickerson, 1994).

Individual irrigation requirements vary with plant size, wind, sunlight, temperature and relative humidity. Paprika plant are shallow rooted and acquire water from the top 30 cm of the soil; irrigation can therefore be scheduled using two moisture probes located in the top 15-20 cm and 50- 60 cm to replenish water in the top and deeper areas of the soil as required (Grattidge, 1993). If visual scheduling is used, plants should be irrigated at the first signs of wilting in the early afternoon (Bosland et al. 1994). Irrigation may not be necessary in the first 3 weeks after emergence, as the roots are growing into the wet soil. Then plants

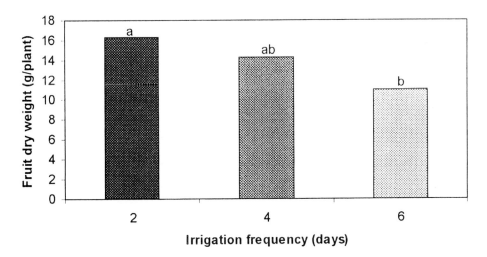

Fig. 2. Effect of water stress after fruit set on marketable yield of paprika fruit. Different letters indicate significant differences at the 5% level (after Klieber, 2000).

are generally irrigated every 5-7 days, with the frequency decreasing according to rainfall and before harvesting to improve quality (Bosland et al. 1994). The most efficient method of water application is drip irrigation, but in some areas sprinkler or furrow irrigation is also practiced. The most important water quality consideration for paprika is the level of salinity. Soil salinity before planting should be less than 1920 ppm (Carter, 1994). Salt injury results in stunted and dead seedlings, burning of root tips, leaf margin necrosis and wilting. To manage salinity, plants are planted in the centre of beds if they are drip irrigated, as salt moves with the water front to the outside of beds. Additionally enough water must be applied at each watering to leach salt below the roots (Goldberg, 1995).

3.2.3. Pest and Disease Management

Paprika is prone to attack by a range of pests that include spider mites, thrips, leafhoppers, aphids, fruit flies, weevils and *Heliothis* (Swaine et al. 1991). These pests reduce yields and quality due to direct damage and through indirect damage as vectors for virus and fungal diseases. 'Stung' fruit rapidly develop fungal diseases due to injury and ubiquitous presence of spores in the environment, resulting in unmarketable fruit as discussed in the quality section above. A range of registered chemical control methods exists for these pest and integrated pest management with for example BT toxin (*Bacillus thuringiensis*) and natural extracts from neem fast becoming common (Walter, 1996).

Weed management can be achieved with mechanical methods, such as hand hoeing, and registered pre-emergence and post-emergence herbicides (Lee and Schroeder, 1995). This aspect of production is important for controlling competition with the paprika plant, but also for controlling alternative hosts for some insects and viruses.

Specific recommendations and regulations for pest and weed control vary greatly between locations and over time; therefore specific local publications must be referred to.

A range of diseases has been well described for paprika by Goldberg (1995), with pictures, and Persley et al. (1989). They include bacterial and fungal root, leaf and fruit rots, viruses and nematodes (Bosland et al. 1994).

Root-knot nematodes result in yield loss, damaging seedlings in particular, and are common in sandy, warm soils (Goldberg, 1995). They spread through soil, transplants and irrigation water and can be controlled by sterilizing potting mixes and soil fumigation; resistance breeding is currently underway.

Paprikas suffer from three soil borne diseases, namely Phytophthora root rot (*Phytophthora capsici*), Verticillium wilt (*Verticillium dahliae*) and Rhizoctonia root rot (*Rhizoctonia solani*) (Goldberg, 1995). Phytophthora root rot occurs where fields become waterlogged due to excessive water from rain or irrigation and drainage is poor. Affected plants die due to root death, with symptoms of discoloured roots, stem bark shedding and straw coloured and defoliated stems. Control with chemicals is limited and as there are no truly resistant cultivars, a 3-4 year rotation with lettuce, brassicas, onions or small grains may be needed. Verticillium wilt is most common in temperate areas and causes yellowing and plant stunting, followed by defoliation as the vascular tissue becomes blocked and discoloured (Goldberg, 1995). There are currently no resistant cultivars, but fumigation and crop rotations as above can help to control this disease. Rhizoctonia root rot reduces yield and can cause plant death as it enters the stem near the soil and then the taproot causing reddish-brown lesions (Goldberg, 1995). Fungicidal treatments as for damping-off achieve best control.

Damping-off prevents seed germination and causes rapid seedling death and is caused by *Rhizoctonia solani, Phytophthora capsici, Pythium* species and *Fusarium* species (Goldberg, 1995). Control is achieved with fungicides or hot water seed treatments, the use of good quality seed and proper drainage.

In addition to the very serious root diseases of paprika, leaf diseases such as bacterial leaf spot (*Xanthomonas campestris* pv. *vesicatoria*), Cercospora leaf spot (*Cercospora capsici*) and powdery mildew (*Leveillula taurica*) are common (Goldberg, 1995). The first two cause crop losses under wet and humid conditions, with symptoms of circular water-soaked lesions that turn purplish-grey with a yellow halo; the latter causes circular lesions that turn grey or white and then dark brown with a reddish edge. Both diseases are best controlled with copper based chemicals, even though some resistant fungal strains exist; these compounds need to be applied before wet weather to be effective. Weed control is important as these diseases can be harboured on many weed species. Powdery mildew prefers warm weather (18-35° C) and can occur at low or high humidity. A powdery white fungal mass covers the lower side of leaves and some discolouration on the top of leaves occurs as this disease develops. Field sanitation and sulphur containing fungicides are the best control options; sprays need to be applied early and thoroughly to achieve control.

Fruit diseases not only reduce yield, but also affect the food safety of the final spice product. Anthracnose (*Colletotrichum* species) is considered a postharvest problem, even though it initially establishes itself in the field during wet conditions. It then lies latent and appears when the fruit ripens as water soaked, dark red to black, lesions. This problem is best controlled with clean seed, crop rotation, and some fungicides. Phytophthora fruit rot infects fruit and seed on the inside and results in shrivelled fruit; control is achieved as with Phytophthora leaf spot. *Erwinia carotovora* pv. *carotovora* causes bacterial soft rot in rainy weather if soil is splashed onto fruit. It enters fruit through minute injury sites, often caused by insects, and liquefies the flesh often turning fruit into water filled bags. This disease is best controlled through insect control in the field and chlorination of wash water after harvest.

Black mould is generally caused by *Alternaria* species, even though other species may also cause these symptoms (Goldberg, 1995). These fungi are of concern as, given the right conditions, they produce mycotoxins, including aflatoxins, the most potent natural carcinogens known. These have been discussed in the quality section above. *Alternaria* occurs due to excessive moisture late in the season. It can be the primary cause of disease or occur secondarily if it infects fruit that already infected by another disease agent. Control methods of black mould include the prevention of other diseases and insects, reducing irrigation and fertilization in the late season, and harvesting ripe fruit as soon as possible. In addition paprika fruit must be kept dry before processing and processing techniques need to be appropriate.

A number of viruses cause disease in chillies; these include beet curly top virus, tomato spotted wilt virus, pepper mottle virus, alfalfa, cucumber and tobacco mosaic viruses and pepper geminiviruses. Symptoms and transmission routes vary somewhat. Beet curly top virus is transmitted by leafhoppers from susceptible plants and results in death of seedlings, stunting of plants and curled, twisted and dimpled leaves with clear veins. Tomato spotted wilt virus is transmitted by thrips and result in green and red fruit with off-coloured spots that never colour red and distortions and chlorotic spots on fruit, while leaves may be distorted and show mosaic and chlorotic ring spots. Pepper mottled virus is transmitted by aphids and results in misshapen leaves that have a mottled and puckered appearance. Alfalfa mosaic virus causes distorted fruit and is transmitted from alfalfa plants that grow in close proximity. Cucumber mosaic virus is transmitted by aphids and causes narrowing of leaves, stunting, yellowish or whitish leaf spots and may cause small, distorted fruit. Tobacco mosaic virus is transmitted mechanically on tools and hands and on seeds; symptoms vary with light and dark green bumps on leaves and possibly small, unevenly coloured fruit. Pepper geminiviruses are transmitted by whitefly and cause stunting, curled or twisted leaves with yellow mosaic patterns and distorted fruit. Control of viruses depends on the type of virus; insect or aphid transmitted viruses are best controlled by controlling the vector and weeds that act as alternate host for the vector and virus. Alfalfa mosaic virus can be avoided by not planting near to alfalfa. Tobacco mosaic virus is hard to control, but control depends on good hygiene practices and clean seed sources. Resistant cultivars are available for tomato spotted wilt virus and tobacco mosaic virus.

Non-microbial disorders, such as nutrient deficiency and toxicity, salt, wind, hail and herbicide damage, blossom end rot and sunburn, also affect paprika fruit (Goldberg, 1995). Blossom end rot is caused by poor irrigation practices that lead to calcium deficiency (Bosland et al. 1994). Fruit develop brown/black spots on the lower portion and ripen prematurely when insufficient water, and with it calcium, is transported into rapidly developing fruit. Sunburn occurs on fruit that are exposed to direct sunlight for extended periods; this may be due to wilting or defoliation from underwatering or diseases. Exposed tissue becomes papery and tan in colour, and often develops secondary fungal infections (Goldberg, 1995).

3.3. Harvesting

Paprika must be harvested at the optimum commercial maturity to have sufficient spice colour levels. Therefore harvest maturity is determined by the colour stage, that is the degree of ripeness of the fruit. Fruit colour changes in a progression from green, to green with some light red colour showing (breaker or chocolate), to light red, to deep red (blood colour), and to deep red with partial drying of fruit. For optimum colour intensity, fruit should be harvested at the deep red colour stage and may or may not be partially dry (Worku et al. 1975; Deli et al. 1996; Klieber, 2000). Processing quality is not affected by partial drying of fruit on the plant, as they will be dried in any case. Delaying harvesting until fruit were fully ripe also enhanced the colour stability of paprika spice (Isidoro et al. 1995). In contrast, light red fruit have an insufficient colour content and therefore dilute the colour of the spice powder.

Timing of paprika harvest is critical, as fruit have a wide range of ripeness on the plant. This is due to the plant architecture (Fig. 3) as the stem repeatedly branches, producing nodes at which the flowers are located. As the plant develops, node 1 produces the first flowers and ripe fruit, node 2 the second and so on. However, fruit are often harvested from 6 or 7 nodes, and therefore it is essential that fruit from the top nodes be allowed to ripen fully. The maximum yield can be achieved by either repeatedly harvesting ripe fruit manually or by strip harvesting in a once-over harvest once most fruit have ripened. The latter is more efficient and is used for mechanical harvesting (Palevitch, 1978). Due to the differing ripeness stages of fruit on the bush, colour (Miles, 1994) and pungency (Zewdie and Bosland, 1996) differences are found between fruit on one bush and between bushes.

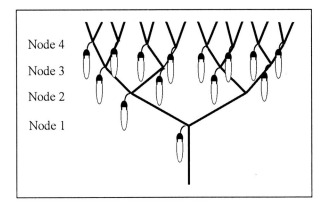

Fig. 3. Schematic representation of a paprika plant.

The plant hormone ethylene is commonly used in horticulture to accelerate ripening. It has been used to synchronise ripening of paprika in the form of ethephon sprays (Indira et al. 1985) and is registered in the US as a preharvest spray in paprika (Kader, 1992). There are reports that it can accelerate (Worku et al. 1975; Batal and Granberry, 1982) or concentrate (Sims et al. 1970) fruit ripeness. However, the effectiveness of this treatment varies depending on rate of application, number of applications, ambient temperature and cultivar (Sims et al. 1970; Batal and Granberry, 1982; Rylski, 1986; Beaudry and Kays, 1988). If the field temperature is below 21° C paprika ripening is not accelerated in response to ethephon, even at concentrations above 3000 µL/L (Knavel and Kemp, 1973). At higher temperatures a concentration range of 1000 to 5000 µL/L was optimal to accelerate fruit ripening in trials where an effect was found (Sims et al. 1970; Worku et al. 1975). However, under these conditions the potential for ethephon to cause defoliation, fruit abscission and sun scald exists (de Wilde, 1971), resulting in serious yield (Fig. 4) and quality losses (Krajayklang et al. 1999). Ethephon application at high temperatures also may cause fruit damage (Krajayklang et al. 1999) (Fig. 5). This fruit injury was not due to sun scalding, as plants were grown in a shade house. In this study and others by Cooksey et al. (1994) and Kahn et al. (1997), spice colour of marketable fruit was not improved by ethephon treatments. Some workers have applied calcium sprays to ethephon treated plants to minimize fruit abscission (Cooksey et al. 1994), but economically it is more sensible to maintain fruit on the plant until the top nodes have matured. The only exception may be locations where

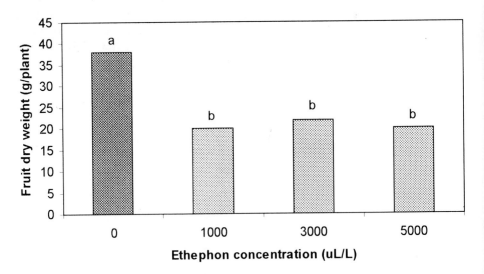

Fig. 4. Effect of ethephon spray on yield of paprika fruit. Different letters indicate significant differences at the 5% level (after Krajayklang et al. 1999).

Fig. 5. Fruit and leaf injury caused by ethephon spray.

paprika has to be harvested as early as possible due to the risk of frosts or excessive rains at the time of harvest.

3.4. Postharvest Handling

3.4.1. Postharvest Ripening and Storage

After harvest, paprika are normally directly dried and ground into powder. However, fruit that are not fully red may continue to ripen after harvest.

Fresh *C. annuum* is often considered to be non-climacteric (Saltveit, 1977; Kader, 1992), that is, they do not ripen with associated increases in metabolism or ethylene production after harvest. However, Gross et al. (1986) suggest that some undergo colour changes and produce increased levels of ethylene. The stage of ripeness at harvest largely determines whether fruit will continue to ripen after harvest or not. Fruit that did not at least have some red colour at harvest did not ripen to a completely red colour (Klieber, 2000). Fruit that were partially red ripened to a red colour, but the extractable colour content of postharvest ripened fruit was inferior to fruit that ripened on the bush (Fig. 6). Application of 100 ppm ethylene does not affect the outcome of postharvest ripening (Klieber, 2000).

If fruit are stored before processing, for example if the processing capacity is not sufficient to process all fruit that are harvested at any given time, care must be taken to minimize deterioration. Especially during mechanical harvesting, physical damage to fruit ensues during harvesting. This allows pathogens to proliferate due to the breakdown of barriers and release of nutrients from the wounds (Kays, 1991). Also, if fruit that are diseased in the field are not removed before storage, there

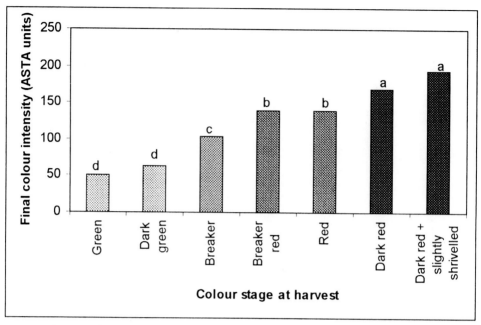

Fig. 6. Final colour intensity of ripened paprika fruit that were harvest at different ripeness stages. Fruit were ripened at 22° C. Different letters indicate significant differences at the 5% level (after Klieber, 1999).

may be a considerable inoculum level present during storage. Quality loss can be reduced by careful handling and storage of fruit at low temperature and humidity. Suitable conditions for processing paprika are about 7° C, with storage for more than 24-36 hours above 15° C causing quality deterioration (Wall, 1996). A relative humidity below 70% reduces mould growth and mycotoxin production (Seenappa et al. 1980); as shrivelling of fruit is not a concern before processing this low RH is preferred. However, overall it is preferable to delay harvesting rather than storing fruit (Wall, 1996).

3.4.2. Processing

After harvest, paprika fruits are generally directly washed to remove debris, dirt and residues. Diseased fruit must be removed at this stage to minimize mycotoxin contamination. Fruit are then dried. Dried paprika powder is normally of an intense red colour, but if processing conditions are wrong the powder becomes brown and this reduces quality (Miles, 1994). The red colour pigments of paprika are masked (Bosland, 1993) by enzymic browning (Mayer and Harel, 1979) and, especially during drying, by non-enzymic Maillard reactions. These accelerate at higher temperatures (Roos and Himberg, 1994) and at intermediate moisture levels of the fruit (Wedzicha, 1984). To prevent browning, high temperature and slow drying must be avoided.

Paprika fruit have been sun dried in some production areas in the world that have sufficient sunshine (Shirvastava et al. 1990) with solar driers improving efficiency (Shirvastava et al. 1990; Tiris et al. 1995). Solar drying, like hot air drying, utilizes high temperatures and does not quickly dry paprika to a low moisture content, thus resulting in brown discolouration of the spice. Industrial driers produce a better and a more uniform product (Minguez-Mosquera et al. 1994b). Industrial processes often use a higher temperature of 60° C initially until the moisture content of the fruit drops. Then after about 6 hours the temperature is reduced to about 45° C to avoid browning during final drying. Heat pump dryers operate at about 40-45° C and low relative humidity, are economical to run and result in a good quality product (Klieber, 2000). However, if a low humidity is not maintained, due to overloading of the dryer, mould growth may occur reducing the quality of the spice.

Drying of paprika is a slow process due to the very slow movement of water through the waxy cuticle of the fruit. To accelerate drying, fruit are often cut into small segments. Using drying oils that disrupt the cuticle, as is common practice for drying of grapes, also accelerates drying but to a lesser extent than cutting (Klieber, 2000). The drying oil contains mostly ethyl ester of fatty acids ($C_{14}/C_{16}/C_{18}$) and potassium carbonate (K_2CO_3) as active constituents. Some workers have dried paprika fruit in about 20 hours (Minguez-Mosquera et al. 1994a), but in commercial operations longer periods are common. The drying time and conditions also have implications for the microbiological safety of the resulting powder as discussed in the quality section above.

The final moisture content of paprika fruit should be about 8%. A moisture content below 4% accelerates colour degradation, above 11% fungal growth (Wall and Bosland, 1993), and above 14% allows aflatoxin accumulation (Anonymous, 1994). As the spice powder is hygroscopic it will quickly absorb moisture in humid environments. Therefore it may be necessary to re-dry spice after grinding or packaging the powder in low RH rooms into moisture proof containers.

Following drying paprika fruit are ground to produce the spice powder. The final quality of paprika is influenced by whether the whole fruit or only the flesh, excluding the seed, is used for manufacturing. Seeds are colourless and their inclusion will dilute the colour (Purseglove et al. 1981; Almela et al. 1991), but the omission of seeds and of the associated placental tissue will reduce the hotness of the powder (Fujiwake et al. 1982; Iwai et al. 1977; Purseglove et al. 1981). Seeds also contain antioxidants that protect carotenoids from autoxidation and spice from loss of red colour during storage (Okos et al. 1990; Biacs et al. 1992; Klieber and Bagnato, 1999). Overall, seed addition reduces the shelf life of spice powder, as it reaches a minimum colour standard quicker due to the colour dilution effect (Klieber and Bagnato, 1999). However, seed exclusion will reduce the spice yield by up to 40% (Klieber and Bagnato, 1999).

The flesh to seed ratio may be varied by separating them after drying and mixing them in a set ratio during grinding. Seed separation is an additional cost, but may be worthwhile where a very high colour intensity of powders is desired.

Two types of mills may be used to grind paprika fruit commercially. Traditionally stone mills are used, but in industrial manufacturing hammer mills predominate. These methods vary in their action. Stone milling is thought to crush and mix seeds more thoroughly with the flesh, resulting in increased gloss and less colour deterioration during storage (Biacs et al. 1992; Minguez-Mosquera et al. 1993). However, Klieber and Bagnato (1999) found no difference in spice quality or shelf life between paprika powders ground with a hammer mill or a plate grinder, which simulated stone milling.

3.4.3. Storage of Powder

Storage conditions of paprika spice powder influence its colour stability. The red carotenoid pigments are easily oxidised, reducing the colour intensity of the spice powder and making it more orange. To minimize the instability of carotenoids different cultivars, drying and storing techniques have been evaluated (Okos et al. 1990; Biacs et al. 1992, 1994; Minguez-Mosquera et al. 1993, 1994b; Osuna-Garcia and Wall, 1996). Natural antioxidants effectively reduce colour loss, with fat soluble α-tocopherol (Vitamin E) being more effective than water soluble ascorbic acid (Vitamin C) (Biacs et al. 1992). Synthetic antioxidants, such as ethoxyquin, also reduced colour loss (Osuna-Garcia and Wall, 1996). Commercially, however, the addition of natural or synthetic antioxidants may not be permissible as, for example, in Australia (Australian and New Zealand Food Authority, 1996). However, different cultivars may contain varying levels of antioxidants in their seeds, giving seeds varying degrees of colour protection (Klieber and Bagnato, 1999). Therefore, breeding cultivars with high antioxidant levels in the seed offers a useful opportunity for producing more shelf-stable spice powders.

Light, air and temperature during storage also have a profound effect on colour retention in paprika spice powders (Malchev et al. 1982). Exclusion of light and oxygen and storage at low temperatures all reduce the rate of colour loss. Oxygen can be excluded by nitrogen storage (Bunnell and Bauernfeind, 1962), and this results in the best possible storage life (Klieber and Bagnato, 1999) (Table 1).

Table 1. Initial extractable colour, rate of extractable colour loss and shelf life at 37° C for paprika spice powder. Vitamin E had 0.3 g of vitamin E added to 100 g of powder, and the N_2 atmosphere powder was held in an atmosphere of nitrogen; neither treatment contained seed. Shelf life was measured as time to drop below an ASTA value of 180

	Spice including seed	Vitamin E	N_2 atmosphere
Initial colour (ASTA units)	208[b]	286[a]	281[a]
Weekly colour loss (ASTA units)	23[a]	25[a]	14[b]
Shelf-life (weeks)	1.2[c]	4.2[b]	7.0[a]

Different letters indicate significant differences at the 5% level (after Klieber and Bagnato, 1999).

4. Conclusion

Producing paprika spice requires the integration of all aspects of production, harvesting, processing and storage if good yields and quality are to be achieved. In particular, yield is sensitive to selection of the correct site and seed, plant establishment, irrigation and fertilization, pest and disease control, and harvest practices. Plant establishment and harvesting become more critical for mechanical harvesting where fruit are harvested in an once-over operation. The main quality aspects of concern are colour intensity and hue, pungency level and freedom from mycotoxins. Colour is affected by cultivar, nutrition, ripeness at harvest, drying conditions and conditions during storage. Pungency is mainly affected by cultivar, temperature during growth of fruit and position on the plant. Mycotoxins can accumulate if the wrong conditions in the field, during processing or during storage of the spice occur. Such conditions include mechanical damage, for example from insects, excessive moisture content of the spice, and poor processing conditions.

5. Perspectives

Paprika spice demand is increasing around the world as consumers are becoming more familiar with international cuisine. Retailers are also becoming more globally oriented, are sourcing product anywhere in the world and are demanding supply chain quality assurance. To stay competitive, spice growers, processors and traders must ensure that they meet the retailers' and consumers' needs. These are, not necessarily in that order, food safety, quality, value for money and environmental sustainability. Therefore the paprika spice industry faces the challenge of continual improvement of its practices. Sustainable irrigation and pest control practices are needed and an integrated quality system needs to be developed that minimises microbial contamination and accumulation of mycotoxins. Colour quality needs to be high at harvest as well as after storage of the powder. However, an area of quality improvement that has not been addressed previously is nutritional improvement. By increasing natural levels of antioxidants, powder colour can be improved, as well as improving health benefits for the consumer.

6. Acknowledgments

Many thanks go to two of my students who added to the understanding of the paprika spice system. Mayuree Krajayklang undertook a PhD study examining the paprika spice system, and Nancy Bagnato examined paprika colour stability in a honours project. Thanks also go to my wife Julia, who first awakened my interest in this spice.

References

Adachi, Y., Hara, M., Kumazawa, N.H., Hirano, K., Ueno, I. and K. Egawa. 1991. Detection of aflatoxin B1 in imported food products into Japan by enzyme-linked immunosorbent assay and high performance liquid chromatography. *Journal of Veterinary and Medical Science* **53**(1): 49-52.

Almela, L., Lopez-Roca, J.M., Candela, M.E. and M.D. Alcazar. 1991. Carotenoid composition of new cultivars of red pepper for paprika. *Journal of Agricultural and Food Chemistry* **39**: 1606-1609.

American Spice Trade Association. 1985. Official analytical methods of the American Spice Trade Association. 2nd ed. American Spice Trade Association, Englewood Cliffs, New Jersey.

Anonymous. 1994. Mycotoxin diagnostic catalogue. Neogen Corporation, Lansing Michigan, 7 pp.

Asian Vegetable Research and Development Centre. 1993. Annual Report for 1992. Taiwan.

Ath-Har, M.A., Prakash, H.S. and H.S. Shetty. 1988. Mycoflora of Indian spices with special reference to aflatoxin producing isolates of Aspergillus flavus. *Indian Journal of Microbiology* **28**(1-2): 125-127.

Australian and New Zealand Food Authority. 1996. Food Standards Code, National Food Authority, Canberra.

Batal, K.M. and D.M. Granberry. 1982. Effects of growth regulators on ripening and abscission of Pimiento and Paprika peppers. *HortScience* **17**: 944-946.

Beaudry, R.M. and S.J. Kays. 1988. Effect of ethylene source on abscission of pepper organs. *HortScience* **23**: 742-744.

Biacs, P.A., Czinkootai, B. and A. Hoschke. 1992. Factors affecting stability of colored substances in paprika powders. *American Chemical Society* **40**(3): 363-367.

Biacs, P.A., Daood, H.G. and M.A. Dakar. 1994. Storage stability of carotenoids from new hybrids produced from Hungarian and Spanish (sic) paprika cultivars. *Acta Horticulturae* **368**: 177-183.

Bosland, P.W. 1993. Breeding for quality in *Capsicum*. *Capsicum and Eggplant Newsletter* **12**: 25-31.

Bosland P.W., Bailey, A.L. and D.J. Cotter. 1994. Growing chiles in New Mexico. New Mexico Cooperative Extension Services Circular, Las Cruces New Mexico, H-230.

Bosland, P.W., Bailey, A.L., and J. Iglesias-Olivas. 1996. *Capsicum* pepper varieties and classification. New Mexico Cooperative Extension Services Circular, Las Cruces New Mexico, p. 530.

Britton, G. and D. Hornero-Mendez. 1997. Carotenoids and colour in fruit and vegetables. In: Phytochemistry of Fruit and Vegetables. F.A. Tomas-Barberan, and R.J. Robins. (Eds), Clarendon Press, Oxford, 11-29.

Bunnell, R.H. and J.C. Bauernfeind. 1962. Chemistry, uses, and properties of carotenoids in foods. *Food Technology* **16**: 36-43.

Carter, A.K. 1994. Stand establishment of chile. New Mexico Cooperative Extension Services Circular, Las Cruces New Mexico. H-238.

Clapham, D.E. 1997. Some like it hot: spicing up ion channels. *Nature* **389**: 783.

Collins, M. and P.W. Bosland. 1994. Measuring chile pungency. New Mexico Cooperative Extension Services Circular, Las Cruces New Mexico. H-237.

Cooksey, J.R., Kahn, B.A. and J.E. Motes. 1994. Calcium and ethephon effects on paprika pepper fruit retention and fruit colour development. *HortScience* **29**: 792-794.

Cotter, D.J. 1980. Review of studies on chile. *New Mexico Agricultural Experiment Station Bulletin* **673**: 1-29.

de Wilde, R.C. 1971. Practical applications of (2-chloroethyl) phosphonic acid in agricultural production. *HortScience* **6**: 364-370.

Deak, T., Fabri, I., King, A.D. (Jr), Pitt, J.I., Beuchat, L.R. and J.E.L. Corry. 1986. Baseline count of molds in paprika. Methods for Mycological Examination of Foods: 200-202.

Deli, J., Matus, Z. and G. Toth. 1996. Carotenoid composition in the fruits of *Capsicum annuum* cv. Szentesi Kosszarvu during ripening. *Journal of Agricultural and Food Chemistry* **44**: 711-716.

Dickerson, G.W. 1994. Growing peppers in New Mexico gardens. New Mexico Cooperative Extension Services Circular, Las Cruces New Mexico, H-240.

FAO. 1999. FAOSTAT database result for chillies and peppers. Downloaded from http://apps.fao.org/cgi-bin/nph-db.pl?subset=agriculture on 6/1/99.
Food and Drug Administration. 1999. Action levels for poisonous or deleterious substances in human food and animal feed. Downloaded from http://vm.cfsan.fda.gov/~lrd/fdaact.html on 7/8/99.
Finoli, C. and M. Ferrari. 1994. Aflatossine in spezie ed erbe aromatiche. *Industrie Alimentari* **33**(328): 732-736.
Frank, H.K. 1977. Occurrence of patulin in fruit and vegetables. *Annales de la Nutrition et de l'Alimentation* **31**: 459-465.
Fuentes, R.G. and W.C. Mora. 1988. Preliminary survey of chilli cultivars (*Capsicum* spp.). *Capsicum Newsletter* **7**: 47-48.
Fufa, H. and K. Urga. 1996. Screening of aflatoxins in shiro and ground red pepper in Addis Ababa. *Ethiopian Medical Journal* **34**(4): 243-249.
Fujiwake, H., Suzuki, T. and K. Iwai. 1982. Capsaicinoid formation in the protoplast from the placenta of capsicum fruits. *Agricultural and Biological Chemistry* **46**: 2591.
Goldberg, N.P. 1995. Chile pepper diseases. New Mexico Cooperative Extension Services Circular, Las Cruces New Mexico. p. 549.
Grattidge, R. 1993. Growing capsicums and chillies. Queensland Department of Primary Industries, Townsville.
Gross, K.C., Watada, A.E., Kang, M.S., Kim, S.D., Kim, K.S. and S.W. Lee. 1986. Biochemical changes associated with the ripening of hot pepper fruit. *Physiologica Plantarum* **66**: 31-36.
Haigh, A.M., Aleemullah, M. and Sumardi. 1996. Water stress reduces yield in chilli by reducing the success of pollination. *In:* Proceedings of the National Pepper Conference. D.N. Maynard (Ed.), December 1996, Naples, Florida, 90-91.
Hertog, M.G.L., van Poppel, G. and D. Verhoeven. 1997. Potentially anticarcinogenic secondary metabolites from fruit and vegetables. *In:* Phytochemistry of Fruit and Vegetables. F.A. Tomas-Barberan and R.J. Robins. (Eds), Clarendon Press, Oxford, 313-331.
Hibberd, A., Finlay, G. and P. Lavrijsen. 1992. Hot chilli varieties for Queensland. Queensland Fruit and Vegetable News, June 18: 11-13.
Hodges, L., Menghua Fu, and J.R. Brandle. 1996. Preliminary report on the effect of wind on pepper (*Capsicum annuum*). *In:* Proceedings of the National Pepper Conference. D.N. Maynard (Ed.), December 1996, Naples, Florida; 86-87.
Indira, P., Gopalakrishnan, P.K. and K.V. Peter. 1985. Response of chilli genotypes to ethephon whole plant sprays. *Agricultural Research Journal of Kerala* **23**(2): 163-167.
Isidoro, E., Cotter, D.J., Fernandez, C.J. and G.M. Southward. 1995. Color retention in red chile powder as related to delayed harvest, *Journal of Food Science* **60**(5): 1075-1077.
Iwai, K., Lee, K.R., Kobashi, M. and T. Suzuki. 1977. Formation of pungent principles in fruits of sweet pepper, *Capsicum annuum* var. grossum during post-harvest ripening under continuous light. *Agricultural and Biological Chemistry* **41**(10): 1873-1876.
Jaffar, M., Saleem, M., Najeeba-Saleem, Maqsood-Ahmed, Saleem, N. and M. Ahmed. 1994. Screening of various raw food commodities for aflatoxin contamination. Part II. *Pakistan Journal of Scientific and Industrial Research* **37**(12): 547-548.
Kader, A.A. 1992. *Postharvest Technology of Horticultural Crops.* 2nd Edition. University of California, Oakland, California.
Kahn, B.A. 1992. Cultural practices for machine-harvested paprika pepper. *Acta Horticulturae* **318**: 239-44.
Kahn, B.A., Motes, J.E. and N.O. Maness. 1997. Use of ethephon as a controlled abscission agent on paprika pepper. *HortScience* **32**: 251-255.
Kays, S.J. 1991. *Postharvest physiology of perishable plant products.* Van Nostrand Reinhold, New York.
Klieber, A. 2000. Chilli spice production in Australia. Rural Industries Research and Development Corporation final report. RIRDC, Canberra, Australia.
Klieber, A. and A. Bagnato. 1999. Colour stability of paprika and chilli powder. Food Australia. (*in press*).

Knavel, D.E. and T.R. Kemp. 1973. Ethephon and CPTA on colour development in bell pepper fruits. *HortScience* **8**: 403-404.

Krajayklang, M., Klieber, A., Wills, R.B.H. and P. Dry. 1999. Effects of ethephon on fruit yield, colour and pungency of cayenne and paprika peppers. *Australian Journal of Experimental Agriculture* **39**(1): 81-86.

Lease, J.G. and E.J. Lease. 1956. Factors affecting the retention of red colour in peppers. *Food Technology* **10**: 368-373.

Lee, R.D. and J. Schroeder. 1995. Weed management in chile. New Mexico Cooperative Extension Services Circular, Las Cruces New Mexico. p. 548.

Levy, A., Palevitch, D. and O. Shoham. 1989. Effect of genetic and environmental factors on the capsaicin content in the fruits of pungent and sweet cultivars of pepper, *Capsicum annuum* L. *In:* European Association for Research on Plant Breeding, The Proceedings of the 7th Meeting on Genetics and Breeding on Capsicum and Eggplant. Smederevska, Yugoslavia, pp. 81-85.

Lorenz, O.A. and D.N. Maynard. 1980. *Knott's Handbook for Vegetable Growers*. 2nd Edition, Wiley, New York.

Macdonald, S. and L. Castle. 1996. A UK retail survey of aflatoxins in herbs and spices and their fate during cooking. *Food Additives and Contaminants* **13**(1): 121-128.

Malchev, E., Ioncheva, N., Tanchev, S. and K. Kalpakchieva. 1982. Quantitative changes in carotenoids during the storage of dried red pepper and red pepper powder. *Die Nahrung* **26**(5): 415-420.

Matta, F.B. and D.J. Cotter. 1994. Chile production in North-Central New Mexico. New Mexico Cooperative Extension Services Circular, Las Cruces New Mexico. H-225.

Mayer, A. and E. Harel. 1979. Polyphenol oxidases in plants. Phytochemistry 18: 193-215.

Meena-Nair, Peter, K.V. and M. Nair. 1990. Organic, inorganic fertilizers and their combinations on yield and storage life of hot chilli. *Vegetable Science* **17**(1): 7-10.

Miles, K.A. 1994. Market development potential for hot chilli peppers. Report Q094008, Queensland Department of Primary Industries, Brisbane.

Minguez-Mosquera, M.I. and D. Hornero-Mendez. 1993. Separation and quantification of the carotenoid pigments in red peppers (*Capsicum annuum* L.), paprika and oleoresin by reversed-phase HPLC. *Journal of Agricultural and Food Chemistry* **41**: 1616-1620.

Minguez-Mosquera, M.I., Jaren-Galan, M. and J. Garrido-Fernandez. 1993. Effect of processing of paprika on the main carotenes and esterified xanthophylls present in the fresh fruit. *Journal of Agricultural and Food Chemistry* **41**: 2120-2124.

Minguez-Mosquera, M.I., Jaren-Galan, M. and J. Garrido-Fernandez. 1994a. Competition between the processes of biosynthesis and degradation of carotenoids during the drying of peppers. *Journal of Agricultural and Food Chemistry* **42**: 645-648.

Minguez-Mosquera, M.I., Jaren-Galan, M. and J. Garrido-Fernandez. 1994b. Influence of the industrial drying processes of pepper fruits (*Capsicum annuum* cv. *Bola*) for paprika on the carotenoid content. *Journal of Agricultural and Food Chemistry* **42**: 1190-1193.

Okos, M., Csorba, T. and F. Szabad. 1990. The effect of paprika seed on the stability of the red colour of ground paprika. *Acta Alimentari* **19**: 79-93.

Osuna-Garcia, J.A. and M.M. Wall. 1996. Use of antioxidants and adjusted moisture content for colour retention of stored paprika. *In:* Proceedings of the National Pepper Conference. D.N. Maynard (Ed.), December 1996, Naples, Florida: 116-117.

Palevitch, D. 1978. Cultural practices and cultivars for once-over harvested sweet paprika. *Acta Horticulturae* **73**: 255-262.

Patel, S., Hazel, C.M., Winterton, A.G.M. and E. Mortby. 1996. Survey of ethnic foods for mycotoxins. *Food Additives and Contaminants* **13**(7): 833-841.

Persley, D.M., O'Brien R. and J.S. Syme. 1989. Vegetable crops: a disease management guide. Queensland Department of Primary Industries, Information Series QI88019, Brisbane.

Purseglove, J.W., Brown, E.G., Green, C.L. and S.R.J. Robbins. 1981. *Spices, vol. 1*. Longman Inc, New York.

Putzka, H.A. 1994. Unerwünschte Stoffe in Futter- und Lebensmitteln. *Angewandte Botanische Berichte* **5**: 112-116.

Quinones-Seglie, C.R., Burns, E.E. and B. Villalon. 1989. Capsaicinoids and pungency in various Capsicums. *Lebensmittel Wissenschaft and Technologie* **22**(4): 196-198.
Roos, Y.H. and M.J. Himberg. 1994. Nonenzymic browning behaviour, as related to glass transition, of a food model at chilling temperatures. *Journal of Agricultural and Food Chemistry* **42**(4): 893-898.
Rylski, I. 1986. Pepper (Capsicum), *In:* CRC Handbook of Fruit Set and Development. S.P. Monselise (Ed.), CRC Press, Boca Raton, Florida, 341-354.
Saito, M, Singh, R.B. and M. Saite. 1976. Reports on study of mycotoxins in foods in relation to liver diseases in Malaysia and Thailand. Institute of Medical Science, University of Tokyo, Tokyo, 85 pp.
Saltveit, M.E. 1977. Carbon dioxide, ethylene, and color development in ripening mature green bell peppers. *Journal of the American Society for Horticultural Science* **102**: 523-525.
Scott, P.M. and B.P.C. Kennedy, 1973. Analysis and survey of ground black, white, and capsicum peppers for aflatoxins. *Journal of the Association of Official American Chemists* **56**(6): 1452-1457.
Seenappa, M., Stobbs, L.W. and A.G. Kempton. 1980. Aspergillus colonization of Indian red pepper during storage. *Phytopathology* **70**(3): 218-222.
Shirvastava, M., Ngomidir, M. and P.H. Pandey. 1990. Design and development of a low cost solar drier. *In:* Proceedings of the International Agricultural Engineering Conference and Exhibition, Bangkok. Asian Institute of Technology, Bangkok: 505-515.
Sims, W.L., Collins, H.B. and B.L. Gledhill. 1970. Ethrël effects on fruit ripening of peppers. California Agriculture 24: 4-5.
Smith, A. 1982. Selected markets for chillies and paprika. Publication of the Tropical Products Institute No. G155.
Somos, A. 1984. The Paprika. Akademiai Kiado, Budapest.
Swaine, G., Ironside, D.A. and R.J. Corcoran, 1991. Insect pests of fruit and vegetables. 2nd Edition. Queensland Department of Primary Industries, Information Series QI91018, Brisbane, 126 pp.
Szucs, K. 1975. [Spice capsicum growing and processing.] Mezogazdasagi Kiado, Budapest, Hungary.
Tabata, S., Kamimura, H., Ibe, A., Hashimoto, H., Iida, M., Tamura, Y. and T. Nishima, 1993. Aflatoxin contamination in foods and feedstuffs in Tokyo: 1986-1990. *Journal of the Association of Official Analytical Chemists International* **76**(1): 32-35.
Tewari, V.P. 1987. Selection of promising lines in perennial chilli (Capsicum frutescens). *Capsicum Newsletter* **6**: 45-46.
Tiris, C., Tiris, M. and I. Dincer. 1995. Investigation of the thermal efficiencies of a solar drier. *Energy Conversion Management* **36**(3): 205-212.
Udagawa, S. 1982. Detection and occurrence of mycotoxin-producing fungi in foods imported to Japan. *Korean Journal of Mycology* **10**(4): 225-231.
van Uffelen, J.A.M. and R. Elgersma. 1990. Chilli pepper. Standard cultivar Torito is the most satisfactory. *Groenten en Fruit* **46**(14): 43.
Wall, M. 1994. Color analysis for dehydrated capsicums. New Mexico Cooperative Extension Services Circular, Las Cruces New Mexico, p. 546.
Wall, M. 1996. Managing preharvest and postharvest quality of dehydrated capsicums. *In:* Proceedings of the National Pepper Conference. D.N. Maynard (Ed.), December 1996, Naples, Florida: 118-119.
Wall, M.M. and C.L. Biles. 1994. Maturation and ripening-related changes in New Mexican type chile fruit. Agricultural Experiment Station Bulletin 769, New Mexico State University, Las Cruces New Mexico, 19 pp.
Wall, M.M. and P.W. Bosland, 1993. The shelf-life of chiles and chile containing products. *In:* Shelf Life Studies of Food and Beverages. G. Charalambous, (Ed.), Elsevier, Amsterdam, 487-500.
Walter, J.F. 1996. The use of botanical pesticides on pepper pest and disease control. *In:* Proceedings of the National Pepper Conference. D.N. Maynard (Ed.), December 1996, Naples, Florida. 52-54.

Wedzicha, B.L. 1984. *Chemistry of Sulphur Dioxide in Foods*. Elsevier Applied Science Publishing, Barking, England.

Wien, H.C. 1997. Peppers. *In:* The Physiology of Vegetable Crops. H.C. Wien (Ed.), CAB International, Oxon, UK, 259-293.

Woodbury, J.E. 1990. Spices and other condiments. *In:* Official Methods of Analysis of the Association of Official Analytical Chemists. K. Helrich, (Ed.), 15th Edition, Arlington, Virginia: 999-1001.

Worku, Z., Herner, R.C. and R.L. Carolus. 1975. Effect of stage of ripening and ethephon treatment on colour content of paprika pepper. *Scientia Horticulturae* **3**: 239-245.

Wright, R. and I. Walker. 1991. Capsicum and chilli varieties. *Queensland Fruit and Vegetable News* **62**(14): 11.

Zewdie, Y. and P.W. Bosland. 1996. The effect of node position in chile on fruit pungency. *In:* Proceedings of the National Pepper Conference. D.N. Maynard (Ed.), December 1996, Naples, Florida, p. 115.

7

Factors Associated with Peach Yield, Quality and Postharvest Behavior

Salvador Pérez González

*Facultad de Química/Area Agrícola, Universidad Autónoma de Querétaro,
CU, Querétaro, Qro. Mexico CP 70100*

1. Introduction

Growing temperate fruit crops in the subtropics represents a challenge and has relied on the implementation of 'imported' cultivars and management practices, which have failed due to vast differences in climate and consumer expectations. The key to successful fruit production is determined mainly by market demand, technical support and natural resources available. Most of the fruit production from subtropical and tropical regions is characterized by:

a. Offseason production and high local demand, both associated with very attractive prices
b. Introduced varieties originally designed to suit other climates and market needs
c. Lack of local experience in temperate fruit growing
d. Large differences among production systems in terms of climate (associated with altitude, mainly temperature and rainfall, experience and cultural background).

This has resulted in a striking heterogeneity among orchards in terms of yield efficiency, harvest season and fruit quality.

Peaches were introduced in Mexico by the Spaniards during the 16th century (Hedrick, 1917), and rapidly spread to the humid sierras due to a wide acceptance of the fruit by the Indians. Over the next 400 years peaches were grown only in family orchards associated with other introduced and native fruit trees. Actually, peaches grow naturally and are cultivated on a very wide range of ecosystems, from sea level to 2900 m altitude, and from the sierras in the tropical regions in the south to the mountains in the semiarid north. Therefore peach is the most important temperate fruit species in Mexico, due to its wide distribution and the large number of growers involved.

Traditional peach production was based on relatively small seedling orchards, while more intensive production systems were initiated in the late 1950s, using local selections in central Mexico and introduced varieties in the north. Both systems have evolved replacing old varieties, increasing planting densities and production, but in many cases increasing production costs associated with protection against frost and hail.

Recently, subtropical ecosystems have been incorporated to peach production because of higher water availability and lower frost risk. The relative success of peach production in these environments is based on:
  a. Application of budbreaking chemicals to concentrate blossom and harvest periods in some introduced varieties already under cultivation
  b. Breeding medium to low chilling requirement varieties, designed to satisfy consumer expectations in terms of quality and length of the marketing season.

This chapter describes the main factors associated with peach production in México and analyzis the possibilities to improve yield, fruit quality and shelf-life, and also provide a background to design future orchard development in the subtropical production systems.

2. Preharvest Factors Influencing Production and Postharvest Behavior

2.1. Climate

Differences in temperature (frost risk, chilling and heat units, daily and seasonal fluctuations), light (quality, daylength, and intensity), rainfall (quantity and distribution during the year), and soil fertility, are considered to be the most important factors required for temperate fruit growing in the tropics and subtropics (Pérez, 1997). However, additional variables related to socioeconomical aspects such as local experience accumulated by growers and official support in terms of research and extension, are definitely required for the development of a solid fruit growing industry.

Mexico is located in the subtropics, but due to large differences in latitude and altitude, tropical, subtropical and temperate fruit species can be cultivated over a wide range of ecosystems in relatively short distances. Satisfactory environmental conditions for growing many temperate fruit species are common at high altitudes in the Mexican subtropical regions, where six regions lead in temperate fruit production (Table 1).

Table 1. Description of main temperate fruit growing regions in Mexico.

Region	Altitude (m)	Chill units (2 to 8° C)	Rainfall (mm)	Fruit species*
I. Chihuahua	1000-2000	600-1000	300-800	A, P, G, Pec
II. Zacatecas	1400-2300	400-700	300-600	P, A, G, Ap
III. Gto.-Ags.	1700-2500	100-600	300-800	P, A, G, Pec
IV. Sierras-1	2100-2600	450-600	1000-2500	P, Pe, Pl
V. Sierras-2	1700-2100	250-450	1000-2500	P, Pl
VI. Subtropics-1	1400-1700	50-250	1000-2000	P, Pl
V11. Subtropics 2	1400-1700	50-250	300-450	P, Pl

* P = peaches, A = apples, G = grapes, Pec = pecans, Pe = pears, Pl = japanese plums, W = walnuts and Ap = apricots, listed in order of importance in each region.

Even though there are some exceptions, it is well known that grapes, apricots and pecans grow well at high light intensities and low relative humidities typical of the semiarid regions in the north, while pears, plums and walnuts prefer the cooler and humid climates of the mountains.

Peaches exhibit the widest distribution and are cultivated in all subtropical climates, from north to south and from high elevations to sea level. Actual surface area is close to 40,000 ha, and yields range from extremely low to 35 t/ha. Differences among production systems are based mainly on climate (temperature and rainfall), cultivated varieties, degree of experience and/or quality of technical support provided.

Temperature is the most important variable to define temperate fruit growing regions in the subtropics and affects both yield and fruit quality mainly by the factors discussed below.

a) Daily fluctuations are well known to influence fruit set, shape and colour (Ryugo, 1988).
b) Heat during blossom increases the proportion of 'double pistils' and reduces fruit set, and induces fruit elongation in most introduced peach varieties (Color Plate 7.1), as well as 'black heart disease' during fruit development in peach and other temperate fruit species, due to increased respiration of the fruit (Ryugo, 1988).
c) The length of the growing season (days between the last spring frost and the first one in the fall) shows a wide range in the Mexican subtropics, from only 180 days in the north to completely frost-free regions in the subtropics. This information allows variety selection and determines the possibilities to shift blossom and harvest season (Pérez, 1997). Most local peach varieties require a long growing season, due to their very long fruit development period (from 160 to 250 days from bloom to harvest).
d) Information on chilling accumulation during tree rest (temperatures between 2° and 8° C) is fundamental for variety recommendation. Chill units accumulated in the Mexican subtropics range from less than 100 at lower elevations (less than 1600 m) to 900 in the northern sierras. However, there are drastic differences between years and ecosystems, 50 to 200 units above or below the mean values.
e) Heat units (of daily mean temperature minus 10° during fruit growth) have a direct influence on the fruit development period. When some peach varieties are grown at higher elevations (2500 m), harvest season is delayed from 20 to 40 days. And when these fruits are transported to regional markets at lower elevations, fruit senesces more rapidly (Pérez, 1997). Heat also influences insect development and this information provides an efficient tool for preventive pest control practices in the orchard.

Rainfall occurs mainly during the summer, but there are large differences among production systems. From the sunny semiarid regions in the north with less than 500 mm of annual rainfall, to the cloudy mountain ecosystems in the slopes facing the oceans with more than 1800 mm of rainfall per year.

2.2. Cultivated Varieties

Peaches were introduced by the Spaniards probably not only from temperate regions in Spain but also from subtropics in the South Pacific, as there are seedling trees with contrasting characteristics and degree of adaptation, now spread over a wide range of ecosystems.

Peach distribution is observed from the tropical lowlands, growing side by side with mangoes, bananas and vanilla (Color Plate 7.1), to the high sierras at 2200-2900 m of elevation (Pérez, 1997).

By the turn of the century, peaches were only grown in family orchards and were used for self consumption or to satisfy local demand. However, during the the past 40 years peaches were commercially cultivated using both local selections and introduced varieties (Table 2). In the northern temperate regions, introduced varieties from California breeding programs are widely grown, while at lower subtropical regions, very low chilling requirement cultivars from Florida and Brazil were initially cultivated commercially, and now are being gradually replaced with higher quality varieties (Table 2, Color Plate 7.2), developed by national breeding programs aimed to satisfy consumer and market standards (Pérez, 1997).

Most of the actual production is derived from seed propagated selections, as peach is an autogamous species and local populations are highly homogeneous (first 6 varieties in Table 2). Introduced freestone cultivars with high chilling requirements (numbers 9 to 13) are important in the northern regions, while medium to low chilling requirement varieties with non-melting flesh (11 to 13), are now being planted in subtropical ecosystems. National market prefers non-melting clingstone flesh types, and pays higher-prices for them than for imported freestone fruit types.

Unfortunately, the most important introduced varieties with low chilling develop an odd pointed fruit shape and are not very firm, such as 'Diamante'. This represents problems during packing and handling (Color Plate 7.3).

2.3. Management Practices

Traditional peach growing in Mexico is based on low input-rainfed seedling orchards (Table 3). However, in the dry-cold environments in north and central Mexico, protection systems against frost and hail are required in order to ensure annual yields of high quality fruit. However, production costs are higher and tend to reduce growers possibilities to compete in the market.

Drip irrigation integrated with a fertilization system is common among peach growers in the north and central part of the country. While rainfed orchards are widespread in the more humid regions in the sierras receiving more than 1000 mm of rainfall per season, and even in dryer regions (500-600 mm of rainfall) located in the mountains (> 2100 m of altitude), where yields are lower and the fruit smaller. Local consumers prefer

Table 2. Main peach cultivars grown commercially in Mexico during the past 20 years

Cultivar	Cr^1	R^2	FDP^3	HS^4	$Weight^5$	$Color^6$	$Flesh^7$	AdditionalComments
Zacatecas[8]	400	II	180	9-10	80-140	o-r	c	Seedling populations
Ags[8]	350	III	170	8-9	80-160	y-r	c	Seedling populations
Michoacán[8]	300	V	170	5-8	80-150	y	c	Seedling populations
Tlaxcala[8]	500	IV	180	10	80-180	r	c	Local selection
Lucero[8]	350	III	130	5	100-200	o	c	Local selection
San Gabriel[8]	300	III	130	5	100-200	o	c	Local selection
Redhaven[9]	900	I	90	4	100-200	r	f	Introduced from US
Springcrest[9]	700	I,II,IV	85	4	80-180	r	f	Introduced from US
Flavorcrest[9]	800	I	100	5	100-200	r	f	Introduced from US
O'Henry[9]	800	I	120	6	120-250	y-r	f	Introduced from US
Flordaprince[10]	200	VI,VII	90	2	80-140	r	f	Introduced from US
Flordagold[10]	250	V	80	3	80-180	r	f	Introduced from US
Magno[11]	300	IV,V	140	5	80-170	y-r	c	Introduced-Brazil
Diamante[11]	250	VI,VII	120	1-4	80-200	o	c	Introduced-Brazil
Diam. Esp.[12]	250	VI,VII	120	1-4	80-160	o	c	Mexican-CP
Regio[12]	200	V-VII	130	1-4	80-250	o	c	Mexican-INIFAP
Toro[12]	300	IV,V	130	3-4	80-140	o	c	Mexican-INIFAP
Acuitzio[12]	250	IV-VII	140	2-4	80-250	o	c	Mexican-INIFAP
San Juan[12]	250	IV-VII	140	2-4	80-180	o	c	Mexican-INIFAP
San Carlos[12]	350	III	160	6	100-200	o	c	Mexican-INIFAP
Amar. Junio[12]	300	III	120	5	80-170	o	c	Mexican-INIFAP
Amiranda[12]	500	II,III	130	6	90-190	o	c	Mexican-INIFAP

[1]**Chilling requirement** in hours between 2-8°C during rest;
[2]**Climatic region**: I: dry, with 700-1000 hours of chilling (h); II: dry, 500-700 h; III: dry, 400-500 h; IV: humid, 450-650 h; V: humid, 350-450 h; VI: humid, 50-250 h; VII: dry, 50- 250 h.
[3]**Fruit development period**: days from bloom to harvest;
[4]**Harvest season**: from extremely early = 1 (February-March), to extremely late = 10 (October)
Fruit characteristics: [5]Mean fruit size in grams;
[6]**Outside colour**: red (r), yellow (y), orange (o);
[7]**Flesh type**: free (f) or clingstone (c);
[8]Local selections
[9]Introduced cultivars with high chilling requirement
[10]Introduced cultivars with low chilling requirement
[11]Introduced from Brazil
[12]Hybrid cultivars recently developed in Alexico

Table 3. Representative peach production systems in Mexico

Region	Representative states	Limiting factors	Level of technology[1]	Production costs (US$/ha/year)
I. North-1	Chihuahua	Dry, frost, hail	4-5	2000-5000
II. North-2	Zacatecas	Dry, frost, hail,	2-3	300-1000
III. Central	Aguascalientes	Dry, frost, hail	4-5	1000-3000
IV. Sierras-1	Michoacán	Summer rains, hail	3-4	500-2000
V. Sierras-2	Michoacán, Puebla	Summer rains, hail	1-3	300-1200
VI. Subtropics-1	Michoacán	Summer rains, hail	2-3	700-2000
VII. Subtropics-2	Guanajuato, Mich.	Dry, low chilling	2-3	800-2000

[1]Based on a scale from 1 = very low, to 5 = excellent: well organized and experienced groups of growers.

smaller peaches (100-180 g) with a high sugar content produced in dry-sunny climates, against those irrigated prior to harvest (Crisosto et al. 1994; Kader et al. 1982).

2.3.1. Nutrition

Nutrition of peach trees is generally supplied through the drip irrigation system and is based on urea and ammonium sulfate during the first growing season (100 to 150 units), supplemented with manure and compost or organic fertilizers derived from humus.

In subtropical ecosystems receiving more than 2000 mm of rainfall per year where deep and fertile soils are common, no supplemental irrigation and even fertilization is supplied. It is common to see highly vigorous trees, reaching heights of 4 to 5 m after the first two growing seasons, making it difficult to control tree size and shape, and thus affecting fruit quality. In mature orchards under these circumstances, heavy N applications are avoided at least one month before harvest, and calcium sprays are practiced by some growers to increase fruit firmness and shelf-life.

In most varieties, fruits ripening in the shady parts of the tree canopy, tend to be insipid and demand careful training and pruning. Early training is particularly important for vigorous varieties growing in highly fertile soils. Most training is based on cup-shaped open systems, but Tatura types (Y) are under trial with leading growers, while the more isolated growing regions rely on very slight pruning and are generally hesitant to thinning.

2.3.2. Budbreaking Chemicals

Rainfall occurs from June to September in subtropical climates, just when local peach varieties are ripening, affecting fruit quality and making harvest difficult. Low chilling requirement varieties tend to bloom after the cool rainy summer, but blossom season is wide and low due to large daily temperature fluctuations, with chilling temperatures at night and 18° to 25° C at noon. Under these circumstances most peach varieties exhibit a very wide harvest season, making it necessary to shift it and concentrate it (Color Plate 7.4)

Means to solve the problem include the use of budbreaking chemicals, as well as breeding and selection for very short fruit development period. The first approach is widely used commercially with medium to low chilling cultivars. Sprays start with induced tree defoliation during late fall-early autum (using zinc sulfate or urea). Two to three weeks later trees are sprayed with different products to induce budbreak, including oils, hydrogen cyanamide and TDZ. This way, harvest season is shifted to winter and spring, and it can be concentrated over a shorter period of two to three weeks (Table 2). However, fruit development and maturity during the cool winter decreases size, soluble solids and general flavor.

Breeding low and medium to high chilling requirement germplasm is underway with very promising selections that defoliate naturally during late summer, and have a short blossom and fruit development period. This approach is aimed at reducing budbreaking chemical sprays, residues and high production costs involved (Pérez, 1997).

The most important diseases in peach orchards decreasing yield and quality of peach fruits are, in order of importance, powdery mildew or oidium (*Sphearoteca pannosa*), brown rot (*Monilinia fructicola*), spot (*Coryneum beikjerinckii*), and to a lesser extent rust (*Tranzchelia discolor*). Losses due to brown rot range from 10 to 35%, depending on cultivar, harvest region and postharvest management.

Breeding for the first problem has led to development of new varieties (Regio, Diamante, San Juan, Acuitzio and Diamante Mejorado). New genotypes derived from intercrossing these varieties with new selections are now under screening for *Monilinia fructicola* (Color Plate 7.5), expecting to identify multiple disease resistant genotypes (Pérez, 1997). However, these genotypes should maitain their high quality standards, as reported for other fruit crops (Granger et al. 1992).

Due to costs of imported chemicals and health risks associated with their use, in most production systems, pesticide use is restricted to wettable sulfur, copper, Captan, Zineb, oils and budbreaking chemicals. Many growers are now using biological pests or disease control practices such as the use of *Trichoderma* to control some soilborne diseases.

2.4. Harvest Season

Traditional peach production was restricted to temperate dry regions, and was harvested during the summer. Recent development of low chilling varieties, with a shorter fruit development period, use of budbreaking chemicals, extensive plantings in subtropical climates with a wide frost free period, and more water available for irrigation , are responsable for the present wide range in the harvest season, which now covers a ten month period, from mid winter to early fall (Table 4). In the mountains, where heavy rainfalls occur during the summer, peach trees bloom from early to late rainfall, and harvest is restricted to winter months at lower elevations (1300 to 1600 m), or to spring and early summer at higher elevations. While in the northern temperate regions where late frost risks are common and summers are relatively dry, harvest starts in June and extends to early rainfall.

Table 4. Peach blossom and harvest season in representative growing regions in Mexico

Region	Blossom season	Harvest season	Representative cultivars
I	March	June to October	FlavorCrest, O'Henry,
II	Mid February-early March	July to October	Local seedlings, s135, s21,
III	Mid February-early March	July to mid September	Lucero, s166, and seedlings
IV	February	May-June	Diamante, Regio, Diam. Esp.
V	January-February	April-August	Diamante, Diam. Esp., San Juan,
VI	November-February	April-August	Local seedlings, Diamante, 2P15
VII	November-February	April-August	Local seedlings, Diamante, Regio

3. Harvest Handling and Marketing

3.1. Fruit Quality Standards

Good fruit quality is the end product aimed to satisfy a particular market, and it can only be defined in terms of consumer expectations associated with a particular species. Highest quality is generally achieved on tree ripened fruit. However, under modern marketing systems it is not possible due to the large distances from production regions to urban markets. Postharvest handling and storage facilities can only maintain fruit quality and delay deterioration to extend the marketing process.

Potential consumers are first attracted by external traits such as fruit size, color, shape and uniformity of the fruit. But once attractiveness has played its role, internal factors come into play: aroma, texture, juice content, soluble solids/acidity, and complex interactions associated with flavor. When both of these phases are satisfactory to the consumer, future preference is largely assured.

Understanding variables associated with fruit quality demands a complex model based not only on individual traits (Arpaia, 1994; Monselise, 1987), but on their interaction. Factors interacting to influence fruit yield, quality and shelf-life in a subtropical production system can be grouped as follows:

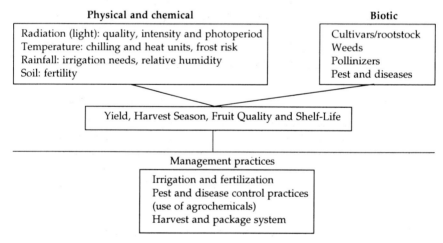

The large number of factors influencing yield, fruit quality and shelf life, make it difficult to predict the possibilities for successful temperate fruit production in subtropical ecosystems. Most experiences are derived and/or have a very strong support from countries at higher latitudes, which promote temperate fruit production outside its 'natural' range.

Peach fruit quality, either expressed as external (attractiveness) or internal factors is directly or indirectly influenced by pre- and postharvest practices (Table 5).

Table 5. Primary and secondary factores influencing fruit quality traits in peach

Attractiveness	
Directly influenced by:	
Secondary factors associated	
Size	Cultivar, tree age and vigour
	Total leaf area, fruits/tree, nutrition, pruning and training system
Shape	Cultivar, pollination, frost, temperature fluctuations,
	Growth regulators,
Color	Cultivar, fruit location,
	Nutrition, training system, radiation, fruit position, growth regulatros,
Scars	Pest and diseases, hail,
	Cultivar, climatic region,
	Extreme temperatures, Protection devices,
	Harvest system
	Experience and economic resources,
Uniformity	
	Cultivar, fruit thinning, packing
	Chilling and heat accumulation,
Internal traits:	
Skin thickness	
	Cultivar, temperature and humidity,
	Nutrition,
	Seed size
	Cultivar,
	Temperature, relative humidity,
Juiciness	Cultivar, irrigation,
	Temperature and humidity
Acidity and	Cultivar, temperature, light
	Total leaf area and fruits/tree,
soluble solids	intensity, thinning
Firmness	Cultivar, maturity, fertilization,
	Irrigation,
Aroma	Cultivar, harvest season,
Flavor	Cultivar, harvest season,
	Chemical residues
Pest/disease control practices and postharvest management	

In the national market, peach fruit attractiveness is associated with a yellow orange background color and slight red blush, medium to low size (70 to 150 g), very round shape, firm and non-melting flesh with high soluble solids to acid ratio. Most of these traits affect flavor, a complex trait that must be defined by consumers, and is influenced mainly by such as aroma, juiciness, texture, soluble solids/acidity, etc.

3.2. Harvest

The maturity index of peaches is based mainly on colour. Fruits are hand picked directly from the tree when they change from green to pale yellow (values from slightly above 0 to 8), deposited in plastic buckets and transported to the shade in large wood boxes. In the dryer sunny regions, most growers avoid picking from noon to 16:00 when days are hot (above 30° C).

Mechanically damaged fruit or fruits with brown rot spots is removed from the boxes before entering the packing house or during the hydrocooling process.

3.3. Packing Facilities and Transportation

Large wood boxes with fresh fruit are first hydrocooled at 3-5° C, then brushed, waxed, air dried and finaly sorted by size and color. Selected fruits are deposited in cardboard boxes with a capacity of 10 to 12 kg. Only fruits that are produced and transported from large distances (> 500 km) are refrigerated from the packing house to the wholesale market.

Fruit weight losses after harvest may be as high as 30% during the 7-9 day period, but there are large differences among varieties and climate where the fruit is produced (Perez, 1996).

Unfortunately, most cultivated varieties from subtropical regions are susceptible to bruising and turning brown as a response to cold water treatment. This has lead to a breeding program aimed to identify sources for brown rot and bruising resistance with good preliminary results (Pérez, 1997).

3.4. Marketing Strategies

Mexican peach production ranges from 180,000 to 250,000 t/year, and all satisfy national demand, either as fresh (70 to 90%) or processed. Most of the production is sold directly to retailers in the packing houses located in each one of the main production regions, but some of the leading growers either pack and sell their fruit to supermarkets or retailers. Others integrate their production on a single packing house and transport it to wholesale/retail markets located in Mexico City, Guadalajara and Monterrey.

There are no long-term contracts between brokers, processors and growers. Every year prices are set according to demand : offer ratio, ranging from 0.30 to 1.4 US$/kg, as influenced by harvest season, fruit type and quality.

4. Perspectives

The future of temperate fruit growing in the Mexican subtropical highlands will be strongly determined by the capacity of growers to organize themselves into well integrated groups. Specific needs, with emphasis on economical and marketing aspects, as well as research and extention, should be addressed, considering short- and long-term goals. The leading edge in this respect has already been set by peach growers unions in the north during the last 15 to 20 years.

Traditional research centers are located on federal or state grounds and are generally isolated, away from representative growing regions. This approach is not only expensive, but makes it difficult to transfer the information generated.

Small family orchards in central and southern Mexico are being promoted, particularly in low-income regions. Agroforestry studies are under way using a combination of local and introduced fruit species and varieties in preliminary trials. These modules are being established with small and highly motivated farmers, expecting them to become 'raw models' and extension centers located in representative ecosystems. In these plots, a large proportion of research trials could be carried out, including variety/rootstock evaluation, pruning-training as well as low input management programs related with use of 'safer agrochemicals', like organic fertilizers and pest-disease control programs.

References

Arpaia, M.L. 1994. Preharvest factors influencing postharvest quality of tropical and subtropical fruits. *HortScience* **29**: 982-985.

Crisosto, C., Johnson, R.S., Luza, J.G. and G.M. Crisosto. 1994. Irrigation regimes affect soluble solids concentration and water loss of O'Henry peaches. *HortScience* **29**: 169-171.

Granger, R.L., Khanizadeh, S., Fortin, J., Lapsley, K. and M. Meheriuk. 1992. Sensory evaluation of several scab resistant apple genotypes. *Fruit Var. J.* **46**: 75-79.

Hedrick, W.H. 1917. The peaches of New York. New York Agr. Exp. Station, N.Y.

Kader, A.A., Heintz, C.M. and A. Chordas. 1982. Postharvest quality of fresh and canned clingstone peaches as influenced by genotype and maturity at harvest. *J. Amer. Soc. Hort. Sci.***107**: 947-951.

Monselise S.P. 1987. Preharvest growing conditions and postharvest behavior of subtropical and temperate zone fruits. *HortScience* **22**: 1185-1189.

Pérez, G.S. 1996. Dynamics of peach fruit color and weight after harvest in the subtropical regions of Central Mexico. Presented at: Vth International symposium on "Temperate fruits in the tropics and subtropics"–ISHS, May 29-June 1,1996, Adoma, Turkey.

Pérez, G.S. 1997. Mejoramiento genético del duraznero. II Reunión Nacional de productores de durazno. Mayo 10-15, Aguascalientes, México.

Ryugo, K. 1988. *Fruit Culture: its Science and Art*. John Wiley and Sons.

Factors Associated with Peach Yield, Quality and Postharvest Behavior 169

Color Plate 7.1. Peach seedlings growing in subtropical environments.

Color Plate 7.2. New peach varieties and selections with non-melting flesh to attend national market expectations.

Factors Associated with Peach Yield, Quality and Postharvest Behavior 171

Color Plate 7.3. Undesirable shape common in most introduced varieties in subtropical climates.

Color Plate 7.4. Twigs showing poor bloom after a warm winter (upper) and three rows sprayed with Hydrogen Cyanamide (Dormex) at the center of the orchard in November.

Factors Associated with Peach Yield, Quality and Postharvest Behavior **173**

Color Plate 7.5. Genetic differences among genotypes during the screening process searching for brown rot resistance (*Monlinia fructicola)* in local germplasm.

8

Influence of Calcium Nutrition on the Quality and Postharvest Behaviour of Apples

Ramdane Dris
Department of Applied Biology,
P.O. Box 27, FIN-00014, University of Helsinki, Finland

Abbreviations

' '	cultivar
N	Newton
kg	kilogram
ha	hectare
SE	standard error of the mean
N	nitrogen
P	phosphorus
K	potassium
Ca	calcium
Mg	magnesium
$CaCl_2$	calcium chloride
$Ca(NO_3)_2$	calcium nitrate

1. Introduction

The nutritional status of apple trees is highly significant in the maintenance of fruit quality (Dris, 1998a; Dris and Niskanen, 1998a, 1999b, 2000b; Dris et al. 1997, 1999). There is also a close relationship between the storage ability of apples and fruit calcium content (Fallahi et al. 1985; Tomala et al. 1993; Tomala, 1997a). The right balance of calcium, magnesium, nitrogen, phosphorus and potassium concentrations in the fruit is essential in preventing physiological storage disorders (Tomala, 1997b). Low calcium concentration causes apples to become susceptible to a number of physiological disorders during storage (Dris, 1998b; Dris and Niskanen, 1997b; Dris et al. 1998a, b). Recently calcium deficiency has become a common problem in apple orchards with the increasing use of fertilizers, which do not contain enough of calcium (Fallahi et al. 1997; Tomala, 1997b). Preharvest calcium treatments have been successfully used to maintain the postharvest quality of apples (Peryea, 1991; Dris et al. 1999; Dris, 1998b; Dris and Niskanen, 1997a, b).

2. Calcium Nutrition of Apple Trees

In the beginning of the 1930s, Delong (Delong, 1936) and Smock later in the 1950s were among the first researchers to find that fruit calcium deficiency causes physiological disorders such as bitter pit (Fallahi et al. 1997; Dris and Niskanen, 1998a, 1999a). It was indicated that a low calcium content in apples caused cork spot which is due to competition between the leaf and the fruit tissues for calcium uptake during the early stage of growth (Fallahi et al. 1997). Thereafter, finding ways to combat bitter pit was recognized as a high priority, and Garman and Mathis were first to use preharvest calcium nitrate sprays to reduce the incidence of bitter pit in apples (Fallahi et al. 1997). During the last decade, calcium nutrition of apple trees has received increasing attention (Tomala, 1997b). Low levels of calcium in apple flesh have been associated with many disorders such as senescence breakdown, Jonathan and lenticel spots, water core, bitter pit and rotting (Fallahi et al. 1997; Tomala, 1997b).

Reduced calcium availability leads to unfavorable nutritional status in apple trees and under adverse weather conditions deficiency symptoms appear (Faust, 1989). Moisture supply, mineral nutrition and vegetative growth influence calcium concentration in the fruit tissue. Calcium is important because it regulates fruit ripening, activates certain enzymes and affects the respiration and ethylene production rates (Ferguson, 1984). Increasing the amount of calcium is a mean of enhancing natural resistance to diseases and maintaining fruit quality (Fallahi et al. 1997).

2.1. Uptake and Translocation of Calcium

The calcium requirement of a mature apple tree is remarkably high in comparison to the requirement of other nutrient elements. In 14 year-old apple trees, calcium comprised 80% of the total content of inorganic nutrients in the above ground portion and 35% of that in roots. The estimation of annual net uptake of calcium is 25 kg ha^{-1} by 9- to 12-year-old 'Cox's Orange Pippin' trees with an average annual yield of 29 tons ha^{-1}; of which 12 kg ha^{-1} was incorporated in the framework of the trees, 2 kg ha^{-1} removed in fruits and 11 kg ha^{-1} in prunings. In a mature heavy-cropping (45 tons ha^{-1}) apple orchard of variety Delicious, the annual net uptake of calcium was 50 kg ha^{-1} (4 kg ha^{-1} in fruits and 46 kg ha^{-1} in the framework of trees). In addition, the amount of calcium returned to the soil as leaf-fall, dropping blossoms, fruitlets and prunings can total 118 kg ha^{-1}.

Calcium uptake is influenced by both the root system and the environment (Color Plate 8.1). The density of the root system affects calcium absorption so that increased root growth is associated with enhanced shoot growth and increased calcium uptake.

It is important that the soil water level is adequate because lack of water decreases calcium uptake. In acidic soil, Al^{3+} ions may disturb calcium uptake. Also other cations like ammonium, potassium and magnesium can compete with calcium in nutrient uptake by the plant. Adequate availability of calcium ions can be met by calcium concentrations ranging from 5 to 40 mg l^{-1} at the root surface (Gerasopoulos and Richardson, 1997).

The xylem is considered to be a major pathway of calcium translocation (Fallahi et al. 1997), although a small amount of calcium moves laterally into the phloem cells. Calcium movement in the xylem is dependent on the transpiration rate; whereas, phloem transport is primarily affected by metabolic activity. Calcium uptake and transfer to the xylem is restricted to the younger parts of roots. In young apple trees, it took about 3 days for calcium to move 30 cm and over 14 days to move 70-80 cm up the stem (Hanger, 1979). Large percentage of the calcium becomes immobile after deposition in a specific tissue (Himelrick and McDuffie, 1983). As a consequence of this lack of redistribution, organs with a high metabolic rate, such as growing fruits or shoots, are dependent on a continuous supply of calcium. Furthermore, rapidly growing spur shoots compete with the developing fruits for calcium. Thus, any cultural practice that stimulates early shoot growth can draw needed calcium away from the young fruits at its crucial period of demand (Fallahi et al. 1997).

2.2. Fruit Calcium

Apple fruits high in calcium have a low respiration rate and potentially longer storage life than fruits with low level of calcium (Fallahi et al. 1997). High calcium also effectively counteracts the detrimental metabolic effects of high nitrogen in the fruits (Fallahi et al. 1997; Dris et al. 1998a, b). Calcium preserves cellular organization, while deficiency levels induce disintegration of cytoplasmic membranes (Gerasopoulos and Richardson, 1997). At low calcium level, fruit breakdown occurs on the tree even before maturity (Shear and Faust, 1975). Apple fruits lose their chlorophyll and are subjected to lenticel breakdown, resulting in deep cracks when exposed to high internal turgor pressures (Fallahi et al. 1997). Calcium maintains adequate protein synthesis, which supplements its effect in preserving cell membranes (Gerasopoulos and Richardson, 1997). Calcium deficiency at an early stage of fruit development restricts cell division and causes disturbances in the structure of cell walls and plasmalemma.

The relatively low calcium content of fruits compared to the leaves is due to the differences in translocation rates (Marschner, 1974). Apple fruits receive the initial supply of calcium via xylem, and later input of nutrient enters via phloem where calcium is relatively immobile. Apple fruit accumulates 90% of its total calcium in the first 6 weeks of growth. Young, developing apples have a large surface area, a very permeable cuticle, a high rate of transpiration and active photosynthesis rate.

During this period, fruits have a high water requirement and a relatively low photosynthate need, with water and calcium supplied primarily through the xylem. As the fruit grows, transpiration and photosynthesis decrease and the surface area to volume ratio falls (Ferguson and Watkins, 1983). The fruit receives assimilates from the leaves in a greater quantity, and the major route of water supply apparently shifts to phloem transport (Ferguson and Droback, 1986). The gain in calcium content later in the season is determined by intake of calcium via phloem at night, when growth occurs, and export during the day when water may be withdrawn due to the contraction of the fruits. During hot, sunny days when transpiration rate is high, water stress may lead to withdrawal of water and calcium from the developing fruits. The movement of water-soluble calcium from fruits to the adjacent leaves before harvest may be responsible for the development of some physiological disorders (Tomala and Dilley, 1990). Heavy rainfall and superfluous watering towards the end of the growth period can enhance the decrease of calcium content of apples through increase in size of the fruits (Bramlage et al. 1980). Similarly, all factors that influence the fruit growth rate, e.g. temperature, the presence of other nutrients, and leaf/fruit ratio have an indirect effect on the calcium nutrition of fruits (Tomala and Dilley, 1990).

Calcium is evenly distributed throughout the fruit early in the growing season. As the season progresses, concentration gradients develop within the fruit, where calcium is highest in the skin, lowest in the flesh and intermediate in the core (Ferguson and Droback, 1986). Faust et al. (1967) found that the apple skin contained about 4 times as much calcium as the underlying flesh. The calcium rich area consists of a layer of epidermal and underlying collenchyma cells. The distribution of calcium around the fruit axis depends initially on xylem transport and thus, on the disposition of vascular tissue and the proximity of an evaporative surface on the fruit skin. Such secondary distribution will be slow and may depend on further water input from both the phloem and xylem as well as on the concentration of other cations such as magnesium and potassium (Poovaiah et al. 1988).

There is a relationship between fruit position on the tree at harvest and incidence of physiological disorders after storage (Tomala, 1997b). Apples located near the tree top are lower in calcium than those farther down (Tomala, 1997b). On the other hand, fruits developed from terminal buds have a significantly higher calcium concentration than those from lateral buds (Tomala, 1997b). Terminal fruits show higher calcium concentration, suggesting an enhancement of calcium flow to fruits in that position during their development (Tomala, 1997b). Calcium in apples increases with a high number of seeds (Bramlage et al. 1980). Calcium accumulation in developing apple fruits is positively linked to polar auxin transport; and seeds may play an important role as a source of auxin (Marschner, 1995; Tomala, 1997b).

2.3. Physiological Functions of Calcium

Calcium is involved in the regulation of many aspects of metabolism including ion uptake, membrane permeability, cell division and the assembly of microtubules (Fallahi et al. 1997). It helps bind together neighboring cell walls and maintains integration and semipermeable properties of the membranes (Ferguson and Droback, 1986). Calcium appears to serve as an intermolecular binding agent that stabilizes pectin-protein complexes of the middle lamella of the cell wall (Poovaiah et al. 1988). It may cause the localized crystallization of the acidic phospho-lipids in the structure of cell membranes (Ferguson and Droback, 1986). Calcium ions also cause temporary neutralization of electrostatic charges (Poovaiah et al. 1988).

Calcium bound to pectates increases their stability against degradation by pectinases which are synthetized by fruits during ripening. Calcium also affects the activity of many enzyme systems and metabolic sequences in plant tissues, including a potential role in the initiation of normal fruit ripening processes (Poovaiah et al. 1988). Changes in membrane function can markedly alter respiration and ethylene evolution. Calcium alters intracellular and extracellular processes and has a pronounced effect on membrane-associated changes which are intimately associated with senescence (Poovaiah et. al. 1988). Studies on membrane leakage (Gerasopoulos and Richardson, 1997) and fruit ripening (Fallahi et al. 1997; Tomala, 1997a) have indicated that the rate of senescence often depends on the calcium status in the tissue. Also a number of senescence processes can be delayed by increasing calcium levels. For example, calcium delays softening of apples by virtue of its ability to delay degradation of the cell wall polymers (Fallahi et al. 1997).

3. Interaction of Calcium with other Nutrients

3.1. Impact of Nitrogen

Nitrogen requirement of deciduous fruits is much less than that of many other crops. However, blossom quality, ovule longevity and fruit set are influenced by nitrogen (Fallahi et al. 1997). Nitrogen fertilization decreases total soluble solids concentrations of fruits and phosphorus level in apple tissue. Furthermore, it increases fruit acidity and flesh firmness at harvest and after storage, and reduces fruit weight loss during storage. High nitrogen level stimulates apples to respire at an accelerated rate, an effect overcome by increased levels of calcium or phosphorus (Letham, 1969).

The form, as well as the total amount of nitrogen used, influences calcium uptake and fruit quality (Fallahi et al. 1997). When compared with leaves, the relative concentration of calcium in fruits is much lower under ammonium nutrition than under nitrate nutrition (Shear and Faust, 1975). Nitrate, as a nitrogen source, increases the movement and accumulation of calcium into mature leaves; while ammonium increases movement into new leaves. As a cation, ammonium reduces calcium uptake and soil pH but increases the availability of metallic ions such as manganese and aluminium, which compete with calcium uptake (Fallahi et al. 1997). The use of fertilizers containing ammonium nitrogen can exacerbate calcium deficiency in apples. The use of ammonium rather than nitrate nitrogen increases the ratio of potassium to calcium in apples by reducing calcium accumulation (Fallahi et al. 1997). Since ammonium is antagonistic to calcium uptake, the use of calcium nitrate as a nitrogen source is recommended and fertilizers containing ammonium should not be applied to apple trees before or soon after bloom.

A high level of nitrogen in the tree favors vegetative growth rather than fruiting leading to poor fruit quality (Dris, 1997; Fallahi et al. 1997). High nitrogen content in apples increases variability of fruit size and maturity, reducing fruit quality and storage life (Fallahi et al. 1997). At harvesting time, fruits with a high nitrogen content tend to be larger, greener and subjected to more pre-harvest drop than fruits with little nitrogen (Fallahi et al. 1997). Apple fruits from orchards that produce fruits with high nitrogen and low calcium concentration are generally not recommended for long-term storage (Fallahi et al. 1997). Many physiological disorders such as scald, bitter pit, internal browning, senescence breakdown and lenticel spot are accentuated under high levels of applied nitrogen (Fallahi et al. 1997).

3.2. Impact of other Nutrients

Adequate fruit phosphorus content is desirable because it maintains fruit firmness and resistance to low temperature breakdown in storage (Sharples, 1980). In apple trees, phosphorus deficiency impairs fruit setting and development of fruits and seeds (Mengel and Kirkby, 1987). Often small fruits with poor quality and delayed maturity are produced as a result (Mengel and Kirkby, 1987). Phosphorus deficiency may contribute to greater deterioration of apples during storage and exacerbate calcium deficiency, since these elements interact intimately in the cell (Bramlage et al. 1980). The increase in cell size of low-phosphorus fruits is a consequence of a decreased phospholipid level in the cell membranes (Letham, 1969).

Magnesium and potassium are also associated with physiological storage disorders in apples (Tomala, 1997b). The involvement of these elements could be due to the synergistic or antagonistic effect on calcium (Fallahi et al. 1997). Potassium and magnesium can exchange with calcium and reduce its uptake into fruits. Apples affected with senescence breakdown often contain more potassium than a sound fruit (e.g. high fruit potassium concentration and K/Ca ratio favor bitter pit). However, brighter apples are obtained with high potassium and low nitrogen (Sadowski et al. 1988). Tomala and Dilley (1990) indicated that physiological disorders could be limited by maintaining leaf potassium content below 1.2% of dry matter. Organic acids and potassium ions accumulate in the affected spots of the apple disorder Jonathan spot, where the potassium-acid accumulation complex is the cause of the disorder. High potassium fertilization in the spring and summer increases vegetative growth, water consumption, flower production, yield and the level of potassium in the fruit, but decreases fruit calcium (Dris, 1997; Sadowski et al. 1988). In apple fruit, high potassium in relation to calcium content induces calcium deficiency and increases susceptibility to cork spot, bitter pit, senescence breakdown, low temperature breakdown and superficial scald. It also tends to reduce fruit size (Bramlage et al. 1980).

Similar to potassium, high level of magnesium in fruits is antagonistic to calcium. This may induce calcium deficiency and, thus, adversely affect postharvest quality (Shear, 1975). In fact, the initiation of storage disorders such as bitter pit is dependent on the calcium/magnesium balance in fruit. When calcium is low, magnesium may be partially incorporated into cell membranes, but when calcium is high, magnesium remains in the cell sap (Letham, 1969). The replacement of calcium by magnesium brings about membrane deterioration and predisposes fruits to disorders.

Finally, boron affects calcium nutrition of apples when applied either in the soil (Tomala, 1997b) or as foliar application. It increases the transport of calcium into mature leaves, but maintains the amount of calcium in the fruit (Sharples, 1980). The effectiveness of early boron sprays in reducing corking disorders may be caused by an enhancement of calcium mobility.

4. Physiological Storage Disorders of Apples

Physiological disorders (Table 1) develop in response to adverse environmental conditions or nutritional deficiency during growth and development (Fallahi et al. 1997; Juan et al. 1997). Susceptibility to disorders is dependent on maturity at harvest, cultural practices and climatic conditions during the growing season, as well as harvesting and handling operations (Fallahi et al. 1997; Dris and Niskanen, 1998b; Dris et al. 1998). Since the early movement of calcium into the developing fruit is essential, water stress during this stage of fruit development may be the critical factor in keeping good quality (Shear and Faust, 1975; Fallahi et al. 1997). However, in some instances, symptoms only appear or worsen during a period of storage after harvest (Fallahi et al. 1997; Tomala, 1997a). As the symptoms of calcium deficiency develop, there is often a stage at which the tissues are water-soaked. Another stage involves cellular breakdown with loss of turgor leading to an internal breakdown of apples.

Bitter pit disorder is caused by a mineral imbalance in apple flesh (Tomala, 1997a). Due to abnormal metabolic activity in apples of low calcium and relatively high potassium and magnesium, cell walls collapse, plasmolysis occurs and pit cavities are formed (Fallahi et al. 1997).

Table 1. Storage conditions and physiological disorders recorded during storage (Dris et al. 1998a)

Cultivar	Storage conditions		Storage period, months	Physiological disorders recorded during storage
	Temperature (°C)	Relative humidity (%)		
Melba	2	90-95	6	Bruising, internal browning
Åerö	2	90-95	6	Bitter pit, bruising
Lobo	4	> 95	6	Senescence, internal and low temperature breakdown (rotting)
Raike	2-4	85-95	3-6	Core browning and breakdown
Red Atlas	2-4	85-95	3-6	Jonathan spot, Core browning and breakdown
Aroma	2-4	85-95	3-6	Bitter pit

Scald is caused by high nitrogen levels, but also associated with low calcium and potassium levels. Also contributing to this disorder are titratable acidity in fruits and low temperature before harvest. After several months of cold storage, the disorder may appear as a diffuse browning of the skin of apples (Dris and Niskanen, 1997a, b; Fallahi et al. 1997; Tomala, 1997a, b).

Senescence breakdown is expressed as softening of the flesh, which becomes mealy and brown. In the advanced stage the skin becomes dull (Blanpied, 1981). Jonathan spots appear on the cuticule, increase in size and coalesce to produce irregular outlines on the fruits during storage. Senescence breakdown and Jonathan spot are associated with low calcium, advanced maturity, high storage temperature and relative humidity, and prolonged storage (Snowdon, 1990; Fallahi et al. 1997; Dris and Niskanen, 1999a, 2000a).

Other physiological disorders such as chilling injury are associated with advanced maturity, occurring when apples are exposed at low storage temperature for an extended period of time (Snowdon, 1990). Bruising injury is a physical damage which occurs during harvesting, handling, packing and transport (Snowdon, 1990). Also climatic factors such as hail storms may cause bruising injury (Snowdon, 1990).

5. Calcium Treatments of Apples

Often the occurrence of physiological disorders of apples result from inadequate or poor calcium distribution in the plant rather than low calcium uptake (Fallahi et al. 1997). There are supplementary techniques such as prehavest calcium sprays and postharvest infiltration, to increase apple calcium content and reduce the incidence of physiological disorders such as bitter pit, internal breakdown, senescence breakdown, lenticel spots and low temperature breakdown (Fallahi et al. 1997; Tomala, 1997a). Calcium treatments delay fruit maturation and softening, increase resistance to ripening, decrease respiration and ethylene production rates and loss of titratable acidity (Dris, 1998b; Dris and Niskanen, 1999a, c, 2000a;

Peryea, 1991; Fallahi et al. 1997). Collectively this leads to an overall improved fruit quality (Table 2). Preharvest foliar sprays containing either calcium nitrate or calcium chloride may be applied as a short-term measure to prevent calcium deficiency (Fallahi et al. 1997; Tomala, 1997a). Sometimes calcium nitrate has been less effective than calcium chloride in reducing bitter pit and senescence breakdown because it also supplies nitrogen (Fallahi et al. 1997).

Table 2. The effect of preharvest calcium chloride and calcium nitrate spraying treatments and calcium nitrate supplied to soil on fruit firmness (N) and soluble solids (%) in apple juice of 'Lobo' at harvest and after two months storage (Dris and Niskanen, 1997a)

Treatment	Firmness harvest	Firmness two month storage	Soluble solids harvest	Soluble solids two month storage
Control	725	500	11.2	10.9
$CaCl_2$ sprays	755	415	11.0	10.4
$Ca(NO_3)_2$ sprays	796	476	11.4	9.9
$Ca(NO_3)_2$ to soil	742	476	11.6	10.2
Mean	754	467	11.3	10.3

Preharvest calcium sprays can increase the calcium content of fruits (Dris, 1998a; Dris and Niskanen, 1998a, 1999a, c, 2000a; Dris et al. 2000a). However, postharvest treatments with calcium chloride solutions may be more effective in increasing fruit calcium content (Table 3) and reducing physiological storage disorders and decay caused by several postharvest pathogens (Fallahi et al. 1997; Juan et al. 1997). Dipping apples in calcium chloride solution can increase tissue calcium, but active vacuum or pressure infiltration procedures that force solutions into fruits are more effective. Pressure infiltration may result in increases in calcium content which are two to three times greater than with spraying and dipping. It has also been shown to be effective in decreasing disorders (Conway et al. 1992).

Table 3. The effect of preharvest calcium chloride and calcium nitrate sprays and soil application of calcium nitrate on nutrient contents (g kg^{-1} dry matter) in leaves of branches bearing fruit (BF) and fruit nutrients (mg kg^{-1} fresh weight) of 'Lobo' apples at harvest (Dris et al. 1997)

	Treatment	N	P	K	Ca	Mg
Leaf	Control	20.4	1.42	15.8	10.7	1.90
	$CaCl_2$ sprays	21.5	1.65	17.1	14.6	2.10
	$Ca(NO_3)_2$ sprays	21.2	1.43	16.1	13.2	2.03
	$Ca(NO_3)_2$ to soil	20.6	1.52	16.7	12.7	2.04
	Mean	20.9	1.50	16.4	12.8	2.02
Fruit	Control	478	87	1307	41	52
	$CaCl_2$ sprays	532	92	1323	43	52
	$Ca(NO_3)_2$ sprays	506	75	1263	50	49
	$Ca(NO_3)_2$ to soil	477	91	1253	46	50
	Mean	498	86	1286	45	51

The amounts of calcium taken into apple fruits from either preharvest sprays or postharvest treatments vary from year to year, by fruit maturity and by cultivar (Fallahi et al. 1997). Calcium enters primarily through lenticels, but cracks in the cuticle and the epiderm also provide entrances especially with late picked fruit (Peryea, 1991; Fallahi et al. 1997). The calcium will have to pass an intact skin containing wax in which strongly lipophylic compounds such as paraffins and esters are found. Ions such as Ca^{2+}, which are strongly hydrophylic, will pass through with difficulty. The solubility in lipophylic substances determines the transport rate. The process of calcium transport through the skin is a combination of diffusion, dissolution in the organic layer, exchange phenomena with the negatively charged groups in the cuticle as counter-ions and transport through lenticels and small cracks. In view of these mechanisms it is reasonable to assume that there will be a positive relationship between calcium transport and its concentration on the skin surface (Poovaiah et al. 1988; Tomala, 1997a; Fallahi et al. 1997).

6. Conclusions

Maintaining good fruit quality during storage remains a challenge. Optimum harvesting time, careful handling and adequate regulation of the storage conditions are of great importance in achieving this goal. The storability of apples depend on the growing seasons. Fruit mineral analysis at harvest is a reliable tool in predicting the postharvest quality and the incidence of physiological storage disorders of apples. Apples with low calcium and phosphorus, and high nitrogen or potassium in fruit flesh are not recommended for long-term storage. Fruits with calcium below 45 mg/kg fresh weight are susceptible to different disorders depending on the cultivar. Preharvest calcium chloride treatments could be considered as a good alternative in fruit quality maintenance, since it reduces the incidence of physiological disorders. Also calcium nitrate spraying retarded senescence breakdown and rotting. The ineffectiveness of soil applied calcium nitrate in controlling disorders in apple fruits can be largely due to drought during the growing season. Therefore, improvement of irrigation systems would be beneficial. It is recommended that there is a need to spray some apple cultivars such as Åkerö or Aroma with a solution containing 2.0 g l^{-1} of calcium chloride combined with 0.2% of surfactant immediately after flowering (end of May or beginning of June). The number of sprays can range from 9 to 11 with a 7 to 10-day interval. There is no need of calcium chloride spraying treatments 2 to 3 weeks before commercial harvest because that will damage the fruits. The benefit of calcium treatments outweighs the input cost as the growers can store apples for a longer time and can get a price premium.

References

Bramlage, W.J., Drake, M. and W.J. Lord, 1980. Influence of mineral nutrition on the quality and storage performance of pome fruits grown in North America. *In:* Mineral

Nutrition of Fruit Trees D. Atkinson, J.E. Jackson, R.O. Sharples and W.M. Waller (Eds), Butterworths, London, pp. 29-39.

Blanpied, 1981.

Conway, W.S., Sams, C.E., McGuire, R.G. and A. Kelman. 1992. Calcium treatment of apples and potatoes to reduce postharvest decay. *Plant Diseases* **76**: 329-334.

Delong, W.A. 1936. Variations in the chief ash constituents of apples affected with blotchy cork. *Plant Physiology* **11**: 453-457.

Dris, 1997. Effect of NPK Fertilization on Clementine. Acta Horticulturae **448**: 375-381.

Dris, R. 1998a. Variation in the storage life of 'Lobo', 'Aroma', 'Red Atlas' and 'Raike' apples during three years. *Acta Horticulturae* **485**: 133-137.

Dris, R. 1998b. Postharvest quality of apples grown in the Åland Islands. *Acta Horticulturae* **466**: 35-40.

Dris, R. and R. Niskanen. 1997a. Postharvest performance of apples grown in the Åland Islands 1993-1995 with reference to preharvest calcium treatments. Department of Plant Production, Horticulture Section, University of Helsinki, Helsinki, 99 p.

Dris, R. and R. Niskanen. 1997b. Effect of calcium on the storage quality of apples grown in Finland. *Acta Horticulturae* **448**: 323-327.

Dris, R. and R. Niskanen, 1998a. Quality changes of 'Lobo' apples during cold storage. *Acta Horticulturae* **485**: 125-131.

Dris, R. and R. Niskanen. 1998b. Nutritional status of commercial apple orchards in the Åland Islands. *Acta Agriculturae Scandinavica, Section B, Soil and Plant Science* **48**: 100-106.

Dris, R. and R. Niskanen. 1999a. Calcium chloride sprays decrease physiological disorders following long-term cold storage of apple. *Plant Foods for Human Nutrition* **54**: 159-171.

Dris, R. and R. Niskanen. 1999b. Nitrogen and calcium status in leaves and fruit of apple trees grown in the Åland Islands. *In:* Agri-Food Quality II. Quality management of fruits and vegetables. M. Hägg, R. Ahvenainen, A.M. Evers and K. Tiilikkala (Eds), The Royal Society of Chemistry Special Publication No. 229. Cambridge, UK, pp. 128-129.

Dris, R. and R. Niskanen. 1999c. Effect of calcium and prestorage heat-treatment on the storage behaviour of 'Raike' and 'Red Atlas' apples. *In:* Agri-Food Quality II. Quality management of fruits and vegetables. M. Hägg, R. Ahvenainen, A.M. Evers and K. Tiilikkala (Eds). The Royal Society of Chemistry Special Publication No.229. Cambridge, UK, pp. 125-127.

Dris, R. and R. Niskanen. 2000a. The impact of calcium chloride and heat-treatments on the quality maintainance of 'Lobo' apples after storage. *Acta Horticulturae (in press)*.

Dris, R. and R. Niskanen. 2000b. Leaf and fruit macronutrient composition during the growth period of apples. *Acta Horticulturae (in press)*.

Dris, R., Niskanen, R. and I. Voipio. 1997. Nutritional status of apple orchards in the Åland Islands 1993-1995. Department of Plant Production, Horticulture Section, University of Helsinki, Publication No. 32, Helsinki, 120 p.

Dris, R., Niskanen, R. and E. Fallahi. 1998a. Nitrogen and calcium nutrition and fruit quality of commercial apple cultivars grown in Finland. *Journal of Plant Nutrition* **21**(11): 2389-2402.

Dris, R., Niskanen, R. and E. Kaukovirta, 1998b. Development of quality characteristics during maturation of apple fruit—mineral composition. *Acta Horticulturae* **466**: 155-159.

Dris, R., Niskanen, R. and E. Fallahi. 1999. Relationship between leaf and fruit minerals and fruit quality attributes of apples grown under northern conditions. *Journal of Plant Nutrition*, **22**(12): 1839-1851.

Fallahi, E., Righetti, T.L. and D.G. Richardson. 1985. Predictions of quality by preharvest fruit and leaf mineral analyses in 'Starkspur Golden Delicious' apple. *Journal of the American Society for Horticultural Science* **110**: 524-527.

Fallahi, E., Conway, W.S., Hickey, K.D. and C.E. Sams. 1997. The role of calcium and nitrogen in postharvest quality and disease resistance of apples. *HortScience* **32**: 831-835.

Faust, M. 1989. *Physiology of Temperate Zone Fruit Trees*. Wiley, New York, 335 p.
Faust, M., Shear, C.B. and C.B. Smith. 1967. Investigations of corking disorders of apples. II. Mineral element gradients in 'York Imperial' apples. *Proceedings of the American Society for Horticultural Science* 91: 69-72.
Ferguson, I.B. and B.K. Droback. 1986. Calcium and the regulation of plant growth and senescence. *HortScience* 23: 262-266.
Ferguson and Watkrins, 1983.
Gerasopoulos, D. and D.G. Richardson. 1997. Fruit maturity and calcium affect chilling requirements and ripening of 'd'Anjou' pears. *HortScience* 32: 911-913.
Hanger, B.C. 1979. The movement of calcium in plants. *Communications of Soil Science and Plant Analysis* 10: 171-193.
Himelrick, D.G. and R.F. McDuffie, 1983. The calcium cycle: Uptake and distribution in apple trees. *HortScience* 18: 147-151.
Juan, J.L., Camps, F. Francés, J. and E. Montesinos. 1997. Decay and physiological disorders of 'Golden Delicious' apples in relation to cultural management and fruit quality. *Acta Horticulturae* 448: 273-282.
Letham, D.S. 1969. Influence of fertilizer treatment on apple fruit composition and physiology. 2. Influence on respiration rate and contents of nitrogen, phosphorus, and titratable acidity. *Australian Journal of Agricultural Research* 20: 1073-1085.
Marschner, H. 1974. Calcium nutrition of higher plants. *Netherlands Journal of Agricultural Science* 22: 275-282.
Marschner, H. 1995. *Mineral Nutrition of Higher Plants*. 2nd Edition. Academic Press Ltd, London and San Diego, 889 p.
Mengel, K. and E.A. Kirkby. 1987. *Principles of Plant Nutrition*. 4th Edition. International Potash Institute, Bern, 687 p.
Peryea, F.J. 1991. Preharvest calcium sprays and apple firmness. *Good Fruit Grower* 42(13): 12-15.
Poovaiah, B.W., Glenn, G.M. and A.S.N. Reddy. 1988. Calcium and fruit softening: physiology and biochemistry. *Horticultural Reviews* 10: 107-152.
Sadowski, A., Scibisz, K., Tomala, K., Kokozanecka, T. and M. Kepka. 1988. Negative effects of excessive nitrogen and potassium fertilization in a replanted apple orchard. *Acta Horticulturae* 233: 85-94.
Sharples, R.O. 1980. The influence of orchard nutrition on the storage quality of apples and pears grown in the United Kingdom. In: Mineral Nutrition of Fruit Trees. D. Atkinson, J.E. Jackson, R.O. Sharples and W.M. Waller (Eds), Butterworths, London. p. 17-28.
Shear, 1975.
Shear and Faust, 1975.
Snowdon, A.L. 1990. *A Colour Atlas of Post-harvest Diseases and Disorders of Fruit and Vegetables*. Vol. 1: General introduction and fruits. Wolfe Scientific Ltd, Barcelona, 302 p.
Tomala, K. 1997a. Effect of calcium sprays on storage quality of 'Shampion' apples. *Acta Horticulturae* 448: 59-65.
Tomala, K. 1997b. Orchard factors affecting nutrient content and fruit quality. *Acta Horticulturae* 448: 257-264.
Tomala, K., Araucz, M. and B. Zaczek. 1989. Growth dynamics and calcium content in 'McIntosh' and Spartan apples. *Communications of Soil Science and Plant Analysis* 20: 529-537.
Tomala, K. and D.R. Dilley 1990. Some factors influencing the calcium level in apple fruits. *Acta Horticulturae* 274: 481-487.
Tomala, K., Myga, W. and J. Kobylinska. 1993. Attempts at predicting storage ability of apples. Pre- and postharvest physiology of pome-fruit. *Acta Horticulturae* 326: 149-156.

Color Plate 8.1. Apple orchard (cv. Lobo) in the Horticultural Research Institute, Piikkiö, Finland.

9

Economic Analysis of Postharvest Systems in Root and Tuber Crops

Gerd Fleischer and Stefan Agne

Institute of Economics and *Communication, Faculty of Horticulture, University of Hannover, Herrenhäuser Straße 2, 30419 Hannover, Germany*

Abbreviations

CIP	International Potato Center
CPI	Consumer Price Index
DAC	District Agricultural Committee
FAO	Food and Agriculture Organization of the United Nations
GASGA	Group for Assistance on Grain after Harvest
GNP	Gross National Product
GTZ	Deutsche Gesellschaft für Technische Zusammenarbeit (GTZ) GmbH (*German technical cooperation agency*)
KARI	Kenyan Agricultural Research Institute
KSh	Kenya shillings
MoALDM	Kenyan ministry of agriculture, livestock development and marketing
PH	postharvest
UK	United Kingdom

1. Introduction

The development of the postharvest (PH) sub-sector in agriculture and horticulture, especially in developing countries, is frequently dominated by isolated activities of technology transfer. Postharvest loss reduction has been focused in the research and development efforts in different product chains. However, when developing new technologies, the wider implications on social coherence and economic viability should be taken into account.

New challenges in agricultural development resulting from market liberalization, targeting of government sector responsibilities and providing food security for growing rural and urban populations have arisen (Goletti and Wolff, 1998). The potential for establishing more value-added activities in rural areas which increase productive employment and provide income opportunities needs to be explored (Wheatiey et al. 1995). Therefore, the reorientation of conceptual approaches is being increasingly considered. For example, major international donor and development agencies have started to adopt a system-oriented approach for postharvest sector of development (GTZ, 1996).

Under the changed conditions, should be answered two questions: a) the importance of postharvest operations for overall development, and b) consideration of concepts for postharvest systems development.

A postharvest system can be defined as a set of operations and functions between production and consumption of agricultural and horticultural commodities which are carried out by different actors in order to achieve an effective and efficient food supply (Flach, 1995; Fleischer et al. 1995). When compared to the traditional perspective of optimizing isolated technical operations, the postharvest systems approach provides the following advantages:

(a) It includes all operations from harvesting of the crop to the consumption of the final product in the analysis.
(b) It combines both the analysis of the operations and that of the actors involved. The traditional commodity perspective is linked with the actors' perspective, thus giving way to the broader consideration of economic, institutional and policy dimensions. The separation of the agricultural production subsystem and the market subsystem is often not possible, especially in the traditional sectors of developing countries. The farm-household is the functional unit where both operations take place.
(c) It provides a clear perspective for targeting public and private sector responsibilities and activities in the development process. This is relevant for programs aiming at strengthening the farming and rural household systems, food processing and marketing, and food security programs.
(d) It focuses on increasing the efficiency of the system by lowering the transaction costs.

This chapter presents the main features of the systems approach in addressing postharvest problems. The approach has been tested in several case studies in Africa which were supported by different development organizations (Fleischer et al. 1999; Boxall, 1997; Henckes and Afful, 1997). The specific methodology and the implications from the findings of the case study on root and tuber crops in Kenya are reviewed.

2. The Approach for Analyzing Postharvest Systems

2.1. General Concept

Based on the earlier work on commodity system analysis (La Gra, 1990) the framework for postharvest system analysis was developed by a number of development organizations. The concept includes both the commodity and the actor orientation (Flach, 1995; Fleischer et al. 1995). The commodity orientation is based on the product-centered concept whereby a set of value-added operations are performed on the product on its way from the producer to the consumer and is considered an integrated whole. The actor orientation stems from the farming system concept. Separate but interacting crop, animal, off-farm and household components are merged taking into consideration the socio-economic linkages.

Postharvest system activities start while harvesting the crop. The analysis considers the chain of operations until the final destination, i.e. at the consumer level. The consumer plays a leading role in this system and influences operations through demand and preferences. For many developing countries, the main actors who operate to add value to the product, are small holder producers and small- to medium-scale entrepreneurs. Indirect and supporting actors are mainly in the credit and input supply system, and in the public sector who govern rules and regulations. A closer look reveals that the links and interdependencies of operations and experts involved can be described as a classical system (Fleischer et al. 1995).

A framework for postharvest system analysis has been proposed by Flach (1995). In view of the institutional framework in the developing countries, the relevant major questions of the system analysis are:

(a) Has the system been able to bridge the gap between production and consumption in a cost-effective manner?
(b) Are technological options appropriate and compatible with the system's characteristics, e.g. ability to adapt new technology.
(c) Is the system able to provide equal and fair opportunities to the interested groups, e.g. low entry barriers for participants. This is especially relevant for women and the marginal poor.
(d) Is the provision of public goods adequate to secure the functioning of the system? (i.e. the performance of the public institutions).

From an economic point of view, transaction costs should be as low as possible. However, the system should also be sustainable in the long run, able to absorb external shocks and rely on sufficient diversity in its structure. This requires a thorough analysis of the system's performance.

The main objective of postharvest system analysis is to identify the constraints and bottlenecks that restrict the performance of the system and limit its development. Such constraints may be technological, financial and economic or institutional. If the postharvest system faces problems in the process of growing commercialization of the agricultural sector, high transaction costs on the way between producer and consumer will be the outcome. This means either low product prices for the producer or high consumer prices or both. In this case, the full potential of the agricultural sector will not be achieved.

2.2. The Approach of the Case Study in Kenya

A pilot study for testing the framework for postharvest system analysis was carried out in Kenya where the agricultural sector is of great importance. The growing trend of urbanization calls for an effective and efficient postharvest chain. Among the many commodities that could be considered in postharvest research, roots and tuber crops were selected for the analysis. Participants of an expert workshop in Nairobi acknowledged the importance for local and national food security. Limited knowledge was available and substantial potential for development was assumed. The analysis thus served as an initial step for the conceptualization of planning development activities in the sub-sector (Fleischer, 1996).

The pilot country study was a component of a four-stage process (Fig. 1). The first phase consisted of the conceptualization of the systems approach among the donor and developmental agencies. Especially the Group for Assistance on Grain After Harvest (GASGA), a co-ordination unit among major agencies such as Food and Agriculture Organization (FAO), Deutsche Gesellschaft für Technische Zusammenarbeit (GTZ) and others, played an important role. In the second phase, the theoretical framework was adopted according to the specific conditions of Kenya. This included the creation of awareness and the selection of an appropriate commodity group. Hypotheses on the existing constraints were developed during a workshop with relevant research, extension and private sector organizations. They were taken as guideline references for the survey among institutions and experts involved. Phase 3 covered the field study and the analysis of results followed by a review phase. Phase 3 incorporated several elements which were found crucial for postharvest system studies, i.e. building of a multidisciplinary team of Kenyan and international experts from different affiliations[1], participatory training of team members in methods of system analysis, selection of adapted rapid appraisal methods, involvement of local organizations from policy planning, research and extension organizations in the planning of the study, and awareness raising on the importance of an integrated multidisciplinary approach. Information gaps that appeared after concluding the field study phase were filled by additional studies which were commissioned to national policy analysis institutions.

The rapid rural appraisal comprised of semi-structured interviews with purposely selected heads of farm-households and their wives, informal group discussions with village community members, semi-structured interviews with rural entrepreneurs (brokers, agents, traders), direct observation of postharvest and marketing operations, and informal interviews with key informants. Furthermore, data collection on the structure of the postharvest system at the urban level was made by using rapid appraisal techniques.

Rapid appraisal techniques do not substitute for formal research in specific areas of interest. They are regarded as advantageous for

---

[1]Participating experts came from the Kenya Agricultural Research Institute (KARI), Egerton University, Ministry of Agriculture, Livestock Development and Marketing, Natural Resource Instiute (UK), Deutsche Gesellschaft für Technische Zusammenarbeit (GTZ, Germany) and the University of Hannover (Germany).

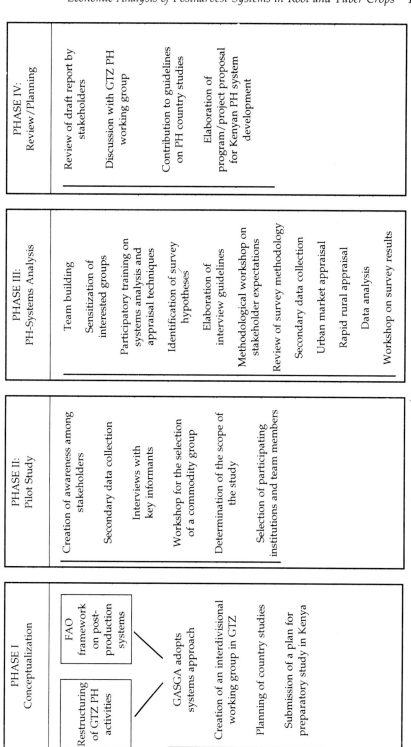

Fig. 1. Overview of the Country Study Approach

exploratory studies (Guyton et al. 1994; Holtzman et al. 1995). Additionally, secondary data was collected from the Ministry of Agriculture, Livestock Development and Marketing, the Central Bureau of Statistics, city councils, the Kenya Agricultural Research Institute (KARI), and the International Potato Center (CIP).

3. Comparative Analysis of the Postharvest System of Potato and Sweet Potato in Kenya

3.1. Importance of Potato and Sweet Potato in Kenya

3.1.1. Food Security

Kenya's development objectives for the agricultural sector include internal self-sufficiency in basic foods according to the comparative advantage in the production of the crops (GOK, 1994). An increased production is needed to cope with population growth. A considerable part of the population growth takes part in urban areas. This requires that postharvest systems to function effectively in meeting the demands.

The population of Kenya increased from 22 million in 1988 to 29.2 million in 1996, indicating 3.9 % annual increase. Over the same period the GNP per capita fell from US$ 410 to just US$ 260 per head—a real decline of more than 5 % per year. The Figure 2 plot changes in the production of major crops from the year 1985/86 up to 1995/96 with changes in population and GNP per capita. Production of major crops decreased in the 1990s. The consumer price index (CPI) for food has risen faster than the general CPI. This is an alarming picture. By the year 2010, the population is expected to reach 44 million and by the year 2025, the population will reach 63 million.

Currently, maize provides three-quarters of the carbohydrates to the fast growing Kenyan population. Since enhancement of maize production is limited due to erratic patterns of rainfall, root and tuber crops have gained more attention in recent years. Especially sweet potato can play an important role in the food security strategy for Kenya since it is considered more nutritious than cassava and more drought resistant than maize. Also, sweet potato is a relatively short-term crop, with a flexible time of harvest allowing a high degree of flexibility in a food security strategy.

3.1.2. Production pattern

In Kenya, potato ranks as the second most important food crop after maize (Guyton et al. 1994). Potato production in Kenya was on average 300,000 metric tons during 1991/92 to 1995/96. The potato crop is grown mainly in the cool, high altitude areas with well distributed rainfall. The most suitable elevation is between 1,500 and 2,500 meters above sea level. Central province produces more than 53 per cent, while Eastern and Rift valley provinces together produce a total of 44 per cent. In Eastern

Economic Analysis of Postharvest Systems in Root and Tuber Crops 195

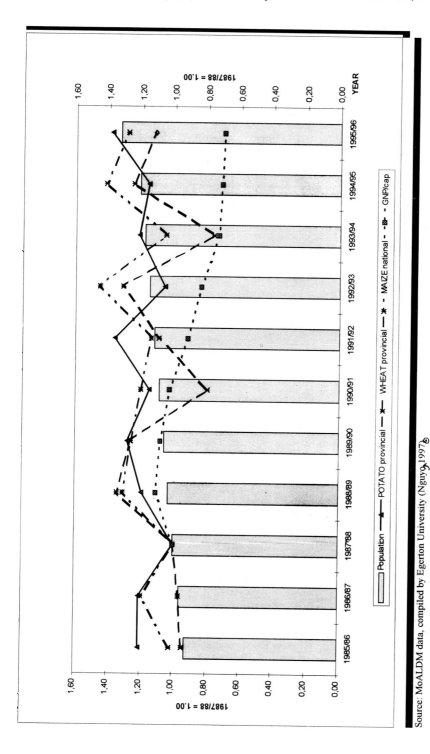

Fig. 2. Changes in production, population and income in Kenya (base year = 1987/88).

Source: MoALDM data, compiled by Egerton University (Nguyo, 1997).

Province, the main growing region is the Meru district in the areas around the slopes of Mount Kenya. In Central province nearly all the districts produce some potatoes, with Nyandarua district being the largest and most diversified potato producing area.

Most producing regions have two growing seasons. Where irrigation is possible, the harvesting time for the irrigated crop is timed to coincide with low market supplies and better prices. The average area planted with potato by each farmer ranges from 0.25 to 2.5 acres.

Sweet potato is mainly grown in western Kenya. There has been a steady increase in the sweet potato growing area from approximately 55,000 hectares in 1988 to about 65,000 hectares in 1996 (FAO, 1997). Average yields are about 10 tons per hectare. Sweet potato matures in a comparatively short period which makes it possible to grow two crops per year. Its cultivation in many parts of the country is enhanced by its ability to adapt to a wide range of climatic conditions including the marginal areas. Although traditionally regarded as a subsistence crop, sweet potato is increasingly being produced for commercial purposes. In some areas, this trend was influenced by the marketing problems of coffee which was the major cash crop. Therefore, sweet potato was seen as an alternative cash crop.

3.2. Characteristics of the Postharvest System

3.2.1. Demand Characteristics

There is heavy demand for potato in the domestic, the restaurant and the institutional sector all-year round. Although potato is usually associated with middle class consumption patterns in developing countries, in Kenya it is also an integral part of the diet of those living in the central area, mainly Kikuyu. It can be expected that the demand will become stronger due to increasing urbanization.

Sweet potato, on the other hand is seen as an inferior food. Sweet potato carries a 'stigma' of being a food of last resort or famine food. On an average per capita consumption of sweet potato is about 24 kg per year with higher proportions consumed in the western parts of Kenya (Scott and Ewell, 1992).

In nutritional terms, sweet potato, particularly the yellow fleshed varieties are good sources of vitamin A, containing 300 micrograms/100 grams fresh weight (Woolfe, 1992). A comparison with other food crops has shown that sweet potato yields more calories per unit area than either maize or potato and nearly as much as cassava, while its protein yield is far higher than the latter.

A major factor for the future development of sweet potato as a cash crop is its uptake by urban consumers as a basic staple food. The following factors are likely to affect further demand:
   a) Demographic changes due to population growth will intensify production.

b) Decreasing income levels, especially among the urban poor, who substitute maize and wheat by the less expensive sweet potato.
c) Uptake of urban processing for sweet potato flour and the incorporation of the flour in traditional and new snacks.

3.2.2. Postharvest Chain

In most growing areas, potato is a commercial crop, both from a production and marketing point of view. The major part of the crop is grown for sale. Potato has a well developed market chain from producers to the wholesale market. Producers harvest, grade and pack the crop themselves. The packaging material is delivered by village level agents. The bulk of potato marketing is done on a regularly scheduled system. Agents or brokers in the villages act for traders by organizing farmers to assemble their crop ready, graded and packed for collection, to fulfil orders made by traders. The lorries deliver the crop directly to the main urban markets (Fig. 3).

The industrial processing of potatoes is restricted to the production of snack type foods, such as crisps and other types of snacks specifically for Asian consumers. These industries are diversifying their product range to include roasted peanuts and spices. There are more than twenty crisps enterprises located in Nairobi (Walingo et al. 1996). However, the number is possibly higher since some processors are operating in private homes and may not be registered. The production depends on the amount ordered by their customers, who are mainly the major supermarkets, hotels and clubs in Nairobi and other major towns. Although the total demand currently does not exceed 5 tons raw product per day, the market shows strong signs of increasing volumes. However, the future growth of the local crisp industry is threatened by the import of products with better packaging, which competes with local products.

At the commercial level, potatoes are mainly consumed as chips served in restaurants and take-away facilities in Nairobi and other major towns in Kenya. The Nairobi City council estimates that there are about 800 restaurants and take-away facilities which sell chips in Nairobi alone. Other major towns in Kenya have similar, but fewer establishments. A study of a few take-aways and restaurants in Nairobi found that each of these facilities sells an average of 3-5 extended bags (each 130 kg) of potato per day and slightly more during weekends and public holidays. This is equivalent to about 2,500 bags per day or 30 seven-ton lorry loads per day.

Some consumers, notably chip manufacturers, have developed direct links to some farms to ensure a regular supply and an assured quality standard. Such contracts represent a very small proportion of the business.

In contrast to the well developed potato postharvest chain, sweet potato is less commercialized. The crop has traditionally been grown on small plots for domestic consumption. In a survey of marketing conditions in

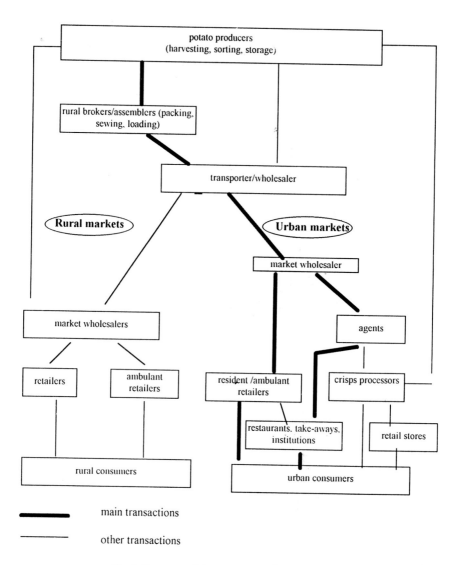

Fig. 3. Structure of the potato post-harvest system.

Kisii district, it has been found that the marketing system for sweet potato responded to a small but significant increase in urban demand (Fleischer et al. 1999). The development of a more sophisticated regular village level buying system helped to stimulate production. This system follows the same pattern as in the case of potato. As the crop is perishable—having a shelf-life of 5 to 7 days—the trader buys the crop within 1 or 2 days after harvest to ensure that they last longer and reaches the final market destinations. Agents buy from farmers who deliver by donkey to the tarmac road, where traders buy on a prearranged schedule. The trader uses a passing truck to carry to rural towns for transhipment to inter-city transports plying between the main towns. The traders are small-size operators and usually do not buy sufficient quantities to fill a lorry on their own. The expansion of the system to other regions depends on the improvement of rural infrastructure, e.g. roads and the further development of urban demand.

On-farm processing of sweet potato is rare. Products made include flour which is mixed with sorghum to make porridge and mild alcoholic beverages from peeled, chopped, pounded and fermented sweet potato. Such processing is only done when the crop has been harvested and there are no other immediate uses for the produce. In many areas of the district, flour production was popular in the 1960s but was abandoned in favor of maize flour.

Sweet potato processing at village level has been promoted in parts of western Kenya by the home economics branch of the agricultural extension service (Owuor, 1996). A majority of the processors are members of women groups who have been trained on the processing technologies for sweet products including chapati, mandazi, crisps, chips and cakes besides other group activities. Products are sold to the neighbouring schools during sports day or at the local markets.

3.2.3. Marketing Margins

The marketing margin can be interpreted as the cost of providing the various types of marketing services. In a perfectly competitive market, the margin on an average and in the long run should be equal to the costs of marketing including costs of capital with a competitive return to labour, management and risk (Tomek and Robinson, 1981).

In the Kenyan study, marketing margins were calculated for the two crops based on the data collected during the survey. Table 1 shows the relative margins received by the different actors in the chain. Potato farmers in Meru receive 69% of the wholesale price both in the local and the Nairobi market, for their product, indicating a high degree of competition and low freight rates. Sweet potato farmers receive 55% on the Nairobi market.

The trader's margin for potato varies considerably, with the highest margins going to the trader in the local production region (Meru). Those traders engaged in transport from Meru to Nairobi gain less per traded volume. This reflects higher costs due to repackaging, but also may reflect a higher price to the farmer due to the enforcement of a more uniform standard of bag in that area.

3.2.4. Labour Requirement and Value Added

The postharvest system of root and tuber crops contributes significantly to employment in rural and urban areas. A comparison of pre- and postharvest labour input shows that more labour is required in the

Table 1. Marketing Margins for Potato and Sweet Potato

Crop	Potato (Ngure)	Potato (Kerr's Pink)	Sweet Potato	Sweet Potato
Origin	Meru district	Meru district	Kisii	Kisii
Destination	District capital	Nairobi	Nakuru	Nairobi
Wholesale price (KSh/bag)	1,300	1,800	1,100	1,200
Margin as % of wholesale price				
Farmer	69%	69%	55%	54%
Rural broker	0%	4%	0%	4%
Transport costs	5%	8%	21%	20%
Handling and packaging	3%	6%	7%	6%
To local govt	2%	2%	2%	2%
Trader:				
Overhead	0%	1%	2%	3%
Trader profit	20%	11%	14%	10%

postharvest chain than in production (Table 2). The model calculation based on crop budget data from Meru and market survey data shows a ratio of 1.9 of pre- to postharvest labour input.

Table 2. Labour input in potato pre- and postharvest operations (model calculation)

Operation	Labour input (man-day/bag)	Comment
Pre-Harvest		
Production (farm family labour)	1	20 days each for sowing, fertilization/ spraying and weeding Yield: 60 bags/acre
Harvesting & grading (casual labour)	0.7	Harvesting 2.5 bags/day, grading 3.5 bags/day
Total	1.7	
Postharvest		
Rural assembling	0.1	10 bags/day
Transporting	0.05	45 bags/day, 3 day per week
Wholesaling	0.04	25 bags/day
Market agent[1]	0.05	10 bags/day
Processing[2]	1	0.5 bag/day
Retailing to consumers	2	0.5 bag/day
Total	3.24	
Ratio (postharvest to preharvest)	1.9	

[1] An estimated 50% of the total amount is marketed by brokers/agents.
[2] An estimated 50% of the total amount is processed to chips and crisps.

Assuming perfect competition in labour markets and nearly equal qualification, returns to labour should be the same in production and postharvest. However, the value-added in the potato postharvest chain is higher than farmers' profit per unit. Returns to labour are 285 KSh/bag in production (only for farm family labour) and 305 KSh/bag in postharvest operation. However, returns on land are not accounted for in the figure of production costs.

In the more intensive farming systems such as found in Meru, all family members participate in most of the farm activities. There is no clear distinction between gender roles in the production of potatoes. In the local markets, there were more women in wholesale and retail of potatoes than men. At the farm level, men and women were active in production, harvesting, sorting and selling. In the marketing chain, the majority of retailers in the local markets were women, while the men did the wholesale and transportation functions.

Where sweet potato is commercialized, both men and women were involved in the production processes and postharvest systems of grading and selling. It was also observed that women form the majority of traders in both the local and distant markets. For the distant markets in major towns such as Nakuru, Nairobi and Mombasa, two or three women or a group of women and men join up and hire a lorry to transport sweet potato to the chosen destination. This may be happening either because one person may not be able to purchase enough produce to fill the whole truck, due to fear of handling large amounts of a highly perishable crop or due to limited resources among the traders, most of whom are women. In locations further away from the market centers, farmers grow sweet potatoes for home consumption and for local markets. As such, it is purely a woman's crop and the men were not interested in the activities related to sweet potatoes.

The entry barriers for women to both the potato and sweet potato trade, may be associated with the initial capital requirements. A beginner requires enough cash to hire a lorry and also to buy the produce. She also has to deal with the cultural expectations as to the role of a woman. She must struggle to penetrate the male dominated business of transportation.

3.3. Performance of the Postharvest System

The performance of the postharvest system can be assessed by its productivity, efficiency, the flexibility to survive and the ability to absorb external shocks. A postharvest system is allocatively efficient if it is able to minimize the transaction costs from production to final destination. It should be able to adopt new technologies easily when the potential for lowering transaction costs arises. Finally, there is the need for public institutions to react appropriately to potential failures of postharvest systems.

3.3.1. Efficiency

Private sector agents dominate both the potato and the sweet potato postharvest system. In the potato sector a well established system of farmers, traders, processors and consumers is found. Key operators such as farmers, traders, transporters, processors and retailers work closely with each other. In general, there are no fixed contractual relations between a specific farmer and a specific trader or processor. Business partners change frequently and depend on supply and demand. As many farmers and many traders are involved there are no cartels.

The effectiveness in bridging supply and demand with special regard to the geographical distance between production and consumption, to seasonal production patterns and product quality is in both systems good. Although the two crops, especially sweet potato, are perishable commodities, there is a constant supply throughout the year. The participants in the postharvest system therefore are continuously receiving market signals down the chain. Potato production has responded to strong urban demand and is a major traded commodity. Steady demand increases due to a shift in urban consumption patterns have led to a constant development of supply chains from the major production regions. In sweet potato marketing, restricted urban demand is guiding the development of the postharvest system. Commercialization has reached only part of its potential.

Generally, informal links between the participants in the chain exist. The daily exchange of information in the main markets means that traders know which way prices will move. Supplies generally balance demand and prices are set to clear the market. There is a fairly stable relationship between regional markets. In the case of potato, middlemen occasionally store the crop to exploit rising prices, but not for very long periods.

The market structure may serve as another indicator for efficiency. Given the number of agents, traders, porters and brokers involved in the system, there are no serious barriers to entry in postharvest activities. Obviously people with sufficient amount of capital can acquire larger consignments, reduce their overhead costs, and achieve scale economies, especially for transport. No serious efforts have been made to set up cooperatives to compete with the private trader and recoup some of the trading margins for the farmer. When farmers were questioned about it, they did not show much interest in the matter.

The postharvest system of both crops involves many participants and creates 'semi' and 'casual' employment for a large section of the poor population. It also provides a training ground in enterpreneurship to persons who may want to start another enterprise or expand up the marketing ladder. Little initial capital and skills are required to take part in the system.

The transaction costs at each stage of the postharvest system can be regarded as reasonable. Excessive margins could not be found. This is mainly due to the involvement of the so called 'agents' who link business partners. In the rural areas they bring together farmers and traders, in the urban markets they are the linkage between the wholesale markets and processors (hotels, restaurants, crisps industries, etc).

There is a difference in the type and number of operations in the postharvest system between a highly commercialized crop (potato) and a crop in the initial stage of commercialization (sweet potato). The structure of the marketing of potato shows more direct links between the farmer and different entrepreneurs. From the farmers' point of view, the market is well developed. The diversity of acting agents and middlemen involved in potato marketing is considerably high. There is effective competition at different levels. From the farmers point of view, there are usually several marketing options including storage at the farm. The system is therefore relatively robust to external shocks.

Although farmers were not so much aware of potato prices in distant urban markets, information about the price level at local markets is good. Farmers know that supply shortages in substitutive commodities like maize boost potato prices. The sweet potato postharvest system is still in the stage of limited commercialization. The postharvest chain operates mainly through a few bottlenecks.

3.3.2. Performance of Public Sector Institutions

The operations in the potato and sweet potato postharvest chains are almost solely executed by private sector agents. At the national level these two crops have not yet been considered as priority crops in comparison with other crops such as maize and wheat. Only to a limited extent public institutions have supported research and extension on potato and sweet potato.

However, there are recent examples where local public entities have intervened. Public research on both crops has been carried out by the Kenyan Agricultural Research Institute (KARI), partly in co-operation with the International Potato Center (CIP). Research on postharvest problems focused on technical issues such as flour processing (Kabira and Imungi, 1991; Kabira, 1992) and postharvest protection. However, links to the extension agencies have been weak. Sweet potato production has been almost completely neglected in extension programs since for a long time it has been considered as a subsistence crop. To a limited extent, sweet potato extension addressed on-farm processing in rural women groups. To date the quantities processed are negligible.

One example for public involvement in the marketing system has been found in Meru district. In 1995, the Meru District Agricultural Committee (DAC) decided to intervene in the potato trade because of the lack of standardization of packaging material. Remarkably, the decision for a standard type of bag was not implemented through a formal law, but through the collaboration of various local stakeholders. At the farm and in the villages, the chiefs and the members of youth clubs made sure that only flat bags were used when packing at the farms, and outside the villages the trucks were checked by officials.

3.3.3. Constraint Factors

Constraints in the postharvest system might place a high burden on economic and social development, thus restraining economic growth. Potato and sweet potato postharvest systems are especially vulnerable to constraint factors because of the bulkiness and perishability of the produce. Constraints in the postharvest system can be: i) technical, e.g. the lack of technologies for improving certain postharvest operations, ii) financial and economic, e.g. input and operational costs are too high, and iii) institutional, e.g. lack of infrastructure and barriers for contractual arrangements.

Constraint hypotheses were revealed in a series of workshops before the start of the field survey. Most of the constraint hypotheses suggested by experts relate to the institutional level (Fleischer, 1996). This led to a hypothesis of major imperfections in the postharvest systems of both crops. The constraint hypotheses were guiding the selection of interview partners and study sites, as well as the analysis of this study (Fleischer et al. 1999).

The analysis of the survey data found less constraints in the system than expected. The postharvest systems are generally able to cope with the specific characteristics of root and tuber crops. The transport system is coping with the bulkiness of both commodities by establishing various lines of transport according to their relative costs. Perishability of the sweet potato is a potential hazard. Although handling is generally rough, the quality of the product that is offered to final destinations is not affected. Market participants see losses generally as negligible.

On the technical level, operations like harvesting, grading and storage require only little improvement. On-farm processing of sweet potato for commercial purposes is still in its initial stage. It seems less a technical constraint but rather lacks economic attractiveness. Lack of credit did not appear to be a major problem for many actors. Although individuals claim the need for it, systematic access barriers to capital could not be found. Entry barriers to markets and operations exist, and are not leading to excessive margins and profits. A high degree of trust among wholesalers was observed, thus indicating the satisfying performance of informal contractual arrangements.

However, major constraint factors with regard to the provision of public goods could be identified. Since transport costs take a considerable share in the marketing margins, the state of the rural road infrastructure is a key factor for lowering transaction costs. In major production districts, feeder roads are inaccessible and thus unable to provide access from distant production areas to the main roads, especially during the rains. The rural infrastructure investment program should consider regions with a high potential of surplus production of root and tuber crops.

The government should consider intervening by setting regulatory standards only in collaboration with the private sector. Market transparency is fairly well developed, even with the farm-households. Improvements, such as the standardization of the packaging on a national scale, would be better achieved in a collaborative effort with the private sector than be enforced. The experience with establishing a marketing board for maize has shown that negative spill-over consequences on other food markets, e.g. root and tuber crops, may occur (Jones, 1995).

The capacity of the existing sweet potato postharvest system is strongly interrelated to the current demand pattern. Urban and rural town demand drives the development. Interventions aimed at increase in production areas in order to stimulate commercialization should therefore not be made without prior analysis of market demand development.

4. Conclusions and Perspectives

The systems approach for analyzing postharvest sector development is a valuable tool for determining performance indicators and eliciting the constraints of further development. Looking on deficiencies in individual operations in the postharvest chain, research and development interventions can be much more targeted, if the linkages of specific operations and functions are well known. For example, in Kenya the potato and sweet potato research organizations had the perception that tremendous losses in the postharvest chain occur. The systems analysis brought about that the private sector agents in the system did not share this view, since incentives exist to avoid economic loss and to use by-products from grading and sorting for secondary purpose, e.g. animal feed.

System analysis cannot be static, but should be done in a historic perspective in order to take into account the inherent dynamism and non-linearities of crucial variables. Thus, a prominent feature for public sector institutions is the regular collection of time-series information. This holds true not only for market price surveys but also for other data on technology and the institutions involved.

Whereas the Kenya case study shows that the private sector is capable of organizing the postharvest chain of an important food commodity under difficult framework conditions of deteriorating infrastructure, there are a number of responsibilities for the government sector. One such intervention is the provision of an adequate legal framework for the private sector activities, for instance, measures of public security and stability, imposition of minimum health and environmental quality standards, etc. For the potato and sweet potato business, for example, the standardization of bag sizes would increase price transparency.

Development activities in the postharvest sector of most of the developing countries are on the cross-roads between an agro-industry and a village-centered approach (Fleischer et al. 1995). Smaller producers will greatly benefit from methods, institutions, and technologies that allow them to compete in markets. This would allow to level some of the imbalances that have been imposed by earlier development strategies which were often systematically biased against rural areas in developing countries.

References

Boxall, R.A. 1997. Constraints analysis of the post-production sector in Zambia. Food and Agricultural Organization (FAO), Rome.
FAO. 1997. Production yearbook. Rome.
Flach, M. 1995. A conceptual framework for analyzing post-production systems. A revision of the document improving the food on the way to the consumer. Food and Agricultural Organization (FAO), Rome.
Fleischer, G. 1996. Postharvest systems analysis—conceptual approach for a country study in Kenya. Deutsche Gesellschaft für Technische Zusammenarbeit, Eschborn, Germany.
Fleischer, G., Waibel, H. and W. Dirksmeyer. 1995. Future priorities in postharvest systems development—the role of donor and development agency support with special reference to GASGA. Report prepared for the group for assistance on grain after harvest on behalf of the Deutsche Gesellschaft für Technische Zusammenarbeit (GTZ), University of Hannover, Germany.
Fleischer, G., Agne, S., Kigutha, H., Oldham, P. and E. Wanjekeche. 1999. Country study on postharvest systems—potato and sweet potato in Kenya. Horticultural systems in the Tropics 2, Institute of Horticultural Economics, University of Hannover, Germany.
GOK (Government of Kenya). 1994. Development plan 1994-1996, Nairobi.
Goletti, F. and C. Wolff. 1998. The impact of postharvest research. International food policy research institute, Washington D.C.
Guyton, B., Sogo, F., Mogire, J. and R. Njuguna. 1994. Kenya's Irish potato subsector—characteristics, performance and participants' information needs. Government of Kenya market information system report No. 94-01, Nairobi, Kenya.
GTZ. 1996. Future priorities in postharvest systems development—the role of the group for assistance on grain after harvest (GASGA). Position paper, Deutsche Gesellschaft für Technische Zusammenarbeit (GTZ), Eschborn, Germany.
Henckes, C. and F. Afful. 1997. Analysis of postharvest systems of tomatoes and yams in Ghana. Ministry of agriculture and .Deutsche gesellschaft für technische zusammenarbeit (GTZ), Accra, Ghana.
Holtzman, J., Lichte, J. and J. Tefft. 1995. Using rapid appraisal to examine coarse grain processing and utilization in Mali. In: Prices, products and people—analyzing agricultural markets in developing countries, G.J. Scott (Ed.), London: Boulder, pp. 43-72.
Jones, S. 1995. Food market reform: The changing role of the state. Food policy 20(6): 551-560.
Kabira, J. 1992. Use of dehydrated potato in composite flours—A case-study of potato flour in Kenya. In: Product development for root and tuber crops, vol. III, Africa, G. Scott, P. Ferguson and J. Herera. (Eds), International potato center and international institute for tropical agriculture, Lima and Ibadan, p. 457-464.
Kabira, J. and J. Imungi. 1991. Possibility of incorporating potato flour into three traditional Kenyan foods. African study monographs 12(4): 211-217.
La Gra, J. 1990. A commodity systems assessment methodology for problem and project identification. Interamerican Institute for Cooperation on Agriculture, University of Idaho.
Nguyo, W. 1997. Statistical data on food crop production. Report for Deutsche gesellschaft für technische zusammenarbeit (GTZ). Policy analysis matrix team, Egerton University, Nairobi.
Owuor, B. 1996. Sweet potato based food enterprises in western Kenya. A report on evaluation of pilot enterprises. International Potato Centre, Nairobi.
Scott, G.J. and P. Ewell. 1992. Sweet potato in African food systems. International Potato Centre, Lima.
Tomek, W.G. and K.L. Robinson. 1981. Agricultural product prices. 2nd Edition, Ithaca, NY, USA: Cornell University Press.

Walingo, A., Alexandre, C., Ewell, P. and J. Kabira. 1996. Potato processing in Nairobi, Kenya—current status and potential for further development. Kenya Agricultural Research Institute, Tigoni, CIRAD, Montpellier and CIP, Nairobi. Draft report, April 1996.

Wheatley, C., Scott, G.J., Best, R. and S. Wiersema. 1995. Adding value to root and tuber crops: A manual on product development. Cali, Colombia.

Woolfe, J. 1992. Sweet potato, a versatile and nutritious food for all. *In*: Product development of root and tuber crops, vol. III—Africa. G. Scott, P. Ferguson and J. Herrera (Eds), Proceedings of the workshop on processing, marketing and utilization of root and tuber crops in Africa, held October 26 - November 2, 1991, at the international institute for tropical agriculture (IITA), Ibadan, Nigeria. International Potato Centre, Lima.

10

Postharvest Handling of Strawberries for Fresh Market

José M. Olías, Carlos Sanz and Ana G. Pérez
Instituto de la Grasa, C.S.I.C. Dept. Physiology and Technology of Plant Products Padre Garcia Tejero 4, 41012-Seville Spain

Abbreviations

CA	Controlled atmosphere
FA	Forced air cooling
HC	Hydrocooling
HPFC	High Pressure Fast Cooling
MA	Modified atmosphere
MAP	Modified atmosphere packaging
PAL	Phenylalanine-ammonio-lyase
PYO	Pick your own
RC	Room cooling
RDA	Recommended daily allowance
RH	Relative humidity
TA	Titrable acidity
TSS	Total soluble solids
UDPGFT	Uridine diphosphate glucose: flavonoid-O^3-glucosyltranferase

1. Introduction

The modern cultivated strawberry (*Fragaria x ananassa* Duch.) is the most widely distributed strawberry crop due to genotypic diversity and a broad range of environmental adaptation. Two other species, *F. vesca*, the most widely distributed of wild species, and *F. moschata* Duch., are also grown commercially, but on a much smaller scale (Hancock and Bringhurst, 1979). The octoploid cultivated strawberry *Fragaria x ananassa* was derived from hybridization between two wild American native octoploid species, *F. virginiana* Duch. and *F. chiloensis* Duch., in the 18th century European gardens. Since then, extensive hybridization between the parent species and their descendents has occurred, making *Fragaria x ananassa* a highly variable and adaptive species with a wide range of morphological and physiological characteristics (Larson, 1994). Due to their hybrid origins, strawberries are adapted to different climates, temperate, moderate, mediterranean, subtropical, and even tropical if grown at high altitudes.

Consumer enthusiasm and positive attitudes about strawberries has coincided with the rise in worldwide production. World strawberry production totaled about 2.30 million tons in 1992, while in the last three years the average world production increased to over 2.65 million tons (FAO, 1998). The major strawberry producing countries are the United States, Spain, Japan, Poland, the former USSR and Republic of Korea (Table 1). Europe and the United States account for more than 70% of the world production. In Europe, Spain is the largest strawberry producer and the main exporter for fresh market, mostly from Western Andalusia. Most (70%) of the strawberry fields in the United States are located in California. Both strawberry producer regions, Andalusia and California, share important similarities in terms of climatic conditions, cultural practices, and strawberry cultivars.

Quality of strawberries depends on their appearance, firmness, and flavour. Strawberries are highly perishable fruits due to their high respiration rate, and high susceptibility to fungal infections. The most important factors in attaining and maintaining good fruit quality are harvesting at the fully ripe stage and providing adequate postharvest handling. A recent study of current consumer attitudes conducted by Retail Management of Santa Clara University (1998) concluded that the frequency with which consumers discard strawberries because they are spoilt has recently improved. From 1982 through 1992, between 56% and 62% of households agreed that they often threw strawberries away because they were spoilt. That number declined to 37% in 1998. Whether this was due to improved packaging, improved strawberry varieties, better temperature control or more everyday consumption is not clear. However, reducing losses due to spoilage has to be encouraged through appropriate handling.

Table 1. Strawberry production in 1997, 1998 and 1999 (from FAO, FAOSTAT Agriculture data, 1997, 1998, and 1999)

Producer	1997 (Mt)	1998 (Mt)	1999 (Mt)
World	2,669,228	2,687,343	2,784,485
Europe	1,143,798	1,117,307	1,033,739
Spain	260,600	315,300	367,200
Poland	162,509	149,858	171,000
Former USSR	125,000	128,000	167,690
USA	738,800	765,900	785,000
Asia	513,394	520,961	520,961
Japan	198,000	198,000	198,000
Korea	151,199	155,521	155,521
South America	65,039	63,365	63,365
Chile	13,650	15,900	15,900
Colombia	15,300	15,300	15,900

2. Morphology

Strawberry is a herbaceous perennial plant in which the stem is compressed into a rosetted crown, with internodes about 2 mm in length. Auxiliary buds in the leaf nodes of the crown either remain dormant or develop into branch crowns or runners depending on environmental conditions (Larson, 1994). Primary flowers are initiated at the terminus of the inflorescence, while secondary and tertiary flowers are initiated from primordia located in the axis of bracts on the inflorescence below the primary bloom. The modern cultivated strawberry, *Fragaria x ananassa* Duch., produces an aggregate fruit that comprises a number of one-seeded fruits or achenes, arranged in a spiral fashion on an enlarged receptacle (Winston, 1902). The achenes are the true fruits of the strawberry, each containing a single embryo. However, the fleshy receptacle constitutes the edible part of the fruit that we commonly call the berry. Arrangement of the achenes in the receptacle affects distribution of growth and, therefore, berry size and shape. Growth of the receptacle is principally a function of cell enlargement in its cortex and pith (Cheng and Breen, 1992).

3. General Physiology

The presence of auxin produced by the achenes is essential for expansion of the receptacle during strawberry fruit development (Nitsch, 1950), and the decline in the concentration of this hormone in the achenes as strawberry fruit matures triggers fruit ripening (Given et al. 1988a; Manning, 1994). Gibberellin, cytokinin and abscisic acid have also been detected in strawberry but appear to have far less direct influence on fruit growth and ripening than auxin (Biale, 1978; Perkins-Veazie, 1995; Martínez et al. 1994). The traditional view that strawberry is a non-climacteric fruit was first confirmed by Knee et al. (1977) who found a continuous decrease in ethylene production during strawberry development. The absence of a climacteric rise in respiration and no effect of exogenously added ethylene to unripe strawberries was further demonstrated by Given et al. (1988a). Strawberries are among the fruits with highest respiration rates. Production of CO_2 at 20° C rises in the first 48 to 72 h after harvest (100-200 mg CO_2/kg.h) followed by a 20% decline (Wills et al. 1998). The ethylene production in a strawberry fruit decreases sharply when the fruit develops from green to white, but changes little with further development, with very low ethylene production 15-80 nL/kgh in ripe strawberries. Although it is clear that ethylene does not play an essential role in strawberry ripening some findings suggest that ethylene may play an indirect role in strawberry fruit physiology by sensitization of the fruit to other hormones (Perkins-Veazie et al. 1987; Luo and Liu, 1994). In this sense, some studies have been published on the possible benefits of removing ethylene during strawberry storage (Kim and Wills, 1998; Wu et al. 1992). In contrast to most non-climacteric fruits that usually lack the pronounced changes in colour and softening associated with the climacteric ones, strawberries continue to increase in size during the ripening process, decrease in titratable acidity, and show different changes in colour and softening (Perkins-Veazie, 1995).

4. Chemical Composition and Quality

The main quality parameters of strawberries include appearance (colour, size, shape, freedom from defects), firmness, flavour, and nutritional value (Kader, 1991). All these quality parameters are related to fruit composition at harvest and change during postharvest handling of strawberries. Due to the great genotypic diversity and broad range of environmental adaptations observed there are large variations in strawberry composition (Table 2).

4.1. Colour

Chlorophyll and anthocyanins are the main pigments responsible for strawberry colour. During fruit development, once the full fruit size is reached, a rapid loss of green surface coloration, associated with chlorophyll degradation, is observed (Martínez et al. 1995). Simultaneously, or shortly thereafter, anthocyanins accumulate related to the induction of de novo synthesis of phenylalanine-ammonio-lyase (PAL) and uridine diphosphate glucose: flavonoid-O^3-glucosyltranferase (UDPGFT) (Given et al. 1988b). The final red colour of the mature fruit is mainly due to pelargonidin-3-glucoside (88% of total anthocyanins) and cyanidin-3-glucoside. The total anthocyanin content of strawberry cultivars varies widely and includes orange-red varieties such as 'Elsanta', and others that overproduce pigment appearing deep-red or purple as 'Camarosa'. Anthocyanin content can change during postharvest life of strawberries, depending on light and storage temperature (Sacks and Shaw, 1993). The two main enzymes related to anthocyanin degradation are polyphenol

Table 2. Chemical composition of strawberry fruit (compiled from Kader, 1991; Perkins-Veazie, 1995; Sanz et al. 1999)

Component	Range
Proteins	0.2-1 g/100 g
Lipids	0.5 g/100 g
Carbohydrates and acids	10-13 g/100 g
Sucrose	0.2-2.1 g/100 g
Glucose	0.8-3.5 g/100 g
Fructose	1.0-3.1 g/100 g
Citric acid	321-1240 mg/100 g
Malic acid	100-680 mg/100 g
Total phenolics	58-210 mg/100 g
Vitamin C	25-120 mg/100 g
Pottasium	164 mg/100 g
Phosphorus	21 mg/100 g
Calcium	21 mg/100 g

oxidase (PPO) which acts indirectly by the formation of hightly reactive quinones (Wesche-Ebling and Montgomery, 1990) and peroxidase (López-Serrano and Ros-Barceló, 1996).

4.2. Firmness

Strawberry softening starts between the green and white ripening stages and continues during fruit colour development. Loss of firmness increases the susceptibility of fruits to physical damage and pathogen infection. There are large variations in firmness, but there is little information as to why strawberry cultivars differ so much in texture. In fact, the mechanism and the enzymes involved in strawberry ripening are not well known. Efforts to reveal the basis of changes in firmness have focused on cell-wall-associated enzymes which contribute to cell wall breakdown. In this sense, endo- and exo-polygalacturonase (Gross and Sams, 1984; Abeles and Takeda, 1990; Nogata et al. 1993), pectinesterase (Barnes and Patchett, 1976), cellulase (Abeles and Takeda, 1990), and pectate lyase (Medina-Escobar et al. 1997) have been described and characterized, but there is not clear evidence as to which enzyme(s) governs the process of softening in the strawberry fruit. Part of the problem relates to the difficulty of extracting active enzymes from strawberries, which have a high polyphenolic concentration and very active polyphenoloxidases.

4.3. Flavour

Flavour is a combination of the tastes and aromas of many different compounds. Strawberry flavour relies upon three main groups of compounds: sugars, organic acids and volatile compounds. All the three have great interest not only as components of flavour but also as indices of strawberry development and ripening.

4.3.1. Sugars

Little is known about the processes involved in the accumulation of sugars in strawberry fruit. Sucrose is the major assimilate transported to the strawberry fruit (Forney and Breen, 1985; John and Yamaki, 1994). However, in ripe fruit, sucrose is accumulated to much lower levels than glucose and fructose, which are in about a 1:1 ratio (Wrolstad and Shallenberger, 1981; Kader, 1991). The absolute and relative amounts of these sugars vary among varieties (Shaw, 1988) and also with the degree of fruit ripeness (Reyes et al. 1982; Pérez et al. 1997), though the soluble solids values (TSS) hardly change during fruit development and ripening, with a mean value around 8° Brix (Maroto et al. 1986). In fact, a very poor correlation has been found for TSS and total sugars (Shaw, 1988; Pérez et al. 1997), and use of TSS in strawberry should be limited to comparative studies in which variations by genotype and enviroment are low.

4.3.2. Organic Acids

Malic, citric, ascorbic, succinic, oxalacetic, glyceric, and glycollic acids are the main organic acids identified in strawberry (Mussinan and Walradt, 1975). Acidity on a per fruit basis increases between the green and white stages but declines during ripening, with malic and citric acids being responsible for most of the differences in acidity between firm-ripe and fully-ripe stages of strawberry maturity. Titratable acidity (TA), a measure of the buffering capacity of the fruit, is generally expressed as a per cent of citric acid, the predominant organic acid in all stages of strawberry development (Pérez et al. 1997). Acids not only determine fruit pH, but they are also implicated in colour quality as was demonstrated by Wrolstad et al. (1970).

4.3.3. Volatile Compounds

The aroma of strawberry is determined by a complex mixture of esters, alcohols, aldehydes, sulphur compounds and furanone derived compounds that have been studied for the past thirty years (Dirinck et al. 1981; Hirvi and Honkanen, 1982; Pérez et al. 1992; Sanz et al. 1994). Although no single volatile compound has been identified as responsible for the aroma of fresh strawberries, some ethyl and methyl esters, and furaneol and its derived compounds seem to be important aroma compounds in strawberry (Schreier, 1980; Pérez et al. 1992; Sanz et al. 1995). In a recent study Ulrich et al. (1997) established that two distinct aroma types can be defined for cultivated strawberries based in the later two groups of compounds. The biosynthesis of these aroma components depends on two main factors: the availability of substrates such as free amino acids, sugars or acyl-CoAs (Pérez et al. 1992; Ueda and Ogata, 1977; Zabetakis and Holden, 1997), and the inherent properties of the involved enzymes. Some of these enzymes related to aroma biosynthesis have been identified and purified alcohol dehydrogenase, alcohol acyltransferase, lipoxygenase, and hydroperoxide lyase—most of them are developmentally regulated and associated with ripening (Mitchell and Jelenkovic, 1995; Pérez et al. 1996b; Pérez et al. 1999a). Since Manning (1994) first extracted strawberry mRNA and studied changes in gene expression during ripening, several molecular studies on strawberry have been initiated with the aim of providing more precise information on the regulation of the main ripening related enzymes.

4.4. Nutritional Value

Strawberry fruits have important contents of natural antioxidant substances such as vitamin C and polyphenols. Strawberries contain larger amounts of vitamin C than citrus fruits, about 25-120 mg/100 g depending on cultivar and climate conditions. This ascorbic acid level reaches a maximum at full ripeness. One serving of eight medium sized mature strawberries contains 140% of the recommended daily allowance of vitamin C, and 20% of the RDA for folic acid. Vitamin C is quite unstable, mainly due to the activity of ascorbic acid oxidase and the reaction with oxygen in the presence of heavy metal ions and light, and thus is taken as an indication of fruit freshness. Despite the interest in vitamin C, polyphenols are probably more involved in the total antioxidant activity of strawberry fruits (Wang et al. 1996). Strawberry fruits are very rich in phenolic compounds such as anthocyanins, flavonols, cinnamic acid derivatives, and simple phenols derived from the shikimate and phenylpropanoid pathways (Cheng and Breen, 1991). The anti-carcinogenic properties of many of these phenols, some of them particularly abundant in strawberry such as ellagic acid (Maas et al. 1991), have important nutritional implications. Strawberries are also a recognized source of potassium and dietary fiber. Like most soft fruits, strawberries are very poor sources of nitrogen substances. Most of the few strawberry proteins have enzymatic activity and free amino acids are considerably lower than those of banana, peach or orange (Kuneman et al. 1988; Pérez et al. 1992).

5. Harvest Maturity

As stated earlier, strawberries are one of the most perishable of all fruits. They are essentially full-ripe at harvest as opposed to many other fruits that are picked 'green' and ripened later. They have a rapid metabolism and last relatively a short time, only 5-7 days at 0° C, 90-95% RH even without the presence of decay micro-organisms (Hardenburg et al. 1986). Strawberries can be harvested at slightly different stages of ripeness, depending on the time and distance to market.

Quality indices recommended for strawberries are generally based on appearance (colour, size, shape, freedom from defects), firmness, nutritional value (vitamin C, polyphenols), and flavour (soluble solids, titratable acidity, aroma volatiles) with a minimum of 7% soluble solids and/or a maximum of 0.8% titratable acidity. However, maturity indices are only based on berry appearance according to the different standards for grades of strawberries.

The United States standards for grades of strawberries (1965) define three different grades: No. 1, No. 2, and Combination, with a fourth denomination, not a grade, named Unclassified for those strawberries which have not been classified in accordance with any of the three foregoing grades.

US No. 1 consists of strawberries of one variety or similar varietal characteristics with the calyx attached, which are firm, not overripe or undeveloped, and which are free from mold or decay and free from damage, caused by dirt, moisture, foreign matter, disease, insects or mechanical or other means. Each strawberry has not less than three-fourths of its surface showing a pink or red colour and a size not less than 19 mm, and a tolerance of not more than 5% for off-size. For this grade not more than 10% of strawberries in any lot can fail to meet the above requirements, of which one-half (5%) is allowed for defects causing serious damage, including no more than two-fifths of this latter amount (2%) for strawberries affected by decay.

US No. 2 is essentially the same than U.S. No. 1 but each strawberry has not less than one-half of its surface showing a pink or red colour, and a diameter not less than 16 mm, and a maximum 5% for off-size. Tolerance for defects specifies not more than 10% for strawberries in any lot which are seriously damaged, including therein not more than three-tenths of this tolerance (3%) for strawberries affected by decay.

US Combination consists of a combination of US No. 1 and No. 2 strawberries. At least 80%, by volume, of strawberries have to meet the requirement of size of US No. 1 grade. Not more than 10% of strawberries in any lot are allowed to be seriously damaged, including therein not more than one-fifth of this tolerance (2%) of strawberries affected by decay. The percentage of US No. 1 strawberries required in the combination and individual containers (cups or baskets) may have no less than 65% US No. 1 strawberries.

California is one of the few states in the United States having quality standards for horticultural crops produced within the state. The standards for fresh strawberry in the California Agricultural Code (1983) defines mature strawberries as those that have not less than two-thirds of fruit surface showing red or pink colour, and free from defect and decay.

The European standards for strawberry (Organization for Economic Cooperation and Development, 1987), establishes four different grades: Extra, First (I), Second (II) and Third (III), depending mainly on external factors such as size, shape (according to variety), uniformity, colour homogeneity, cleanliness, calyx presence, and bruising absence. The varietal factor is highly important in terms of size for grading. These standards establish minimum sizes among the grades: 25 mm for extra, 22 mm for I and II (except for cultivars 'Primella' and 'Gariguette', which are 18 mm), and 15 mm for III. Tolerance for quality and size specifies not more than 5% for strawberries grade Extra provided they fulfil grade I requirements, 10% for grade I strawberries provided they fulfil grade II minimum requirements, 10% for strawberries grade II including therein not more than 2% fruit weight with defects causing serious damage and unfit for consumption, and 15% for grade III strawberries including therein not more than 4% unfit for consumption.

6. Postharvest Problems

6.1. Mechanical Injuries

Strawberry fruit structure presents a thin skin that is easily broken, and a soft flesh allowing the fruit to be easily crushed and bruised. This structure makes this fruit very vulnerable to spoilage, so the quality when the consumer receives it depends on how the job was done by the previous handlers. Therefore, proper supervision of the entire harvest operation is essential. This must include all aspects of the harvest-picking, field handling, and local transport to the cooler facility. Thus, picker damage can nullify all other attempts to maintain fruit quality. Mitchell et al. (1964) found less unmarketable fruit with careful pickers (14.4%) than with less careful pickers (33.7%). On the other hand, over-packed or over-filled berries are very often injured by abrasion against the corrugated crate or cut by basket rims as a result of being packed over the top of the open-mesh plastic basket that are used for packaging in some strawberry producer regions. Injuries from this type of packing cause both a direct loss and subsequent rot losses by different micro-organisms, mainly *Botrytis*. Preharvest mechanical injury should be also taken into account, whose main cause is the hail. Brown coloured scars, often form on unripe fruits hit by hail, and even though healed over, make these fruits unmarketable when they mature.

6.2. Physiological Disorders

6.2.1. Fruit Malformation

There are several factors such as climate, insects, diseases, drought, humidity or frost producing this effect (Garren, 1981). Misshapen strawberry fruits can be caused by damage to the developing receptacle tissues or achenes or by a lack of pollination. Nitsch (1950) found out that the development of the fleshy receptacle is entirely dependent upon the presence of fertilized achenes, and if the achenes are removed, the receptacle immediately ceases to grow. If an achene fails to develop because of a lack of pollen or unsuccessful fertilization or if it is destroyed by any of the factors mentioned above, the receptacle tissue below the achene may not fully expand. If a high proportion of achenes fail to develop in a given part of a berry, growth in that region will be reduced, which results in a malformed fruit.

6.2.2. Shrivelling

Due to the high respiration rate and thin skin, strawberries are subject to rapid water loss by transpiration, which can cause the fruit to shrivel and appear old and deteriorated. Robinson et al. (1975) observed a maximum permissible water loss of approximately 6% before the marketability of strawberries was impaired. On the other hand, since strawberry for fresh consumption should retain the calyx it is important to avoid water loss that would promote the fruit calyx to wilt and dry.

6.2.3. Overripeness

Also due to its high rate of physiological activity, strawberry quickly pass from ripeness to an overripe or senescent state if held at warm temperatures or if held too long.

6.2.4. Nutrient Deficiencies

Lower than normal translocation of sugars to the fruit during maturation is the primary cause of fruit albinism (Ulrich et al. 1980; Albregts and Howard, 1982). Albino fruits are normal in size and general appearance but lack full colour, being mottled pink and white internally. They are also soft and insipid in flavour, and they rot quickly after harvest. Because of their poor colour and transportation qualities, these fruits represent a loss to the grower.

Different mineral deficiencies have been described for strawberry that may affect plant development and berry production (Ulrich et al. 1980), but without apparent effect on the fruit. Among the main mineral deficiencies that affect the fruit are those caused by boron and calcium. Boron-deficient strawberries may cause fruit distortion increasingly with culture time (Ulrich et al. 1980). Typical fruit deformation due to boron deficiency in southwestern Spain is the so called 'cat-face', and in some cases brown discoloration appeared around the core of the berry. Calcium-deficient strawberries are small, hard, seedy or with seedy patches, and sour. Phosphorus deficiency produces soft fruits with insipid flavour. Soft fruits and lighter colour may be symptoms of magnesium deficiency (Maas, 1984).

6.3. Postharvest Diseases

The short postharvest life of strawberries, measured only in days, and the fragility of the strawberry fruit compared to other fruit crops make this fruit very susceptible to postharvest diseases. The major objective of disease control during the postharvest life of the fruit is to retard development of infections already initiated in the field. Although there are lots of diseases described that can potentially affect this fruit (Maas, 1984), the most important postharvest diseases of strawberry are the following.

6.3.1. Grey Mould Rot

Among the various postharvest diseases found in strawberries, grey mould, *Botrytis cinerea*, is the most serious and most studied (Jarvis, 1977). Affected tissue turns dull pink to brown, and the whole fruit may become completely rotted without disintegration, and very little exuded juice. In time, the lesion on the fruit surface exhibits white mycelium that turns grey as the fungus sporulates. The fungus 'nests' when mycelium from a rotting fruit penetrates and colonizes adjacent fruits through wounds or stomata. Occasionally, irregularly shaped, black sclerotia may form on the fruit surface, and under favourable conditions sclerotia germinate and produce spores which can be dispersed by wind, splashing water or insects. Spores require free water to germinate, coming from splashes or water condensation on the fruit, and can do so over a wide range of temperatures with a minimum temperature for growth about $-2°$C. The fungus is able to complete the cycle of infection to spore production very quickly especially under warm and humid conditions. Spores are nearly always present from bud-break to harvest. Flower parts may become infected by air-borne spores. Infections of the calyx end may remain in a latent state and not become active until the fruit is fully ripe (Hennebert and Gilles, 1958; Gilles, 1959; Powelson, 1960; Jarvis, 1964). Ripening fruit also may become infected directly or by contact with other infected fruits, so at the time of harvest, fruits should be moved promptly from the field to a cooler in order to lower fruit temperature at or near 0° C to suppress fungal activity.

6.3.2. Leak or Rhizopus Rot

Leak rot is produced by ubiquitous fungi belonging to the *Rhizopus* genus (*R. stolonifer* and *R. nigricans*) that grow as saprophytes on decaying organic matter. They are capable of infecting berries via wounds, after they are completely ripe. The infection gives rise during storage and transport to fruit softening and the characteristic exudation so that postharvest life is drastically shortened. Appropriate storage and transport temperatures (0-2° C) do not allow *Rhizopus* growth because these are below the minimum temperature required for growth (5° C). On the other hand, germinating spores are killed by these low temperatures, so the rot after removal to ambient temperatures is usually minor.

6.3.3. Hard Rot

Caused by *Rhizoctonia solani*, this rot is easily identified since it occurs on the portion of the fruit touching the bare soil so that soil particles generally adhere to the surface at the rotted part. In this sense, *R. solani* may be controlled during preharvest using any kind of mulch to keep the developing fruit from touching the ground.

6.3.4. Leather Rot

Leather rot is caused by *Phytophthora cactorum*. Infected fruits present a noticeable lighter colour than non-infected fruits. Diseased fruit remains quite firm and leathery, having an unpleasant taste that permits ready identification. Weather conditions are the determining factor. The disease is promoted by free water on the fruit caused by excessive rainfall or fog during periods of moderate temperatures (Wright et al. 1966). Mulching may be the only available method for minimizing or controlling this disease, especially the use of plastic.

6.3.5. Anthracnose

Four different species of *Colletotrichum* (*C. fragariae*, *C. gloeosporioides*, *C. dematium* and *C. acutatum*), have been described as causal agents responsible for the anthracnose diseases of strawberry, although it is difficult to know which of these are essentially fruit rotting fungi. Anthracnose lesions develop as tan or light brown, circular, sunken lesions on ripe or ripening fruit. As sporulation occurs, cream to salmon or pink spore masses appear. The appearance of these fungi is sporadic and seems to be rain-dispersed and developed at temperatures between 15-30° C leading in some cases to serious loss of fruit.

6.3.6. Mucor Rot

This disease resembles Rhizopus rot. *Mucor piriformis* also causes copious leakage of juice from strawberries, but it can be differentiated from *Rhizopus* genus because of its ability to grow at lower temperatures, even below 0° C, and by the much longer sporangiophores than *Rhizopus*.

6.3.7. Tan Brown Rot

This rot is caused by *Pezizella lythri*, which can live saprophytically or parasitically on a large number of host plants, so that inoculum is nearly always present to infect strawberry fruit. It occurs sporadically, but it can be destructive locally (Plakidas, 1964).

7. Harvesting Practices

Strawberries are generally harvested manually due to their tenderness and vulnerability to mechanical damage. Strawberries for fresh market are picked, graded and packed in the field by the harvester, who is also responsible for removing rotten or overripe fruits from plants. This process reduces the time between harvest and cooling (Mitchell, 1992). In Europe, women are preferred for manual harvest and further handling. The pick-your-own (PYO) method of harvesting strawberries has become a substantial way of moving strawberries in the United States, with a peak in the early 1980s, and the consequent labour cost savings. In some states the PYO operation accounted for over 90% of the strawberry operations (Childers, 1983).

The growth and maturation habits of the strawberry and the tenderness and high perishability of the fruit make it an unlikely prospect for mechanized harvest. However, manual harvesting of strawberries accounts for one-half or more of total production costs, and dependable labour for harvesting is increasingly difficult to find. Labour problems in the 1960s and the need for good net financial returns encouraged research in machine harvesting of strawberries mainly for processing. Various harvesters were patented around 1970 (Rasmussen, 1968; Adrian, 1970; Quick, 1971; Hecht, 1972). Denisen and Buchele (1967) developed some prototype harvesting equipment based on Denisen's earlier testing of various cultivars for one-over harvest by using a stone fork passed through the foliage for stripping the fruit. The harvesting machine progressed to the combination of leaf mower, vibrating stripping teeth, and an air elevator to convey fruit to a cross conveyor (Quick and Denisen, 1970). Further refinements involving the placement of flexible netting over the strawberry bed during the dormant season were tested for strawberry fresh market (Stang and Denisen, 1971). Several other harvesting machines were developed and only three types have been used since with harvest efficiencies between 75-94% (Hergert and Dale, 1989; Theusen, 1984; Swinkels and Murray, 1991). All experimental machines in development are based on specially bred strawberry cultivars for this purpose. The major requirements for a mechanically harvested strawberry cultivar are: high once-over yields, i.e. a combination of high total yield and concentrated ripening; easy calyx removal; long, strong, upright peduncles; and firm, resilient, bruise-free fruit with fruit rot resistance (Gooding et al. 1983; Lawrence et al. 1990; Moore, 1983). Mechanized strawberry harvest for the fresh fruit market, however, appears unlikely in the near future.

8. Postharvest Handling

The strawberry fruit perishes easily due to rapid loss of water, high respiration rate and high susceptibility to fungal infections. In order to prolong shelf-life for transport and marketing of fresh fruit, and to meet industry demands for processing, several cooling methods and combinations of handling procedures have been studied and developed.

8.1. Cooling

Temperature has the most marked effect on strawberry postharvest life. At high temperature strawberry respiration increases markedly, about 4-fold for each $10°$ C change in temperature over the range 0-30°C (Hardenburg et al. 1986), leading to a depletion of nutrient reserves, so that fruit senescence is accelerated. Even under proper temperature (0-2° C) and relative humidity (RH 95%) conditions strawberry shelf-life is about seven days. Rapid precooling to remove field heat immediately after harvest is critical for strawberry. Delays of more than 1 hour between picking and cooling greatly accelerates fruit deterioration (Boyette et al. 1989; Talbot and Chau, 1991; Collins and Perkins-Veazie, 1993). Growers are encouraged to schedule frequent delivers of small and partial loads of strawberries to the cooler in order to minimize deterioration. However, immediate precooling is often not attained as strawberries are frequently left in the field under the sun and exposed to high temperatures. Delaying the start of cooling 6 hours (at field temperature about 29-30° C) results in about 50% greater water loss and increased losses of ascorbic acid and sugars (Nunes et al. 1995). The process of removing field heat can be achieved by several different methods such as room cooling (RC), forced air cooling (FA) or hydrocooling (HC). HC could have some potential advantages as a very rapid precooling method (Ferreira et al. 1994) but is not commercially used for strawberries due to concerns that decay might be increased by free water left on the fruits after HC (Boyette et al. 1992). In fact, this cooling technique is indirectly prohibited by European Union quality standards, since strawberry washing is not allowed. Although traditional RC facilities are still used, strawberries are mostly cooled using FA (Kasmire and Thompson, 1992). This method reduces strawberry temperature from 30° C to 5° C in no more than 2 hours, compared to 9 h in traditional RC (Boyette et al. 1989; Ferreira et al. 1994). The most used FA system for strawberries is the forced-air-tunnel in which two adjacent rows of pallets form a tunnel with reinforced canvas placed over the space between the rows. Fans pull air from one end of the tunnel producing cooling as refrigerated air from the cold storage room is pulled through the containers of the product (Talbot et al. 1995). The newest cooling concept derived from the forced air cooling idea developed by Guillou (1960) is high pressure fast cooling (HPFC). In a typical HPFC cell, fans pull air from a false ceiling producing cooling as refrigerated air is pulled through the pallets of fruits (Fig. 1). There are two zones in a HPFC cell, the suction and blow zones, only connected by the product. Using these cells berries can be cooled to 2° C in 45-90 min. The shortening of strawberry cooling times seems to produce better retention of fruit firmness, flavour and colour than fruits traditionally room-cooled (Pérez et al. 1998). A correlation between strawberry cooling time and marketability is inferred from published results, which relates to the same effect observed for precooling delays of strawberries.

In highly condensed strawberry producing regions such as California (USA) and Andalusia (Spain) berries are typically harvested, graded and packed by trained pickers into plastic punnets in the field. Remarkable increases in strawberry packaging options have been developed during the past few years such as the open-mesh ribbed plastic baskets covered with polymeric films, and the thermoformed plastic clamshells are the most frequently used to protect against mechanical damage and to

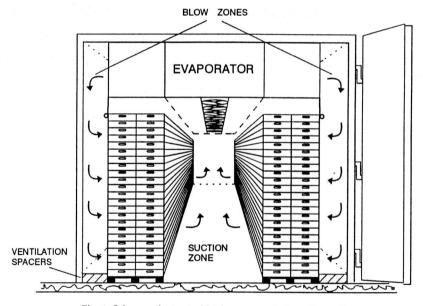

Fig. 1. Scheme of a typical high pressure fast cooling cell.

prevent moisture loss. Special attention for optimizing strawberry precooling and cooling processes are required such as dimensions, vent sizes and locations, and percentage of vent openings of these strawberry containers (Talbot et al. 1995).

In some cases controlled heat treatments could be useful to control decay of strawberry, followed immediately by rapid cooling and storage at 0-2° C for preserving fruit quality. Heating of strawberries at 45° C for 15 min has been described to control Botrytis development during berry shelf-life without affecting sensory quality of the fruit (García et al. 1995).

8.2. Controlled and Modified Atmosphere Storage

Controlled atmospheres (CA) and modified atmospheres (MA) usually involve reduction of O_2 and/or elevation of CO_2 concentration. The use of both CA and MA in strawberries should always be considered as a supplement to proper temperature and relative humidity management (Kader et al. 1998). Rapid removal of field heat immediately after harvest and prompt cooling of strawberries to 0-2° C is the most effective practise to slow undesirable quality changes and increase strawberry shelf-life. Strawberries are among the fruits with higher limits of tolerance to elevated CO_2 concentrations. For this reason, they have been the subject of much research effort on CA/MA storage (Ke et al. 1991; Larsen and Watkins, 1995). Current recommendations for strawberry stored at 0-5° C are 5-10% O_2 and 15-20% CO_2 (Table 3) (Kader et al. 1998). This atmosphere composition reduced respiration rate, retarded softening and slowed various compositional changes associated with strawberry senescence (Ke et al. 1991; Smith and Skog, 1992; Picón et al. 1993). The primary benefit of CA/MA for short-term storage and/or transport of strawberries is the control of decay. Elevated CO_2 levels (15-20%) significantly inhibit development of Botrytis rot on strawberries.

CA and MA differ only in the degree of control. The high degree of gas regulation associated with CA is quite expensive and makes this technology more appropriate for long-term storage commodities (Zagory and Kader, 1988). In contrast, MA is less expensive and commonly used in the postharvest handling of strawberries. Many strawberry loads in Europe and USA are shipped under MA. The usual method, after fruit cooling, is to cover the whole pallet with a plastic film bag that is sealed to the plastic pallet beneath the crates. After sealing, the pallet is injected with CO_2 up to approximately 20% and resealed. This atmosphere can be maintained throughout transit, if properly sealed, with the CO_2 produced by fruit respiration approximately matching CO_2 leakage from the pallet.

An interesting alternative to the use of MA in strawberry pallets is modified atmosphere packaging (MAP). In order to reduce water loss and to help minimize damage during postharvest handling, many fresh strawberries are routinely harvested and packed in open-mesh ribbed plastic baskets which are wrapped with different plastic films (Olías et al. 1994; García et al. 1998). Macroperforated films provide only a mechanical

Table 3. Summary of modified atmosphere requirements for strawberries (from Kader, 1989).

Temperature recommendations	Optimum temperature: – 0.5 to 0° C Expected range: 0 to 5° C	
	Modified atmosphere considerations	
	Reduced O_2	Increased CO_2
Beneficial level	5-10%	15-20%
Benefits	Reduced respiration rate	Firmness retention Reduced decay
Potential for benefits	Good	Very good
Injurious level	< 2%	> 20%
Injury symptoms	Off-flavours	Off-flavours, brown Discoloration of berries
Potential for injury	Slight to moderate	Moderate

barrier to gas exchange but non-perforated semi-permeable or microperforated films offer an additional possibility of creating a modified atmosphere inside the package. In these semipermeable plastic packages a beneficial gas atmosphere is created and maintained by the interaction of fruit respiration, that causes O_2 depletion and CO_2 enrichment of the headspace, and gas diffusion through the packaging film. The use of MAP during strawberry transport and distribution can extend the effective shelf-life of berries by aproximately two more days compared to fruits packaged in normal atmosphere. MAP systems designed to produce optimum O_2 and CO_2 concentrations at suitable temperatures have been mathematically modeled (Exama et al. 1993; Cameron et al. 1995). Despite the high number of plastic films available for packaging most authors conclude that for commodities with high respiration rates, such as strawberry, only combinations of polymeric and microperforated films could provide adequate fluxes of O_2 and CO_2 (Beaudry, 1999). In this sense, the use of microperforated films for MAP of strawberries has the advantage of overcoming temperature changes during storage, transport and distribution that could damage other MAP systems (Edmon and Chau, 1990; Sanz et al. 1999). Application of semi-permeable edible coatings, which change the internal atmosphere of berry represent another approach to MAP of strawberries. In this sense, chitosan coatings on fresh strawberries have proved to be effective in prolonging strawberry quality and reducing decay (El-Ghaouth et al. 1991; Zhang and Quantick, 1998).

Although high CO_2 and/or low O_2 levels in the atmosphere of strawberries extends shelf-life, there are also some detrimental effects related to the use of CA/MA (Li and Kader, 1989). Some described negative effects of MAP in strawberries include delayed cooling of fruits and increased potential for water condensation within the package, which may encourage fungal growth. Prolonged exposure to extremely high CO_2 levels combined with low levels of O_2 can lead to the development of anaerobic metabolism, resulting in the formation of off-flavours and discoloration of the fruit tissue. This can be a particularly serious problem when strawberries are transported or stored at higher than optimum temperature as often occurs in commercial practice. Although strawberries should ideally be kept at 0-2° C after harvest, during commercial transport strawberry temperatures actually range from 2 to 10° C. It must be clearly pointed out that the use of MA should imply a strict control of temperature during strawberry handling. The primary cause of off-flavour appears to be related to accumulation of volatile compounds such as acetaldehyde, ethanol and ethyl acetate, caused by anaerobic respiration (Ke et al. 1991; Shamaila et al. 1992; Larsen and Watkins, 1995). Besides the accumulation of these compounds, the biosynthesis of volatile esters and furaneol derived compounds which contribute to strawberry aroma, are also greatly affected by CA/MA storage (Ke et al. 1994; Pérez et al. 1996a, b). High CO_2 atmospheres ($\geq 20\%$ CO_2) have also some adverse effects on strawberry colour, with an induced paleness or 'bleaching' of the anthocyanins in the internal flesh of the fruit (Gil et al. 1997; Holcroft and Kader, 1999).

Other attempts to preserve strawberry quality by modifying the atmosphere describe the use of ozone, carbon monoxide (CO), and high O_2 atmospheres. Contradictory data have been presented on the efficacy of ozone on Botrytis cinerea, the main pathogen fungus in postharvest life of strawberry (Spalding, 1966; Liew and Prange, 1994). Ozone has been claimed to be effective in preserving some other commodities but its usefulness in controlling decay during strawberry handling seems to be limited, and some detrimental effects on fruit flavour have been described in ozonated strawberries (Pérez et al. 1999b). A further problem with gaseous ozone is the risk of human exposure. CO is a respiration inhibitor that can be useful in reducing respiration without accumulation of off-flavours often originated in normal CA/MA atmospheres. CO was shown to reduce *Botrytis cinerea* development in strawberry (El-Goorani and Sommer, 1979) but its use requires stringent safety precautions since it is an odorless poison. Over the last few years, some research groups have been experimenting with MAP systems containing high O_2 mixtures (70-100% O_2) with surprisingly beneficial results. Although high O_2 MAP has been reportedly effective for extending the shelf-life of soft fruits (Day, 1996) more research needs to be done to evaluate its effect on strawberry quality.

Apart from the effects of these gases, there are a number of biologically active molecules that could have promising application in strawberry storage. Among them are aroma compounds and aroma precursors, such as acetic acid or (E)-2-hexenal, that have antifungal activity (Archbold et al. 1997; Moyls et al. 1996; Ntirampemba et al. 1998). In some cases these applied volatile compounds can alter the flavour of berries, but the effect is usually not permanent, and may even enhance the aroma to some extent.

8.3. Transport

Refrigerated transportation is designed primarily to maintain product temperature, not to reduce temperature after loading. Advances in postharvest biology and technology, improved packaging, and the development of faster, more uniform cooling methods and equipment increased the potential for cooling strawberries to their desired transport temperatures prior to loading for shipment. The recommended transit environment for strawberries is temperature in the range − 0.5-0° C, and RH of 90-95% (Ryall and Pentzer, 1974). Precooling to the proper loading temperature, ensuring that the transport vehicle is thoroughly cooled before loading, and an adequate air flow around and through the entire load is essential to preserve quality of strawberries during and after transport. Research on perishables truck transportation in the 1950s and 1960s demonstrated the need for adequate air flow to absorb heat from the product (sensible and vital heat) and from external sources, mostly conducted across the outer surfaces of the vehicle, to maintain desired product temperatures. Later studies demonstrated the considerable benefit of bottom delivery air circulation over top-air delivery systems.

Center-line stacking patterns for palletized loads of commodities such as strawberries is highly recommended (Kasmire and Ahrens, 1992). A minimum 5 cm gap between sidewalls and load to allow airflow preventing heating or freezing is also recommended. For this purpose, air-filled bags bracing the load away from the sidewalls and wood wedges support the pallet bottom to avoid load shifting are usually used.

Mechanical refrigeration units have progressed from small units with very limited refrigeration and air circulation capacities and crude temperature sensing and controls, to highly sophisticated units with excess refrigeration capacity, both supply and return air temperature monitoring and control, and refrigerant modulation that can control circulating air temperatures within ±1° C of the set point. Data may now be accessed in real time by remote satellite communication.

Surface transport can damage strawberry due to vibration. Vibration damage is caused by the low level vibration in transit, more than occasional impacts because of rough roads, but it is not considered serious for strawberry because of its elasticity (Mitchell, 1992). Trailer manufacturers have improved the design and construction of refrigerated trailers in different aspects including more gentle air-ride suspension systems to minimize vibration damage.

Maintenance of high RH to prevent shriveling can be a problem in mechanically refrigerated vans because of the limited coil surface and relatively large spread between air and coil temperatures, although in well precooled strawberries moisture loss is seldom a problem in a normal haul. Some refrigerated trailers are equipped with humidifiers to increase the RH in the load compartment (Ashby, 1970). On the other hand, dry ice has often been added in rail express or truck shipment of strawberries to increase CO_2 concentration around the berries when modified atmosphere pallets were not used. The effectiveness of this treatment is usually limited to one day or less because of leakage from vehicle, particularly when in motion.

9. Conclusions and Recommendations

Besides its attractiveness and quite high nutritional value, the strawberry is a highly perishable fruit. Therefore, special care should be taken both during harvest and postharvest handling. Training courses and supervision by growers and/or experienced pickers are recommended. Non-serious bruises caused by non-experienced pickers at harvest could become unacceptable after long-term transport. On the other hand, different studies have shown the benefits of rapid cooling to preserve strawberry quality. In order to minimize strawberry deterioration rate, growers are encouraged to schedule frequent deliveries of partial loads to the cooler instead of keeping the fruit waiting in the field for completing the load, especially when the fruit is placed under direct sunlight.

Taking into account the susceptibility of strawberry to pathogen infection, especially if free water is on its surface, no cooling system involving fruit soaking should be used. In this sense, among the minimum requirements for European strawberry grades it is stated that this fruit should not to be washed. Forced air cooling has proved to be the most convenient cooling system for strawberry. Fruit load should be thoroughly and rapidly cooled as close as possible to the recommended temperature, -0.5-$0°$ C. Due to technical limitations in cooling facilities, it is advisable to lower strawberry temperature to 0-2° C in order to avoid freezing risk. The quick cooling provided by the forced air system minimize moisture loss so that humidifiers in this system are not strictly necessary.

Finally, special care should be taken during transport in order to maintain strawberry quality. Besides the importance of the stacking patterns for the load and the different technical possibilities of the refrigeration system, a properly cooled strawberry load, a transport vehicle thoroughly cooled before loading, and an adequate air flow through the load are the main points to be considered in order to assure strawberry quality preservation during and after transport. When strawberry production is intended for long distance markets with longer transport times, the use of MA, either in the whole pallet or using MAP, is recommended. MA has proved to be effective in promoting shelf-life extention and control decay when proper temperature is not achieved during transport.

References

Abeles, F.B. and F. Takeda. 1990. Cellulase activity and ethylene in ripening strawberry and apple fruit. *Sci. Hort.* **42**: 269-275.
Adrian, J.J. 1970. U.S. Patent No. 3,521,438. Strawberry harvester, 21 July 1970.
Albregts, E.E. and C.M. Howard. 1982. Effect of transplant stress on strawberry performance. *HortScience* **17**: 651-652.

Archbold, D.D., Hamilton-Kemp, T.H., Barth, M.M. and B.E. Langlois. 1997. Identifying natural volatile compounds that control gray mold (*Botrytis cinerea*) during postharvest storage of strawberry, blackberry and grape. *J. Agric. Food Chem.* **45**: 4032-4037.

Ashby, B.H. 1970. Protecting perishable food during transport by motortruck. U.S.D. Handbook 105.

Barnes, M.F. and B.J. Patchett. 1976. Cell wall degrading enzymes and the softening of senescent strawberry fruit. *J. Food Sci.* **41**: 1392-1395.

Beaudry, R.M. 1999. Effect of O_2 and CO_2 partial pressure on selected phenomena affecting fruit and vegetable quality. *Postharvest Biol. Technol.* **15**: 293-303.

Biale, J.B. 1978. On the interface of horticulture and plant physiology. *Ann. Rev. Plant Physiol.* **29**: 1-23.

Boyette, M.D., Estes, E.A. and A.R. Rubin. 1992. Hydrocooling. Maintaining the quality of North Carolina fresh produce. North Carolina Agric. Ext. Serv. Circular 414-4.

Boyette, M.D., Wilson, L.G. and E.A. Estes. 1989. Postharvest cooling and handling of strawberries. North Carolina Agric. Ext. Serv. Circular 413-2.

California Department of Food and Agriculture. 1983. Fruit and vegetable quality control standardization. Extract from the Administrative Code of California. Dep. Food Agric., Sacramento, CA.

Cameron, A.C., Talasila, P.C. and D.W. Joles. 1995. Predicting film permeability needs for MA packaging of lightly processed fruits and vegetables. *HortScience* **30**: 25-34.

Cheng, G.W. and P.J. Breen. 1991. Activity of phenylalanine ammonia-lyase (PAL) and concentration of anthocyanins and phenolic in developing strawberry fruits. *J. Am. Soc. Hort. Sci.* **116**: 865-869.

Cheng, G. and P.J. Breen. 1992. Cell count and size in relation to fruit size among strawberry cultivars. *J. Am. Soc. Hort. Sci.* **117**: 946-995.

Childers, N.F. 1983. Strawberry growing. *In:* Modern fruit science. Orchard and small fruit culture. Horticultural Publications, Gainesville, FL, pp. 451-480.

Collins, J.K. and P. Perkins-Veazie. 1993. Postharvest changes in strawberry fruits under simulated retail display conditions. *J. Food Quality* **16**: 133-143.

Day, B. 1996. High oxygen modified atmosphere packaging for fresh prepared produce. *Postharvest News and Information.* **7**: 31N-34N.

Denisen, E.L. and W.F. Buchele. 1967. Mechanical harvesting of strawberries. *Proc. Am. Soc. Hort. Sci.* **91**: 267-273.

Dirinck, P.J, De Pooter, H.L. Willaert, G.A. and N.M. Schamp. 1981. Flavour quality of cultivated strawberries: The role of the sulfur compounds. *J. Agric. Food Chem.* **29**: 316-321.

Edmon, J.P. and K.V. Chau. 1990. Use of perforations in modified atmosphere packaging. *ASAE paper.* **90**: 6512-6517.

El-Ghaouth, A., Arul, J. Ponnanpalam, R. and M. Boulet. 1991. Chitosan coating effect on storability and quality of fresh strawberries. *J. Food Sci.* **56**: 1618-1620.

El-Goorani, M.A. and N.F. Sommer. 1979. Suppression of postharvest plant pathogenic fungi by carbon monoxide. *Phytopathology* **69**: 834-838.

Exama, A., Arul, J. Lencki, R.W. Lee, L.Z. and C. Toupin. 1993. Suitability of plastic films for MA packaging of fruits and vegetables. *J. Food Sci.* **58**: 1365-1370.

FAO. 1998. Production Yearbook. FAO Statistics Series, Roma.

Ferreira, M.D., Brecht, J.K. Sargent, S.A. and J.J. Aracena. 1994. Physiological responses of strawberry to film wrapping and precooling methods. *Proc. Fla. State Hort. Soc.* **107**: 265-269.

Forney, C.F. and P.J. Breen. 1985. Sugar content and uptake in the strawberry fruit. *J. Am. Soc. Hort. Sci.* **111**: 241-247.

García, J.M., Aguilera, C. and M.A. Albi. 1995. Postharvest heat treatment of Spanish strawberry (*Fragaria x ananassa* Cv. Tudla). *J. Agric. Food Chem.* **43**: 1489-1492.

García, J.M., Medina, R.J. and J.M. Olías. 1998. Quality of strawberries automatically packed in different plastic films. *J. Food Sci.* **63**: 1037-1041.

Garren, R. 1981. Causes of misshapen strawberries. *In:* The strawberry. N.F. Childers (Ed.), Horticultural Publications, Gainesville, FL, pp. 326-328.

Gil, M.I., Holcroft, D.M. and A.A. Kader. 1997. Changes in strawberry anthocyanin and other polyphenols in response to carbon dioxide treatments. *J. Agric. Food Chem.* **45**: 1662-1667.

Gilles, G. 1959. Biology of control of *Botrytis cinerea* Pers. on strawberries. Hfchen-Briefe **3**: 141-168.

Given, N.K., Venis, M.A. and D. Griergson. 1988a. Hormonal regulation of ripening in strawberry, a non-climacteric fruit. *Planta* **174**: 402-406.

Given, N.K., Venis, M.A. and D. Griergson. 1988b. Phenylalanine ammonia-lyase activity and anthocyanin synthesis in ripening strawberry fruit. *J. Plant Physiol.* **133**: 25-30.

Gooding, H.J., McNicol, R.J. and J.H. Reid. 1983. Studies of strawberry fruit characters necessary for machine harvesting. *Crop Res.* **23**: 3-16.

Gross, K.C. and C.E. Sams. 1984. Changes in cell wall neutral sugar composition during fruit ripening: a species survey. *Phytochemistry* **23**: 2457-2461.

Guillou, R. 1960. Coolers for fruits and vegetables. *Calif. Agr. Exp. Sta. Bull.* 773.

Hancock, J.F. and R.S. Bringhurst. 1979. Ecological diferentiation in perennial, octaploid species of *Fragaria*. *Amer. J. Bot.* **66**: 367-370.

Hardenburg, R.E., Watada, A.E. and C.Y. Wang. 1986. The commercial storage of fruits, vegetables, and florist and nursery stocks. USDA Agriculture Handbook 66, US Govt Print. Off., Washington, D.C.

Hecht, C.L. 1972. U.S. Patent No. 3,698,171. Mechanical picker for strawberries, 17 October 1972.

Hennebert, G.L. and L. Gilles. 1958. Epidemiologie de *Botrytis cinerea* Pers. sur fraisiers. Meded. LandbHoogesch. *Opzoekstns Gent.* **23**: 864-888.

Hergert, G.B. and A. Dale. 1989. Performance of a mechanized strawberry production system. ASAE/CSAE Metting Presentation, No. 89-1074.

Hirvi, T. and E. Honkanen. 1982. The volatiles of two new strawberry cultivars, 'Annelie' and 'Alaska Pioneer' obtained by backcrossing of cultivated strawberries with wild strawberrries, *Fragaria vesca*, Rugen and *Fragaria virginiana*. *Z. Lebensm. Unters. Forsch.* **175**: 113-116.

Holcroft, D.H. and A.A. Kader. 1999. Controlled atmosphere-induced changes in pH and organic acid metabolism may affect color of stored strawberry fruit. *Postharvest Biol. Technol.* **17**: 19-32.

Jarvis, W.R. 1964. The effect of some climatic factors on the incidence of grey mould of strawberry and raspberry fruit. *HortScience* **3**: 65-71.

Jarvis, W.R. 1977. Botryotinia and Botrytis species: taxonomy, physiology, and pathogenicity. *Can. Dept. Agr. Res. Br. Monogr.* 15.

John, O. and S. Yamaki. 1994. Sugar content, compartmentation, and efflux in strawberry tissue. *J. Am. Soc. Hort. Sci.* **119**: 1024-1028.

Kader, A.A. 1991. Quality and its maintenance in relation to the postharvest physiology of strawberrry *In:* The Strawberry into the 21st Century. J.J. Luby and A. Dale (Eds), Timber Press, Portland OR, pp. 145-152.

Kader, A.A., Singh, R.P. and J.D. Mannapperuma. 1998. Technologies to extend the refrigerated shelf-life of fresh fruits. *In:* Food Storage Stability. I.A. Taub and R.P. Singh (Eds), CRC Press, Boca Raton, Florida, pp. 419-434.

Kasmire, R.F. and M.J. Ahrens. 1992. Handling of hortivariable crops and destination markets. *In:* Postharvest Technology of Horticultural Crops. A.A. Kader (Ed.), University of California, Davies pp. 175-180.

Kasmire, R.F. and J.F. Thompson. 1992. Cooling horticultural commodities. III. Selecting a cooling method. *In:* Postharvest Technology of Horticultural Crops. A.A. Kader (Ed.), University of California, Davies, pp. 63-68.

Ke, D., Zhou, L. and A.A. Kader. 1994. Mode of oxygen and carbon dioxide action on strawberry ester biosynthesis. *J. Am. Soc. Hortic. Sci.* **119**: 971-975.

Ke, D., Goldstein, L., O'Mahony, M. and A.A. Kader. 1991. Effects of short-term exposure to low O_2 and high CO_2 atmospheres on quality attributes of strawberries. *J. Food Sci.* **56**: 50-54.

Kim, G.H. and R.B. Wills. 1998. Interaction of enhanced carbon dioxide and reduced ethylene on the storage life of strawberries. *J. Hort. Sci. Biotechnol.* **73:** 181-184.

Knee, M., Sargent, J.A. and D.J. Osborne. 1977. Cell wall metabolism in developing strawberry fruits. *J. Exp. Bot.* **28:** 377-396.

Kuneman, D.M, Braddock, J.K. and L.L. McChesney. 1988. HPLC profile of aminoacids in fruit juices and their (1-fluoro-2,4-dinitrophenyl)-5-L-alanine amide derivatives. *J. Agric. Food Chem.* **36:** 6-9.

Larsen, M. and C.B. Watkins. 1995. Firmness and concentrations of acetaldehyde, ethyl acetate and ethanol in strawberries stored in controlled and modified atmospheres. *Postharvest Biol. Technol.* **5:** 39-50.

Larson, K.D. 1994. Strawberry. *In*: Handbook of Environmental Physiology of Fruit Crops. Volume I: Temperate Crops. B. Schaffer and P.C. Andersen (Eds), CRC Press Inc., Boca Raton, Florida, pp. 271-297.

Lawrence, F.J., Galletta, G.J. and D.H. Scott. 1990. Strawberry breeding work of the US Department of Agriculture. *HortScience* **25:** 895-896.

Li, C. and A.A. Kader. 1989. The residual effects of controlled atmospheres on postharvest physiology and quality of strawberries. *J. Amer. Soc. Hort. Sci.* **114:** 629-634.

Liew, C. and R.K. Prange. 1994. Effect of ozone and storage temperature on postharvest diseases and physiology of carrots (*Daucus carota* L.). *J. Am. Soc. Hortic. Sci.* **119:** 563-567.

López-Serrano, M. and A. Ros-Barceló. 1996. Purification and characterization of a basic peroxidase isoenzyme from strawberries. *Food Chem.* **55:** 133-137.

Luo, Y. and X. Liu. 1994. Effects of ethylene on RNA metabolism in strawberry fruit after harvest. *J. Hort. Sci.* **69:** 137-139.

Maas, J.L. 1984. *Compendium of Strawberry Diseases.* APS Press, St. Paul, MN.

Maas, J.L., Wang, S.Y. and G.J. Galleta. 1991. Evaluation of strawberry cultivars for ellagic acid content. *HortScience* **26:** 66-68.

Manning, K. 1994. Changes in gene expression during strawberry fruit ripening and their regulation by auxin. *Planta* **194:** 62-68.

Maroto, J.V., Pascual, B., Alagarda, J. and S. López-Galarza. 1986. Estudio sobre las principales características productivas de algunos cvs. de fresón de utilización frecuente en España. *Agrícola Vergel*, **57:** 493-496.

Martínez, G.A., Chaves, A.R. and C. Añón. 1994. Effects of gibberellic acid on ripening of strawberry fruits. (*Fragaria ananassa* Duch.) *J. Plant Growth Regul.* **13:** 87-91.

Martínez, G.A., Civello, P.M., Chaves, A.R. and M.C. Anoy. 1995. Partial characterization of chlorophyllase from strawberry fruit (*Fragaria ananassa* Duch.) *J. Food Biochem.* **18:** 213-216.

Medina-Escobar, N., Cárdenas, J., Moyano, E., Caballero, J.L., J. Muñoz-Blanco. 1997. Cloning, molecular characterization and expression pattern of strawberry ripening-especific cDNA with sequence homology to pectate lyase from higher plants. *Plant Mol. Biol.* **34:** 867-877.

Mitchell, F.G. 1992. Postharvest handling systems: small fruits (table grapes, strawberries, kiwi fruits). *In*: Postharvest Technology of Horticultural Crops, A.A. Kader (Ed.), University of California, Davies, pp. 223-231.

Mitchell, W.C. and G. Jelenkovic. 1995. Characterizing NAD- and NADP-dependent alcohol dehydrogenase enzymes of strawberries. *J. Am. Soc. Hort. Sci.* **120:** 798-801.

Mitchell, F.G., Maxie, E.C. and A.S. Greathead. 1964. Handling strawberries for fresh market. *Calif. Agr. Expt. Sta. Cir.* p. 527.

Mussinan, C.J. and J.P. Walradt, 1975. Organic acids from fresh California strawberries. *J. Agric. Food Chem.* **23:** 482-484.

Nitsch, J.P. 1950. Growth and morphogenesis of the strawberry as related to auxin. *Am. J. Bot.* **37:** 211-215.

Nogata, Y., Ohta, H. and A.G.J. Voragen. 1993. Polygalacturonase in strawberry fruit. *Phytochemistry* **34:** 617-620.

Ntirampemba, G., Langlois, B.E., Archbold, D.D., Hamilton-Kemp, T.R. and M.M. Barth. 1998. Microbial populations of Botrytis cinerea-inoculated strawberry fruit exposed to four volatile compounds. *J. Food Protec.* **61**: 1352-1357.

Nunes, M.C.N., Brecht, J.K., Morais, A.A.M.B. and S.A. Sargent. 1995. Physical and chemical quality characteristics of strawberries after storage are reduced by a short delay to cooling. *Postharvest Biol. Technol.* **6**: 17-28.

Olías, J.M., Sanz, L.C., Verdier, M. and A.G. Pérez. 1994. Effect of the temperature and the material of the basket on the market life of strawberries. *In*: Contribution du Froid a la Preservation de la Qualité des Fruits, Legumes et Produits Halieutiques. A.L. Bennani and D. Messaho (Eds), Actes Editions, Rabat. pp. 267-273.

Organization for Economic Cooperation and Development. 1987. International standardization of fruits and vegetables. OECD, Paris, France.

Pérez, A.G., Olías, R., Espada, J., Olías, J.M. and C. Sanz. 1997. Rapid determination of sugars, non-volatile acids, and ascorbic acid in strawberry and other fruits. *J. Agric. Food. Chem.* **45**: 3545-3549.

Pérez, A.G., Olías, R., Olías, J.M. and C. Sanz. 1998. Strawberry quality as a function of the 'high pressure fast cooling' desing. *Food Chem.* **62**: 161-168.

Pérez, A.G., Olías, R., Sanz, C. and J.M. Olías. 1996a. Furanones in strawberries: evolution during ripening and postharvest shelf-life. *J. Agric. Food. Chem.* **44**: 3620-3624.

Pérez, A.G., Ríos, J.J., Sanz, C. and J.M. Olías, 1992. Aroma components and free amino acids in strawberry variety Chandler during ripening. *J. Agric. Food Chem.* **40**: 2232-2235.

Pérez, A.G., Sanz, C., Olías, R. and J.M. Olías. 1999a. Lipoxygenase and hydroperoxide lyase activities in ripening strawberry fruits. *J. Agric. Food Chem.* **47**: 249-253.

Pérez, A.G., Sanz, C., Olías, R., Ríos, J.J. and J.M. Olías. 1996b. Evolution of strawberry alcohol acyltransferase during fruit development and storage. *J. Agric. Food Chem.* **44**: 3286-3290.

Pérez, A.G., Sanz, C., Ríos, J.J., Olías, R. and J.M. Olías. 1999b. Effects of ozone treatments on postharvest strawberry quality. *J. Agric. Food Chem.* **47**: 1652-1656.

Perkins-Veazie, P. 1995. Growth and ripening of strawberry fruit. *Hort. Reviews* **17**: 267-297.

Perkins-Veazie, P., Huber, D.J. and J.K. Brecht. 1987. Respiration, ethylene production and ethylene responsiveness in developing strawberry fruit. *HortScience* **22**: 1128-1132.

Picón, A., Martínez-Jávega, J.M., Cuquerella, J., Del Rio, M.A. and P. Navarro. 1993. Effects of precooling packaging film, modified atmosphere, and ethylene adsorber on the quality of refrigerated Chandler and Douglas strawberries. *Food Chem.* **48**: 189-193.

Plakidas, A.G. 1964. Strawberry Diseases. Louisiana St. Univ. Studies, Biological Sci. Ser. 5. Louisiana St. Univ. Press, Baton Rouge, LA.

Powelson, R.L. 1960. Initiation of strawberry fruit rot caused by Botrytis cinerea. *Phytopathology* **50**: 491-494.

Quick, G.R. 1971. US Patent No. 3,596,456. Strawberry harvesting device, 3 August 1971.

Quick, G.R. and E.L. Denisen. 1970. A strawberry harvest mechanization system. *HortScience* **5**: 150-151.

Rasmussen, C.E. 1968. US Patent No. 3,589,572. Strawberry harvester, 25 June 1968.

Reyes, F.G., Wrolstad, R.E. and C.J. Cornvell. 1982. Comparison of enzymatic, gas-liquid chromatographic, and high performance liquid chromatographic methods for determining sugars and organic acids in strawberries at three stages of maturity. *J. Assoc. Off. Anal. Chem.* **65**: 126-13.

Robinson, J.E., Browne, K.H. and W.G. Burton. 1975. Storage characteristics of some vegetables and soft fruits. *Ann. Appl. Biol.* **81**: 399-408.

Ryall, A.L. and W.T. Pentzer. 1974. *Handling, Transportation and storage of fruit and vegetables*. Vol. 2. Fruits and Vegetables. AVI Westport, CT.

Sacks, E.J. and D.V. Shaw. 1993. Color change in fresh strawberry fruit of seven genotype stored at 0° C. *HortScience* **28**: 209-210.

Sanz, C., Pérez, A.G., Olías, R. and J.M. Olías. 1999. Quality of strawberries packed with perforated polypropylene. *J. Food Sci.* **64**: 748-752.

Sanz, C., Pérez, A.G. and D.G. Richardson. 1994. Simultaneous HPLC determination of 2,5-dimethyl-4-hydroxy-3(2H)-furanone and related compounds in strawberries. *J. Food Sci.* **59**: 139-141.

Sanz, C., Richardson, D.G. and A.G. Pérez. 1995. 2,5-Dimethyl-4-hydroxy-3(2H)furanone and derivatives in strawberries during ripening. *In*: Fruit Flavors. Biogenesis, Characterization, and Authentication. R.L. Rouseff and L.L. Leahy (Eds), ACS Symposium Series 596, Washington DC, pp. 268-275.

Schreier, P. 1980. Quantitative composition of volatile constituents in cultivated strawberries, *Fragaria ananassa* cv. Senga Sengana, Senga Litessa, and Senga gourmella. *J. Food Sci.* **31**: 487-494.

Shamaila, M., Powrie, W.D. and B.J. Skura. 1992. Sensory evaluation of strawberry fruit stored under modified atmosphere packaging (MAP) by quantitative descriptive analysis. *J. Food Sci.* **57**: 1168-1172.

Shaw, D.V. 1988. Genotypic variation and genotypic correlations for sugars and organic acids of strawberries. *J. Am. Soc. Hort. Sci.* **113**: 770-774.

Smith, R.B. and L.J. Skog. 1992. Postharvest carbon dioxide treatment enhances firmness of several cultivars of strawberry. *HortScience* **27**: 420-421.

Spalding, D.H. 1966. Appearance and decay of strawberries, peaches, and lettuce treated with ozone. USDA Marketing Res. Rept. 756.

Stang, E.J. and E.L. Denisen. 1971. A proposed system for once-over machine harvesting of strawberries for fresh use. *HortScience* **6**: 414-415.

Swinkels, P.M. and R.A. Murray. 1991. Development of the Bragg strawberry harvester. *In*: The Strawberry into the 21st Century. A. Dale and J.J. Luby (Eds), Portland, OR, Timber Press, pp. 266-267.

Talbot, M.T., Brecht, J.K. and S.A. Sargent. 1995. Cooling performance evaluation of strawberry containers. *Proc. Fla. State Hort. Soc.* **108**: 258-268.

Talbot, M.T. and K.V. Chau. 1991. Precooling strawberries. Florida Coop. Ext. Serv. Circular 942.

Ueda, Y. and K. Ogata. 1977. Coenzyme A-dependent esterification of alcohols and acids in separated cells of banana pulp and its homogenate. *Nippon Shokuhin Kogyo Gakkaishi* **24**: 624-630.

Ulrich, A., Mostafa, M.A.E. and W.W. Allen. 1980. Strawberry Deficiency Symptoms: A Visual and Plant Analysis Guide to Fertilization. University of California, Division of Agricultural Sciences, Berkeley, California, CA.

Ulrich, D., Hoberg, E., Rapp, A. and S. Kecke. 1997. Analysis of strawberry flavour, discrimination of aroma types by quantification of volatile compounds. *Z. Lebensm. Unters Forsch.* **205**: 218-223.

USDA. 1965. US Standards for grades of strawberries. 30 F.R. 6711.

Wang, H., Cao, G. and R.L. Prior. 1996. Total antioxidant capacity of fruits. *J. Agric. Food Chem.* **44**: 701-705.

Wesche-Ebling, P. and M.W. Montgomery. 1990. Strawberry polyphenoloxidase: its role in anthocyanin degradation. *J. Food Sci.* **55**: 731-734.

Wills, R., McGlasson, B., Graham, D. and D. Joyce. 1998. *Postharvest. An Introduction to the Physiology and Handling of Fruit, Vegetables and Ornamentals*. 4th Edition. University of New South Wales Press Ltd. Sydney.

Winston, A.L. 1902. The anatomy of edible berries. Conn. Agr. Exp. Stn. Rep. **26**: 288-325.

Wright, W.R., Beraha, L. and M.A. Smith. 1966. Leather rot on California strawberries. *Plant Dis. Rptr.* **50**: 283-287.

Wrolstad, R.E., Putnam, T.P. and G.W. Varseveld. 1970. Colour quality of frozen strawberries: Effect of anthocyanins, pH, total acidity, and ascorbic acid variability. *J. Food Sci.* **35**: 448-452.

Wrolstad, R.E. and R.S. Shallenberger. 1981. Free sugars and sorbitol in fruits—A compilation from the literature. *J. Assoc. Off. Anal. Chem.* **64**: 91-103.

Wu, Y.M., Gu, C.Q., Tai, G.F. and Y. Liu. 1992. Effects of ABA and ethylene on the ripening and senescence of postharvest strawberry fruit. *Acta phytophysiologica Sinica* **18**: 167-172.

Zabetakis, I. and M.A. Holden. 1997. Strawberry flavour: analysis and biosynthesis. *J. Sci. Food Agric.* **74**: 421-434.

Zagory, D. and A.A. Kader. 1988. Modified atmosphere packaging of fresh produce. *Food Technol.* **42**: 70-77.

Zhang, D. and P.G. Quantick. 1998. Antifungal effects of chitosan coating on fresh strawberries and raspberries during storage. *J. Hort. Sci. Biotechnol.* **73**: 763-767.

11

Postharvest Biological Changes and Technology of Citrus Fruit

Giovanni Arras

CNR - Istituto per la Fisiologia della Maturazione e della Conservazione del Frutto delle Specie Arboree Mediterranee, Via Dei Mille, 48 - 07100 Sassari, Italy

Abbreviations

A. alternata	*Alternaria alternata*
C. sinensis	*Citrus sinensis*
C. reticulata	*Citrus reticulata*
C. limon	*Citrus limon*
C. paradisi	*Citrus paradisi*
EU	European Union
IMZ	imazalil
MIC	minimal inibitory concentration
OPP	Orthophenylphenate
POM	Programma Operativo Multiregionale
P. digitatum	*Penicillium digitatum*
P. citrophthora	*Phytophthora citrophthora*
P. parasitica	*Phytophthora parasitica*
P. syringae	*Phytophthora syringae*
RH	relative humidity
SOPP	sodium orthophenylphenate
TBZ	benzimidazole
TSS	total soluble solids

1. Introduction

The problem of storing food has always been important for man and his evolution. The first methods were simply to keep food in naturally cool places like caves and underground stores. Refrigeration systems to preserve fruit during transport by sea were introduced only at the end of the nineteenth century. In Italy, the refrigeration of fruit and vegetables began around 1923 in fruit processing plants which had introduced modern technology and refrigeration and exported fruit outside the country.

In recent years, profound social and economical changes have brought new eating habits, increasing the demand for fresh fruit and vegetables.

Nowadays the interval between harvesting and consumption tends to be longer. Thus the products are liable to decay on account of pathological changes and biochemical processes.

Modern storage systems, besides safeguarding the physiological aspect of the fruit, play an important role in the fruit industry since the marketing of the fruit can be delayed. The products can be put on sale at the right time, eventually leading to a levelling of prices.

Italy holds the first place in the European Union (EU) in the horticultural sector with an annual production of fresh fruit (excluding wine grapes) of about 7 million tonnes and a vegetable harvest (excluding potatoes) of about 12 million tonnes. As regards citrus fruit, production in the last few years has been about 30,000 tonnes. However, the market for citrus fruit in Italy, and also in Europe, is in crisis in comparison to the past. The production cannot keep up with changes in distribution, a rising demand for high-quality products and ever-increasing competition from developing countries which are able to price their products more competitively. In view of this current situation, it would make sense economically to find new high-quality cultivars with an extended period of maturation that are suitable for long-term storage and early degreening.

From the botanical and morphological point of view, citrus fruits belong to the genus *Citrus* which includes many species and varieties of sweet oranges (*C. sinensis*), clementine and mandarins (*C. reticulata*), lemons (*C. limon*), grapefruits (*C. paradisi*), etc. The citrus fruit is composed of the epicarp or flavedo (orange-coloured), rich in essential oils, of the mesocarp or albedo (white and spongy) and the juicy endocarp which is the most important part from a nutritional viewpoint. The latter consists of segments around a central core (medulla or intercapillary axis) each separated by carpellary membranes. Inside the segments are the fusiform vesicles containing the juice and small amounts of essential oils (Fig. 1). The juice is made up of several constituents, including carbohydrates, acids, acid alcohols, amino acids, nitrogenous substances, proteides, lipids, glucosides, vitamins, pigments, aromatic compounds and enzymes.

2. Storage of Citrus Fruits

During postharvest, citrus fruits are subjected to a number of metabolic processes like respiration and transpiration, and the production of heat and volatile compounds causes biochemical transformations. How quickly these last reactions take place depends on the temperature. The transformations affect both the organoleptic characteristics of the fruits and their quality as merchandise, which can be compromised over fairly long periods of time. Suitable refrigeration systems which can guarantee longer storage periods can slow down these processes (Arras and Schirra, 1988). Refrigeration reduces respiration and slows down the metabolic activity of the fruits; this results in less deterioration of the products and consequently they can be on sale far longer than fruits kept at ambient temperature. Thus, refrigeration not only allows the marketing period of

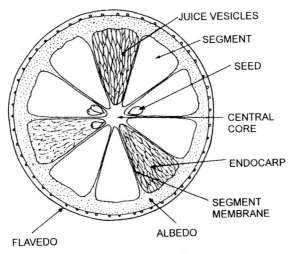

Fig. 1. Cross section of a citrus fruit.

fruit and vegetable products to be extended but also preserves the nutritional value of fruit since it reduces vitamin and weight loss.

Citrus fruits are not climacteric; in fact, the respiratory activity continues to decrease very slowly after harvesting (unlike other fruits, they do not have a climacteric peak) (Baldini and Scaramuzzi, 1980). For this reason, citrus fruits mature on the tree and show a slow, gradual utilization of soluble sugars during their postharvest life.

The subtropical origin makes citrus fruits sensitive to low temperatures. They are easily affected by postharvest deterioration caused by temperature fluctuation, e.g. too-high temperatures lead to increased development of microbiological lesions and greater transpiration. It is, therefore, necessary to find the right balance to reduce to a minimum these diseases, which are discussed in sections 6-9.

The life of the fruit can be divided into three physiological stages: growth, maturation and senescence. However, it is not easy to make a clear distinction between these stages. Growth implies cell multiplication (cytokinesis) followed by cell diffusion. During the maturation process a series of physiological and biochemical changes take place, such as darkening and maculation (change of colour due to the destruction of chlorophyll and the formation of anthocyanins and carotenoids), pH reduction, decrease in acids, increase in sugars and total soluble solids and an increase in the percentage of juice and metabolites (ethanol, acetaldehyde, etc). These processes can be affected by regulating storage conditions to delay senescence (section 2).

The quality of the citrus fruits depends on a series of intrinsic and extrinsic factors which also affect the fruit's potential for storage. The quality of the fruit is judged according to its organoleptic characteristics like smell, sweetness, acidity, juiciness and fibre content. A panel of expert tasters would be one way of evaluating these parameters but since this is not always possible, some chemical analyses are done instead. The most important are as follows:

1. *Total free acidity*, due mainly to the citric and malic acid content, only traces of other acids (tartaric, benzoic, etc) are present. Acidity is expressed as the percentage of citric acid, determined by titration with NaOH 0.1 N up to pH 8.2. This value, which accounts for the agreeable taste of citrus fruits, decreases considerably during storage (Agabbio et al. 1982a).

2. *The total soluble solids* (TSS) in the juice, mainly sugars and to a lesser extent organic acids, mineral salts, vitamin C and pectin (Royo and Perez, 1977). This parameter varies considerably according to storage conditions, so it is important to be monitored, using Abbe's refractometer to read the index of refraction, expressed in Brix degrees or as a percentage.

3. *The maturation index* is the ratio between total soluble solids and total acidity. During the maturation period, the TSS increases while the percentage of acidity decreases, so the maturation index tends to rise even during storage. It is obviously important to monitor the TSS : acidity ratio to determine the peak harvesting time and its evolution during storage (Pennisi, 1973).

4. *Ethanol and acetaldehyde*: An increase in these metabolites occurs during maturation. As regards ethanol, the increase can be correlated with the maturity index of the fruits (Davis, 1970, 1971). It has been observed that during postharvest life these compounds increase more markedly in relation to storage conditions, so they can be held directly or indirectly responsible for the quality deterioration of the citrus fruits due to the physiological disorders that arise during storage (Davis et al. 1974).

5. *Ethylene* (C_2H_4) is a natural product of the plant's metabolism and, along with other hormones (auxin, gibberellin, quinine and ascorbic acid), affects the ripening of the fruit. Ethylene promotes increased respiration, destroys chlorophyll and influences pectinase and cellulase activity by increasing membrane permeability and reducing the cementing effect of the insoluble protopectins: their transformation into water-soluble pectins reduces pulp consistency. In citrus fruits, endogenous ethylene is either absent or else is present in small quantities compared with other fruit species which are climacteric (Arras and Usai, 1991). It can increase by as much as ten times when the fruits are attacked by pathogens or otherwise stressed. The properties of ethylene make it an important indicator to monitor fruit in storage. Too high an increase is a sign either of senescence of the fruits or of biotic or abiotic changes.

It is obvious that it is far from easy to keep the original quality of the fruit unchanged after harvest. Several authors have reported a reduction of total soluble solids and ascorbic acid in the juice of late Valencia cv oranges during refrigeration (Pritchett, 1962; Arras and Schirra, 1988), besides loss of citric acid (Khalifah and Kuykendall, 1965). The reduction was greater when the fruits were ripened on the tree (El-Zeftawi, 1976). These and other variations in the components of the juice cause a change in flavour and in any case the fruit loses its pleasant taste.

One of the most important methods of lengthening the storage life of fruits is fungistat treatment, necessary to reduce to a minimum losses due to rot caused mainly by *Penicillium italicum* Wehmer and *P. digitatum* Sacc. (Eckert, 1997). Rot continues to be widespread, owing to the development of strains resistant to TBZ and IMZ products (Houck, 1977; Dave et al. 1990). This has led researchers to try to find new control techniques, which will be reviewed later (section 7).

Other authors (Lutz and Hardenburg, 1968) have pointed out the need for a more careful assessment of all the interdependent factors that affect the storage of citrus fruits. An understanding of the correlation between genetic factors, environment and the suitability of the fruit for storage is essential for successful refrigeration (Grierson and Hatton, 1977).

To reach the objectives previously described, adequate temperature and humidity level for each species and variety must be used so that fruits can be stored without altering its metabolism. The main parameters affecting the physiological processes are temperature, humidity, composition of the air in the cold store, ventilation and CO_2 treatment.

2.1. Influence of Temperature

In citrus fruits, the temperature has a significant influence on pathological, physiological (respiration, transpiration, ethylene production, aromatic compounds, etc) and biochemical processes (biosynthesis of ethanol, acetaldehyde, methanol, ethyl acetate, etc) affecting the length of the storage life of the fruit (Arras and De Cicco, 1998).

The most suitable temperature varies according to intrinsic and extrinsic factors such as species, cultivar, soil and climate, rootstock, ripening stage, etc which must be taken into account. The optimal minimum temperature varies considerably between species of citrus fruits: some, like the grapefruit and the lemon, are very sensitive to cold and temperatures below 10° C cause physiological changes, while other species like the orange and mandarin are less susceptible and can be kept at 6-8° C for quite long periods (Table 1).

Within each species, the storage potential varies markedly according to the cultivars. Early-ripening oranges such as 'Washington navel' and 'Tarocco' can be stored for periods of between 7 and 8 weeks (Agabbio et al. 1982b), while late-ripening cultivars like Valencia late, Ovale, Biondo comune, Tardivo di S. Vito, etc are less sensitive to cold damage and rot and tolerate 3-4 months storage without injury (Agabbio, 1986; Arras and Schirra, 1988).

The climate and soil have a considerable influence on the morpho-qualitative characteristics of the fruits and therefore also on their susceptibility to mycological and physiological injury. In studies on Valencia late cv. (Khalifah and Kuykendall, 1965) and Washington navel oranges (Arras et al. 1985), it was found that fruits produced in warmer areas with sandy soil seemed to be more resistant to physiological injury during storage than those grown in cooler areas with a heavier soil type. The

Table 1. Optimal temperature and storage time for citrus fruit produced in Italy

Species	Cultivar	Temperature (° C)	Storage weeks
Orange (Citrus sinensis)	Washington navel	7-8	6-8
	Moro	6-8	9-11
	Tarocco	6-8	10-12
	Sanguinello	6-8	8-10
	Belladonna	5-6	12-14
	Biondo comune	5-6	10-12
	Ovale	5-6	12-16
	Valencia late	5-6	15-18
Mandarin (Citrus reticulata)	Avana	5-6	4-8
	Tardico Ciaculli	4-5	7-8
	Fortune	6-7	6-7
	Ortanique	4-5	8-11
Lemon (Citrus limon)	Femminello	10-12	8-10*
	Monachello	10-12	8-10*
	Eureka	13	8-10*
Grapefruit (Citrus paradisi)	Marsh seedless	10-12	8-10

*The storage time indicated refers to ripening lemons in colour change.
The best RH for citrus fruit is about 85-90%.
Air exchange in the cold store should take place every 2-4 h.

Tarocco and Washington navel cultivars grown in clayey ground seemed to be more resistant to *P. italicum* and *P. digitatum* infections than fruit produced in sandy soil (Ibraham, 1968; Arras, 1988). One example reported by Baldini and Scaramuzzi (1980) concerned Valencia fruits which can be refrigerated at 0-1° C if grown in Texas or Florida, at 2-3° C if grown in Spain, at 5-6° C if grown in Italy and at 4-7° C if grown in California.

Rootstock (McDonald and Wutscher, 1974; Grierson and Hatton, 1977) and stage of maturity (Pantastico et al. 1968; Chessa et al. 1982; Arras et al. 1984) influence the number of disorders and diseases caused by cold and microbiological lesions in the grapefruit and orange.

2.2. Relative Humidity

The relative humidity of the air inside the cold room is a decisive factor in storing the products since it directly affects transpiration. This is the process by which the tissues emit water in the form of vapour. The amount depends on the difference in pressure between the cold room and the outside environment.

Weight loss is mostly due to transpiration which is one of the main causes of the dehydration and deterioration of fruits, reducing their commercial and nutritional value. Transpiration can be reduced by increasing relative humidity, lowering the temperature and ventilation, and using packaging like plastic film, wax, etc (Long and Leggo, 1959). When relative humidity is 100%, transpiration is completely inhibited but other disadvantages could arise due to a greater number of microbiological and physiological disorders. As the slight loss in weight is of less importance than these diseases, the optimal RH for citrus fruits is around 85-90%, giving a weight loss of 0.9-1.5% a week, depending on species, cultivar and postharvest treatment (Agabbio et al. 1982b).

Relative humidity is inversely proportionate to the temperature; due to this phenomenon greater transpiration occurs as the temperature rises (Wells, 1962). Humidifiers should be boosted in cold rooms used for storing fruits like grapefruit and lemons which are sensitive to cold. The regulation of RH in cold rooms is important and involves problems of an economic and technical nature. The latter can be resolved satisfactorily in different ways such as the use of vaporizers, regulation of the temperature to avoid condensation, air exchange or alternatively, air cleaning. RH in cold stores can be measured with various instruments like a hair hygrometer (cheap, but not very accurate), a psychrometer (accurate and easy to use) or an electronic hygrometer (expensive, but extremely accurate). When these instruments are connected to the cold room humidifier, RH can be stabilized at values very close to the ones required and graph tracings allow continuous monitoring (Anelli and Mencarelli, 1990).

2.3. Air Exchange in Cold Stores

During storage, citrus fruits emit CO_2, water vapour and heat, besides volatile substances like acetaldehyde, ethanol, limonene, α-cymene, β-myrcene, ethyl butyrate, ethyl formate (Norman, 1977) and ethylene after respiration and transpiration. These compounds, particularly ethylene, can provoke a series of negative phenomena leading to premature aging of the fruit and an increase in microbiological lesions, so the metabolites must be eliminated from the cold store. Several studies have shown that the presence of ethylene is extremely harmful as it is conducive to increased respiratory activity (Kusunose and Sawamura, 1980), opaqueness of the skin and a certain aftertaste in the juice caused by the abnormal presence of some metabolites like ethanol and acetaldehyde (Davis et al. 1974; Smagula and Bramlage, 1977).

Two different techniques can be used to eliminate these volatile compounds from the cold store: air exchange or air cleaning. The first consists of replacing the air inside with air from the outside which has been cooled and humidified. The second allows gaseous products emitted by the fruits to be eliminated by various methods; a particularly effective one is washing counter-current in water. The device used is a tower placed near the refrigerator where the air is made to pass first through a very fine shower of water and then into 'Raschig' type rings designed to increase the surface of air-water contact, thus promoting the absorption of the air's gaseous compounds, like CO_2, aromatic compounds and ethylene (Gorini, 1979, Gorini; et al. 1989).

2.4. Ventilation

In order to guarantee uniformity of temperature, relative humidity and air cleaning inside the cold store, ventilators are needed to provide active circulation of air. The most appropriate circulation of volumes of air is dictated by the amount of water vapour emitted by the conserved fruit, the size of the cold store, the way the fruit is stowed and the length of refrigeration time. Air movement must be moderate, however, to avoid an increase in transpiration which causes weight loss.

Distribution methods are circulation and canalization. The first is used in small or medium-sized cold rooms. The air is insufflated into the room through the entire refrigerant surface, equipped with a powerful ventilator. The second method is employed to make air distribution uniform in very large cold stores. Special channels with slits distribute the air in different directions (Gorini, 1979).

3. Postharvest Treatment with CO_2

The use of atmospheric modification in storage (modification of the gaseous composition, reduction of O_2 and increase in CO_2) to eliminate metabolic activity has usually had disappointing results with regard to citrus fruit (Hatton and Reeder, 1967; Chace et al. 1969; Harding, 1969). Oranges and grapefruit are so sensitive to CO_2 that injuries to the peel and flesh can be seen even at low concentrations, and the fruit has a bitter aftertaste and unpleasant smell (Tomkins, 1938).

Despite this, keeping the fruit for a short time before storage in an atmosphere with a high CO_2 content helps maintain quality. CO_2 slows down the respiratory metabolism and senescence processes of the fruit. The inhibitory action of CO_2 on the biosynthesis of ethylene, pectin hydrolysis and chlorophyll degradation is well known. In some cases, citrus fruits have tolerated massive CO_2 treatment.

Studies by Pratella et al. (1974) found that 'Washington navel' oranges showed no damage with CO_2 concentrations at 7-20° C for 10 days, and neither did lemons or late oranges. The significant slowing down of senescence processes which results from CO_2 treatment makes the technique of interest, from both a commercial and an experimental point of view in the citrus fruit sector postharvest. The lemons showed such good tolerance that no damage was seen after 9 days of treatment with 5-25% CO_2 (Pratella and Tonini, 1984; Cohen et al. 1983).

4. Degreening

Degreening with ethylene is a fairly widespread practice when the flesh of some varieties and species of citrus fruit has acquired excellent organoleptic characteristics but the peel is not yet of a colour suitable for marketing. This is particularly evident in early-ripening fruit and when climatic conditions delay the destruction of chlorophyll.

The process of degreening can be carried out in two different ways:
1. The method of Cohen (1981) which is based on the exposure of lemon and satsuma fruits to a continuous flow of a 5 ppm ethylene concentration at 25° C for 12 h a day, followed by a 12 h interval; the treatment is then repeated at 12 h cycles for 3 days with 90% RH and a 0.6 change in air volume per hour.
2. The method of Kitigawa et al. (1977) in which 'Satsuma mandarin' fruits are exposed to 500 ppm ethylene at 25° C and 90% RH for 15 h.

Naturally, the most suitable method depends on the species and ripening stage of the fruit, keeping in mind the risks involved in the use of ethylene. However, the practice is justified by the good quality of the results and by the economic advantage derived from a reduction in operational costs (Arras and Chessa, 1986).

5. Packinghouse Procedures

Productivity can be increased by introducing efficient handling procedures and by using modern technology to mechanize and computerize processing stages (Beni and Colorio, 1996). To ensure standard quality, morpho-qualitative checks (on the basis of EU regulation no. 929/89 of 10/4/89) should be carried out in the central packinghouse, which would encourage associate producers to keep the quality high.

5.1. Delivery and Sorting

Different machines can be used to feed the line: semiautomatic conveyors or else roller feeders which mechanically invert the boxes of fruit. After harvesting, the fruits can be delivered to the central packinghouse in pallet-boxes or bins which can be mechanically introduced into the sorting line by bin dumps (turnobins) carried by a forklift. The produce is sorted by hand on conveyor belts. Electronic colour sorting, though it cannot completely replace hand sorting, is useful since it identifies rotten or damaged products and separates fruits according to stage of maturity (Beni and Colorio, 1996).

5.2. Washing, Waxing and Antifungal Treatment

The fruits are washed in brushing-washing machines with detergents. Waxing consists of applying resins which reduce water loss besides conferring a shine to the fruits, improving their appearance. Waxing is often done together with antifungal treatment using TBZ or IMZ, the latter proving more efficacious (Arras, 1982; Beni and Colorio, 1996).

5.3. Sizing and Packing

The latest technology consists of electronic systems which sort the fruits by diameter, weight and colour. The mechanical sizers used in central packinghouses (cup belts, expanding sorting belts) have proved to be sufficiently accurate and reliable in relation to their technical features, if regularly serviced.

However, electronic sizers by weight and diameter are more efficient than mechanical sizers and can also separate the fruits by species. More use could therefore be made of machines, with a consequent increase in productivity. Packing can be done either manually or by mechanical means if cardboard boxes are used and the fruits are mostly similar in size.

Other types of packaging (wood, plastic) are not suitable for automatic systems. These materials can be used to package fruits for distribution to wholesalers, while small containers like nets are more practical for retailers (Beni and Colorio, 1996).

6. Diseases of Citrus Fruits

Postharvest microbiological and physiological diseases of citrus fruits are influenced by genetic factors, climate (temperature, thermal variations between day and night, rainfall, relative humidity, duration and force of winds, degree of luminosity) and agronomic features (fertilizer, irrigation, antiparasitic and hormonal treatments, age of the tree, rootstock, pruning, productivity, position of the fruits on the tree, time and methods of harvesting, etc). Other equally important factors include the maturation stage of the fruits, processing methods (sorting, washing, waxing, fungicide treatments, packing, etc) and the parameters adopted during storage (temperature, relative humidity, air exchange and ventilation). All the above play a part in determining the quality of the fruits since they affect the incidence of disease and consequently the suitability of the product for storage.

6.1. Microbiological Diseases

Most losses of fruit and vegetable products are due to microbiological infections or diseases, which can reach high proportions during the postharvest period depending on species, storage conditions and duration, and marketing. As regards citrus fruits, losses are around 15-25% of production on average and 80-90% can be attributed to attacks by pathogenic fungi. In economic terms, the damage is even greater when packing and transportation costs are taken into account, in addition to loss of prestige on domestic and international markets. An incidence of 0.5-1% of rotten fruits depreciates the whole consignment and also the brand image and consequently the company concerned.

Even in those cases in which adequate storage facilities with the most sophisticated technology are available, limitation of postharvest fungal pathogen infections is carried out mainly by the application of plant protection products whose active principles are usually TBZ and IMZ which have become less effective owing to the formation of resistant strains (Dave et al. 1990).

The highest incidence of microbiological disease appearing in citrus fruits postharvest is caused by wound-attacking parasites which enter the fruits though cuts made during harvesting, processing and storing, sometimes penetrating the stem-end or injured skin tissue.

6.1.1. Green and Blue Mould Rot

The two fungi *Penicillium digitatum* (Colour Plate 11.1) and *Penicillium italicum* (Colour Plate 11.2) are the agents which cause the most frequent diseases of citrus fruits. Infection occurs through lesions on the flavedo or diseased skin tissue. External symptoms are well defined soft areas slightly darker than the healthy tissue. A band of white mycelium forms around the lesion, later turning blue or green when spore germination takes place. This stage is critical because infection can be passed to the other fruits by the many very small powdery-type spores which develop even at temperatures of around zero.

The green rot can be distinguished from the blue variety not only by colour but also because of the broad white band surrounding the sporulated area. Rapid and total decay of the tissue of infected fruits occurs, especially in a warm, humid environment. The disease is controlled by carefully sorting the fruits and discarding those which are injured and thus open to infection and ensuring that storage facilities are kept as free as possible of microorganisms.

6.1.2. Grey Mould

The agent of grey mould (*Botrytis cinerea* Pers.) is a polyphagous mycete whose spores are found in storage facilities and on the fruits themselves, especially if they have been grown in damp or wet areas. The disease develops rapidly even at low temperatures (3-6° C). Mandarins are the most susceptible to this type of infection which usually enters through lesions or the stem-end. Symptoms seen on mandarins, lemons (Colour Plate 11.3) and grapefruits are brown or bluish-brown spots characterized by the presence of a halo. As the disease progresses, the spots turn a dark brown or blackish colour and become enlarged, either round or lobed in shape. If the fruit is infected through the rosette, a spot often forms opposite the primary one. The fruits sometimes show symptoms similar to those of brown rot but without the characteristic rancid smell of decay of this disease.

In wet years, the fruits may be infected on the tree and the disease become apparent during storage. This causes serious damage since grey mould is an optimal breeding ground for other saprophytic mycetes like *P. digitatum* and *P. italicum* which inevitably sporulate, thus contaminating the cold room. The mould is controlled by not harvesting in wet weather, keeping the fruits in the packinghouse for 1-2 days and carefully sorting them before storage.

6.1.3. Black Rot of the Carpellary Axis

The disease can be caused by *Alternaria citri* (Ell. & Pierce) or *A. alternata* (Fr.) Keissler. The first is more common and usually infects oranges and mandarins while the second attacks mainly lemons (Colour Plate 11.4). The symptoms are much the same and infection generally occurs in the grove, the spores invading the carpellary axis which turns first brown then a blackish colour because of sporulation of the pathogen. The fungus spreads to the segment membranes and then symptoms appear externally near the rosette. Since external symptoms are not immediately apparent, infected fruits are stored with healthy products, promoting the growth of penicillia, as in the case of grey mould. The disease is particularly prevalent at the end of the fruits' growth cycle if the growing season has been wet. No effective fungicide is known and the disease is controlled by the methods used for grey mould.

6.1.4. Sour Rot

Sour rot disease is caused by *Geotrichum candidum* (Link) and manifests as small yellow spots with dark streaks. The peel first shrivels and then collapses, giving off a sour, putrid smell and emitting a slimy liquid full of spores which passes to the adjacent fruits, spreading the disease. Sour rot is not very common as the pathogen lives in the soil and the spores are not easily carried by the wind; however, the infected fruits, although few in number, can cause a great deal of damage through secondary contamination as they provide an excellent breeding ground for saprophyte fungi. The disease is found particularly in lemon fruits harvested in April (Colour Plate 11.5) in warm, damp areas (20-28° C) and left in boxes on the ground, which are contaminated by contact with the soil. No effective treatment exists. Control measures include preventing injury to the fruit, and disinfecting the packing cases, processing machinery and store rooms.

6.1.5. Brown Rot

Brown rot is caused by different species of the genus *Phytophthora* (*P. citrophthora*, *P. parasitica*, *P. syringae*, etc.). The symptoms are almost identical in each case. Infection usually occurs in the grove, particularly affecting lemons (Colour Plate 11.6). The disease appears as roundish patches of discolouration which turn dark brown as the rot extends, giving off a characteristic pungent, rancid smell. The fruit remains firm at first but then softens. The rot is not always evident at the sorting stage so infected fruits are sometimes stored or packed among healthy fruits with the consequences already mentioned with regard to sour rot. To control this rot it is sufficient to wash the fruits in hot water for 2 min at 46° C or 1 min at 49° C or else apply fungicides in hot water which strengthens their inhibitory effect.

Some diseases, like *Sclerotinia sclerotiorum* (Lib.) Mass. which develops a cottony white mycelium or *Phomopsis citri* (Faw.), *Colletotrichum gloeosporioides* (Penz.), and *Diplodia nataliensis* (Pole-Evans) which cause stem-end rot in fruits, are not often found in citrus fruits grown in Italy whereas they cause serious problems in Florida and Israel. Other rots which appear rarely are *Trichoderma viride* (Pers & Fr.), *Rhizopus* spp., *Dothiorella ribis* (Fuck.) Sacc., and *Fusarium* spp.

7. Control of Fungal Infection

7.1. Use of Postharvest Fungicides

Control at present is generally by means of chemical compounds, mostly containing the active principles IMZ, TBZ and SOPP. Prolonged use of TBZ and afterwards IMZ resulted in resistant strains, with consequent reduced efficacy of the active principle. Ortho-phenylphenate can be used alone and in the compound without sodium (OPP), mixed with wax together with TBZ in order to strengthen the fungitoxic effect and lessen the likelihood of the selection of tolerant strains (Arras and Chessa, 1982; Dave et al. 1990). The fungistatic and/or fungicidal action of these plant protection products is directed at the conidia of the mycetes on the fruit's surface. If the fruit has been infected on the tree and the mycelium is already inside, the treatment has little effect (Table 2).

Table 2. Fungitoxic activity of fungicides against postharvest citrus fruit parasitic mycetes when conidia are present on the epicarp

FUNGI	IMZ	TIABEN-DAZOL	SOPP	TIABENDAZOL + SOPP
Penicillium italicum	3*	2	1	3
Penicillium digitatum	3	2	1	3
Botrytis cinerea	2	2	1	3
Alternaria citri	1	1	1	2
Geotrichum candidum	2	2	1	2
Phytophthora spp.	1	1	1	2
Phomopsis citri	2	2	1	3

*3: excellent fungitoxic activity
2: good
1: unsatisfactory

7.2. Biological Control

Public awareness of the risks of pesticides to human health and the ecosystem in recent years has led to increased research in biological control. In particular, fungicides used to control postharvest pathogens must be of low toxicity because of the very short interval between produce treatment and marketing. It is therefore necessary to find alternatives that can be integrated with or even replace the more widely employed synthetic chemical agents.

Plants have their own natural defense mechanisms at the pre- and post-infectional stages. Microbial antagonism also plays a crucial role in control of diseases.

Pre-infectional: The static resistance due to the constitutional characteristics of the plant such as physical barriers and naturally occurring preformed compounds.

Post-infectional: The dynamic resistance of the plant under abiotic (chemical, physical and thermial treatments) or biotic (host-pathogen interaction) stresses which may determine the induction of necrosis, lignification and biosynthesis of toxic compounds (phytoalexins).

Microbial antagonism: Plays an important role in naturally occurring control of many plant diseases affecting decay development (antagonist-pathogen interaction) either in the field or during postharvest.

The practice of biological control demands an understanding of the local ecosystem and the action mechanisms of the inhibitors. Recent Scanning Electron Microscope (SEM) observations have shown that the presence of large numbers of microbial cells on the leaves and fruits is not accidental but has a specific purpose. The surface of the leaves and fruits provides a habitat for numerous microbial; species adhering so closely to the surface (Colour Plate 11.7) that they form a biological membrane (Spurr, 1994). These membranes contain polysaccharides (95%) which are present to stop the epiphytic microflora from drying up, and to protect them from UV radiation, mechanical injury, viruses, etc.

This being the case, microbial antagonism plays an important role in the natural control of numerous pathogens both in the field and postharvest (Wilson et al. 1993). The isolation and use of antagonist microorganisms is proving to be a reliable method of natural control of pathogens (Spurr, 1994; Arras et al. 1998). The identification of microorganisms with a high level of antagonists capable of both inhibiting disease and stimulating the fruit to the biosynthesis of phytoalexins at the site of the wounds, where pathogens most frequently enter, would be of particular scientific interest (Arras, 1996). Several antagonistic microorganisms with good inhibitory action have been isolated (*Pichia guilliermondii*, *Candida oleophila* and *Rhodotorula glutinis*). During postharvest handling the fruits sustain various kinds of injuries which become points of entry for postharvest pathogens (*P. digitatum*, *P. italicum*, *Botrytis cinerea*, etc). Moreover, the washing and brushing stage removes pathogen-inhibiting microorganisms. Fungicide treatment therefore becomes essential to prevent rot; the alternative would be to colonize the fruits with efficacious antagonists in order to rebuild a set of choice microflora better able to prevent disease.

7.3. Physical Control

Recently there has been renewed interest in thermal treatment (hot water at 50° C for 2-3 min) applied alone or combined with low concentrations of plant protection products (in a dose about 10 times lower than the one usually employed). This treatment is still in the experimental stage but has also produced satisfactory results with regard to costs. Disease control is realized due to the fungitoxic action of the hot water and closure of the entry points for the pathogen in the wounds by means of changing and redistributing the waxes on the fruit's surface (Schirra, 1997). An interesting aspect is the integration between biological control, hot water treatment (D'Hallewin et al. 1999) and possibly low concentrations (100 ppm) of TBZ (Arras et al. 1999).

7.4. Antimicrobial Activity of Some Natural Compounds

Several studies have investigated antifungal properties of essential oils, particularly those distilled from plants of the genera *Thymus and Origanum* and their fractions (Brasseur, 1983; Arras and Grella, 1992), providing important preliminary data for their eventual use in the field.

In particular, *Thymus capitatus* (L.) has biotoxic activity against *P. italicum, P. digitatum, Botrytis cinerea, Alternaria citri, Phytophthora tracheiphila*, etc. and two strains of *Pseudomonas syringae* isolated from orange and grapefruit fruits (Arras and Picci, 1984). The minimal inhibitory concentration (MIC) was between 0.2 and 0.3 g/l for fungi and 0.4 and 0.6 g/l for the bacteria tested (Arras, 1988).

However, the highly effective fungicidal activity observed in vitro was not found in vivo on account of the volatile nature of the constituents of the oil. Recent research has overcome this problem by combining physical and chemical methods in the treatment of citrus fruits (Arras et al. 1994), in particular by volatizing thyme oil in subatmospheric conditions.

7.5. Effect of Acetaldehyde against Fungal Pathogens

The inhibitory effect of acetaldehyde against pathogenic fungi has been known for some time. Prasad (1975) found that 10% (v/v) acetaldehyde vapour concentration at 21° C for 10 h completely inhibited *P. digitatum* and *P. italicum* spore germination in vitro. Subsequent work carried out in vivo on grapes showed that ≥ 0.5% acetaldehyde application significantly reduced rots caused by the following fungi: *Botrytis cinerea, Rhizopus stolonifer, Aspergillus niger, Alternaria alternata* (Avissar and Pesis, 1991). Acetaldehyde produced an increase in CO_2 production and a reduction in ethylene in grapes (Pesis and Marinansky, 1992).

Recent studies in vitro aimed at determining the effect of acetaldehyde on pathogens have shown that this probably destroys the cell membrane of *Rhizopus stolonifer* and *Botrytis cinerea* fungi. A loss of electrolytes, reducing sugars and aminoacids, was observed in these fungi treated with the volatile compound. Spore germination, germ tube elongation and spore formation was inhibited (Avissar et al. 1990).

7.6. Effect of Ultraviolet Rays

Among the methods used to elicit the defense mechanisms of citrus fruits, UV rays (254 nm) play an important role since they induce the production of phytoalexins (scoparone) even without pathogen infection (Rodov et al. 1994). The amount of phytoalexin produced depends, within certain limits, on the UV dose. However, the UV ray method has the disadvantage of causing a deterioration in the quality of the fruit. Research is now directed at improving the technique to obtain the benefits of the positive effects of UV rays and reduce the undesirable consequences.

8. Chilling Injury

8.1. Introduction

Citrus fruits are susceptible to a series of physiological disorders of the epicarp and endocarp when they are stored at temperatures between 0° and 15° C for longer than 3-4 weeks. The degree of sensitivity to chilling injury varies according to species and cultivar (section 2.1).

The etiology of the disorders is still not clearly understood; however, a likely explanation is the cytological effect of chilling stress, causing an increase in cell and sub-cell membrane permeability until the membranes split, with loss of electrolytes and metabolic distress due to changes in some enzymatic processes. The altered biochemistry gives rise to an accumulation of intermediate metabolites including ethanol and acetaldehyde. Over a certain concentration these become toxic and damage the cell structure, first reversibly, then irreversibly (Lyons, 1973; Lyons and Breidenbach, 1987). Chilling injury is thus correlated with the duration of the undesirable temperature, which would explain the reduced incidence of these injuries in fruits subjected to temperature cycles (see section 2.1).

8.2. Symptoms of Chilling Injury

This is one of the most common injuries of the peel and the one which most limits the length of time the citrus fruits can be stored. It disfigures the appearance of the fruit and therefore its market value. Symptoms of cold damage can appear after about 3-4 weeks of cold storage, varying according to species and stage of the disease, as follows:

- Cold pitting, pox manifests with round or irregular spots, usually sunken, light brown becoming darker, of a size varying between a few mm to 1-3 cm (Colour Plate 11.8). An example is shown of histological and cytological changes produced by cold pitting in an orange fruit (Colour Plate 11.9). It is most frequently seen in oranges and grapefruits but often appears also on other citrus fruit species.
- Browning of the oil glands is most commonly found in oranges and grapefruits. The area between the oil glands at first remains healthy but later the disorder covers a large part of the fruit, producing an unpleasant smell.
- Blotched is the last stage of cold pitting, with the same symptoms but affecting large areas of the fruit. The blotch outlines are well defined and coloured from dark brown to black.
- Scald, brown staining is characterized by large, poorly-defined areas, not sunken; at first the tissues are of normal consistency, then they soften and collapse. The colour is at first pale yellow, then light brown. The injury is sometimes associated with a watery breakdown and occurs most frequently in mandarins and grapefruits.
- Albedo browning appears as darkening of the white part of the peel (albedo) which shows through to the fruit's exterior as light brown and opaque. It is mainly found in lemon.

- Red blotch is limited almost exclusively to lemons harvested green during the colour-changing stage of maturation, in damp and cold areas. Symptoms are many tiny reddish specks on the peel, scattered at first, then in clusters, which tend to join up during cold storage, giving the fruit a darker colour. Most of the fruit becomes affected, particularly in cold rooms with insufficient changes of air or poorly ventilated.

Some diseases can appear in citrus fruits both on the tree and during cold storage. The most common are the following:
- *Oleocellosis (rind oil spot):* This disorder, manifesting as darkening of the interglandular tissue, is found frequently in oranges and lemons and is caused by the phytotoxic action of oils extruding from the oil glands after the fruit has been cut or bruised. The disease can be controlled by not harvesting on wet days and delaying the processing of the fruits for at least 12-24 h to allow the oil utricles to lose some turgidity.
- Membranous stain appears inside lemon fruits with dark spots in the carpellary membrane covering the segments. Sometimes it also develops in the carpellary axis and the albedo (Colour Plate 11.10). The disorder is usually found in fruits refrigerated at temperatures lower than 10° C, and can reach an incidence of 80% in fruits of the cultivar Femminello. However, susceptibility is correlated with the cultivation environment and stage of maturation: late fruits showed less sensitivity (Pratella, 1978).

9. Control of Chilling Injury

As reported in section 8, the disorders can be controlled by selecting the right temperature, relative humidity, air exchange and duration of cold storage according to species and variety, and taking into account the environmental factors mentioned in the introduction. The incidence of chilling injury is closely linked not only to temperature and length of storage but also to the plant variety and fungicide treatments. In this regard, Chessa et al. (1986) compared the sensitivity to cold stress of 4 orange cultivars with some structural components of the epicarp (pectins and minerals) and showed that resistance to disease varies according to when each variety typically reaches maturity. Early-ripening orange cultivars like Washington navel and Tarocco, proved more susceptible to chilling injury than late varieties like Belladonna and Valencia late which could be kept longer in cold storage. The same study found that wax treatments containing thiabendazole (0.12% TBZ) and orthophenylphenate (0.4% OPP) markedly reduced damage in 4 orange cultivars.

Resistance to chilling injury can be improved by the induction or use of compounds produced by microorganisms present in the fruit. Some authors (Nordby and McDonald, 1990) demonstrated a positive correlation between the squalene present in the grapefruit's epicuticular layer, and a reduction in chilling injury. The biosynthesis of this compound can be aided by 7 days' heat treatment at 15° C (McDonald et al. 1991).

Since chilling injuries are reversible if the temperature is raised in time (Lyons, 1973), several studies have been conducted on the effects of different temperature cycles. The results have generally been positive with regard to 'Marsh' and 'Ruby Red' grapefruits (Hatton and Cubbedge, 1982) at different stages of maturation (21 days at 1° C + 7 days at 10°, 16°, 21° C), 'Villa Franca' lemons (Cohen et al. 1983) harvested before maturation and stored for 6 months (21 days at 2° C + 7 days at 13° C) and 'Thompson navel' oranges (Arras and Usai, 1992) (6 days at 2° C + 1 day at 14° C). On the other hand, mixed results were obtained with 'Murcott' mandarins (Arras and Usai, 1991). This type of refrigeration is already in use in commercial storage of citrus fruits in Israel, and has the added advantage of reducing energy consumption by about 10%.

The soil and climate have a considerable influence on the quality of the fruits and also the fruits' susceptibility to disorders and diseases (see section 2.1). Trials conducted on 'Marsh seedless' grapefruits (Chessa et al. 1982) showed that maturation stage affected chilling injury which was infrequent between the end of January and the beginning of March but reached a high level in May. A significant correlation has been detected between cold damage, time of maturation, acidity, soluble solids, maturation index and ascorbic acid.

In conclusion, optimal conditions for storage are a compromise between the need to inhibit the development of fungi by keeping the temperature very low, and the need to prevent cold damage by maintaining higher temperatures during cold storage. At the same time, the metabolic characteristics of the fruits must be taken into account.

References

Agabbio, M. 1986. Scelte varietali e agrotecniche nell'ampliamento del calendario di maturazione. *In:* Proceedings of the 'Il Recente contributo della ricerca allo sviluppo dell'agrumicoltura italiana', Cagliari, 29 April-3 May 1986, pp. 327-346.
Agabbio, M., Schirra, M., and I. Chessa. 1982a. Test morfoqualitativi e chimici applicabili ai frutti di agrumi frigoconservati. Atti *'Attivit' Scientifica e Tecnica'* dell'Istituto Arboricoltura Mediterranea del Consiglio Nazionale delle Ricerche. p. 107-123.
Agabbio, M., Chessa, I., Schirra, M. and G. Arras. 1982b. Frigoconservazione degli agrumi. Prime esperienze in Sardegna sulla conservazione delle cultivar di arancio 'Washington navel', 'Tarocco' e 'Valencia late'. 'Studi Sassaresi'. Sez. III, Ann. Facoltá di Agraria Universitá di Sassari, 29: 3-24.
Anelli, G. and F. Mencarelli. 1990. Conservazione degli ortofrutticoli-Tecnologie e aspetti fisiologico-qualitativi. REDA, Roma, pp. 119-127.
Arras, G. 1982. Controllo dei *Penicillium* spp. sui frutti di agrumi dopo la raccolta con l'impiego di prodotti a diverso meccanismo di azione. Atti *'Attivit' Scientifica e Tecnica'* dell'Istituto Arboricoltura Mediterranea del Consiglio Nazionale delle Ricerche. p. 221-229.
Arras, G. 1988. Influenza dell'ambiente pedoclimatico sulla suscettibilitá dell'arancio 'Tarocco' e 'Washington navel' al *Penicillium italicum* ed al *P. digitatum* dopo la raccolta. 'Notiziario Orto floro frutticoltura Italiana' **3**: 95-101.

Arras, G. 1996. Mode of action of an isolate of *Candida famata* in biological control of *Penicillium digitatum* in orange fruits. *Postharvest Biology and Technology* **8**(3): 191-198.
Arras, G. and I. Chessa. 1982. Prove su alcuni trattamenti fungicidi antipenicillium postraccolta sui frutti degli agrumi. Atti 'Giornate Fitopatologiche' 1982, **2**: 253-260.
Arras, G. and V. Picci. 1984. Attivitá fungistatica di alcuni oli essenziali nei confronti dei principali agenti di alterazione post-raccolta dei frutti di agrumi. 'Rivista di Ortoflorofrutticoltura Italiana' **5**: 361-366.
Arras, G. and I. Chessa. 1986. Influenza dell'etilene e dell'ossigeno sullo sverdimento dei limoni 'Eureka' in relazione all'epoca di raccolta. In: Proceedings of the 'Il recente contributo della ricerca allo sviluppo dell'agrumicoltura italiana', Cagliari, 29 April-3 May 1986, p. 347-356.
Arras, G. and M. Schirra. 1988. Modificazioni fisiologiche e biochimiche delle arance 'Valencia late' conservate in differenti condizioni termiche. *Il Freddo* **42**(2): 211-216.
Arras, G. and M. Usai. 1991. Response of 'Murcott' mandarins to storage temperature. *Advances in Horticultural Science* **5**(3): 99-103.
Arras, G. and G.E. Grella. 1992. Wild thyme, *Thymus capitatus* (L.), essential oil seasonal changes and antimycotic activity. *Journal of Horticultural Science* **67**(2): 197-202.
Arras, G. and M. Usai. 1992. Reduction of chilling injury by intermittent warming during cold storage of 'Thompson Navel' oranges. *Agricoltura Mediterranea* **122**: 90-96.
Arras, G. and V. De Cicco. 1995. 'Washington navel' and 'Valencia late' orange fruit storability as regards some metabolic changes and post-harvest diseases. In: Proceedings of The Sixth Int. Symp. Cost 94 The Post-Harvest Treatment of Fruit and Vegetables Oosterbeek, The Netherlands, 19-22 October 1994, pp. 201-208.
Arras, G., Agabbio, M. and I. Chessa. 1984. Studi sui danni da raffreddamento dell'arancio 'Valencia late' in relazione allo stadio di maturazione dei frutti e alle temperature di conservazione. 'Studi Sassaresi'. Sez. III, Ann. Facoltá di Agraria **31**: 15-22.
Arras, G., Chessa, I. and M. Schirra. 1985. Studi sulla suscettibilitá alle fisiopatie da raffreddamento dell'arancio 'Washington navel' in relazione all'ambiente di produzione. *Studi Sassaresi* **32**: 33-44.
Arras, G., Piga, A., and G. 1994. Effectiveness of *Thymus capitatus* aerosol application at subatmospheric pressure for postharvest control of green mold. In: Proceedings of the Int. Symp. on Postharvest Treatment of Hortultural Crops, Kecskemét (Ungheria), 30/8-3/9 1993. *Acta Horticulturae* **368** (1): 382-386.
Arras, G., De Cicco, V., Arru, S. and G. Lima. 1998. Biocontrol by yeasts of blue mould of citrus fruits and mode of action of an isolate of *Pichia guilliermondii*. *The Journal of Horticultural Science* **73**: 413-418.
Arras, G., Dessi, R., Sanna, P. and S. Arru. 1999. Inhibitory activity of yeasts isolated from fig fruits against *Penicillim digitatum*. In: Proceedings of the Int. Symp. Preharvest Postharvest Effects on Fruits and Vegetables, Warsaw, 3-7 August 1997. *Acta Horticulturae* **485**: 37-46.
Avissar, I. and E. Pesis. 1991. The control of postharvest decay in table grapes using acetaldehyde vapours. *Ann. Appl. Biol.* **118**: 229-237.
Avissar, I., Droby, S., and E. Pesis. 1990. Characterization of acetaldehyde effects on *Rhizopus stolonifer* and *Botrytis cinerea*. *Ann. Appl. Biol.* **116**: 213-220.
Baldini, E. and F. Scaramuzzi. 1980. Gli agrumi. ED Reda, Roma, pp. 251-256 and p. 273.
Beni, C. and G. Colorio. 1996. Tecnologia del condizionamento per gli ortofrutticoli freschi. Stampa e servizi - Roma.
Brasseur, T. 1983. E'tudes botaniques, phytochimiques et pharmacologiques consacrées au thym. *J. Pharm. Belg.* **38**(5): 261-272.
Chace, W.G., Davis, P.L. and J.J. Smoot. 1969. Response of citrus fruit to controlled atmosphere storage. In: Proceeding of the XII Congress of Refrig., Madrid **3**, pp. 383-391.
Chessa, I. and G. Arras. 1986. Influenza del portinnesto sulla conservabilitá della cultivar di arancio 'Washingtong navel' sel. Frost. Notiziario de ortofrutticoltura italiana 5.
Chessa, I., Arras, G. and M. Agabbio. 1982. Osservazioni sulla resistenza alle fisiopatie da raffreddamento del pompelmo 'Marsh seedless' sel. Frost in relazione all'epoca di raccolta. Atti I.A.M. **1**: 193-204.

Chessa, I., Schirra, M. and G. Arras. 1986. Sensibilitá varietale dell'arancio ai danni da freddo in relazione ad alcuni componenti strutturali della buccia. *In:* Proceedings of the Congress Il recente contributo della ricerca allo sviluppo dell'agrumicoltura italiana, Cagliari, 29 aprile-3 maggio 1986, **1**: 407-417.

Cohen, E. 1981. Methods of degreening Satsuma mandarin. *Proc. Int. Soc. Citriculture* **2**: 748-750.

Cohen, E., Shuali, M. and Y. Shalom. 1983. Effect of intermittent warming on the reduction of chilling injury villa franca lemon fruit stored at cold temperature. *J. Hort. Sci.* **58-4**: 593-598.

Dave, B., Sales, M. and M. Walia. 1990. Resistance of different strains of *Penicillium digitatum* to imazalil treatment in California citrus packinghouses. *In: Proc. Florida State Hort. Soc. For.* **102**: 173-179.

Davis, P.L. 1970. Relation of ethanol content of citrus fruits to maturity and to storage conditions. *Proc. Fla. State Hort. Soc.* **83**: 294-298.

Davis, P.L. 1971. Further studies of ethanol and acetaldehyde in juice of citrus fruits during the growing season and during storage. *Proc. Fla. State Hort. Soc.* **84**: 217-222.

Davis, P.l., Hofmann, R.C. and T.T. Hatton (Jr). 1974. Temperature and duration of storage on ethanol content of citrus fruit. *Hort. Sci.* **9**: 376-377.

D'Hallewin, G., Arras, G., Dessi, R., Dettori, A. and M. Schirra. 1999. Citrus green mold control in stored 'Star Ruby' grapefruit by the use of a bio-control yeast under curing conditions. *In:* Proceedings of the Int. Symp. Preharvest Postharvest Effects on Fruits and Vegetables, Warsaw, 3-7 August 1997. *Acta Horticulturae* **485**: 37-461.

Eckert, J.W. 1997. Past and present in the post-harvest treatment of citrus fruits to control decay during storage and marketing. *In:* Proceedings of the Congress Acireale, 13 October, pp. 129-135.

El-Zeftawi, B.M. 1976. Cool storage to improve the quality of Valencia oranges. *Journ. Hort. Sci.* **51**: 411-418.

Gorini, F. 1979. Frigoconservazione dei prodotti ortofrutticoli. Reda, Roma, p. 35-39.

Gorini, F., Rizzolo, A. and A. Polesello. 1989. Risultati delle ricerche sulle tecniche di depurazione dell'etilene. Atti MACFRUT, Cesena, 28 April 1988, pp. 103-126.

Grierson, W. and T.T. Hatton. 1977. Factors involved in storage of Citrus fruits: a new evalutation. *In:* Proceeding of the Int. Soc. Citriculture **1**: 227-231.

Harding, P.R. 1969. Effect of low oxygen and low carbon dioxide combination in controlled atmosphere storage of lemons, grapefruit and oranges. *Plant. Dis. Reptr.* **53**: 585-588.

Hatton, T.T. and W.F. Reeder. 1967. Quality of Persian limes after different packinghouse treatments and storage in various controlled atmospheres. *Proc. Trop. Reg. Amer. Soc. Hort. Sci.* **11**: 23-32.

Hatton, T.T. and R.H. Cubbedge. 1982. Conditioning Florida grapefruit to reduce chilling injury during low-temperature storage. *J. Amer. Soc. Hort. Sci.* **107**: 57-60.

Houck, L.G. 1977. Problems of resistance to citrus fungicides. *Proc. Int. Soc. Citriculture* **1**: 263-269.

Ibraham, R. 1968. The effect of type of soil and time of picking on the storage of Navel oranges. *Agr. Res. Rev.* **46**: 57-71.

Khalifah, R.A. and J.R. Kuykendall. 1965. Effect of maturity, storage temperature, and prestorage treatment on storage quality of Valencia oranges. *Proc. Amer. Soc. Hort. Sci.* **86**: 288-296.

Kitagawa, H., Kawada, K. and T. Tarutani. 1977. Degreening of 'Satsuma' mandarin in Japan. *Proc. Int. Soc. Citriculture* **1**: 219-223.

Kusunose, H. and M. Sawamura. 1980. Ethilene production and respiration of postharvest acid citrus fruits and wase Satsuma mandarin fruit. Agricultural and Biological Chemistry 1980.

Long, J.K. and D. Leggo. 1959. Waxing citrus fruit. *CSIRO Food Preserv. Quart.* **19**: 32-37.

Lutz, J.M. and R.E. Hardenburg. 1968. The commercial storage of fruits, vegetables and florist and nursery stocks. *USDA, Agr. Handbook* **66**: 94.

Lyons, J.M. 1973. Chilling injury in plants. *Annu. Rev. Plant Physiol.* **24**: 445-466.
Lyons, J.M. and R.W. Breindenbach. 1987. Chilling injury. *In*: Postharvest Physiology of Vegetables, J. Weichmann (Ed.) Marcel Dekker, New York, pp. 305-326.
Monzini, A. 1979. Interventi ed operazioni prima e dopo la raccolta ai fini della conservazione dei prodotti ortofrutticoli. *Frutticoltura* III-IV: 37-42.
McDonald, R.E. and H.K. Wutscher. 1974. Rootstocks affect harvest decay of grapefruit. *Hort. Sci.* **9**: 455-456.
McDonald, R.E., Mc Collum, T.G. and H.E. Nordby. 1991. Surface treatments and temperature conditioning affect weight loss and chilling injury on grapefruit. *HortScience* **26**: 698 (Abstr.).
Nordby, H.E. and R.E. McDonald. 1990. Squalene in grapefruit wax as a possible natural protectant against chilling injury. *Lipids* **25**: 807-810.
Norman, E.B. 1977. The role of volatiles in storage of citrus fruit. *Proc. Int. Soc. Citriculture* **1**: 238-242.
Pantastico, E.B., Soule, J. and W. Grierson. 1968. Chilling injury in tropical and subtropical fruit II. Limes and grapefruit. *Proc. Trop. Reg. Amer. Soc. Hort. Sci.* **12**: 171-183.
Pennisi, L. 1973. La caratterizzazione del succo delle arance nel corso della maturazione. *Ann. Ist. Sper. Per l'Agrumicoltura Acireale* **6**: 257-274.
Pesis, E. and R. Marinansky. 1992. Carbon dioxide and ethylene production by harvested grape berries in response to acetaldehyde and ethanol. *J. Amer. Soc. Hort. Sci.* **117**(1): 110-113.
Prasad, K. 1975. Fungitoxicity of acetaldehyde vapour to some major post-harvest pathogens of citrus and subtropical fruits. *Ann. Appl. Biol.* **81**: 79-81.
Pratella, G.C. 1978. Le fisiopatie da raffreddamento degli ortofrutticoli refrigerati. *Not. CRIOF* **8**(3): 1-32.
Pratella, G.C. and G. Tonini. 1984. Recenti acquisizioni sperimentali sulla protezione dei prodotti ortofrutticoli per mezzo dell'anidride carbonica. *Ortofrutticoli freschi* **2**: 175-248.
Pratella, G.C., Tonini, G. and A. De Nardo. 1974. Fisiopatologia dell'impiego dei gas criogenici. *Not CRIOF* **5**(3): 1-20.
Pritchett, D.E. 1962. Changes in Valencia orange composition during marketing. *Calif. Citr.* **48**: 29-30.
Rodov, V., Ben-Yehoshua, S., Fang, D.Q., 'Hallewin, D.G. and T. Castia. 1994. Accumulation of phytoalexins scoparone and scopoletin in citrus fruits subjected to various postharvest treatments. *Acta Horticulturae* **381**: 517-523.
Royo, I. and T.M.J. Perez. 1977. Relationship between the Brix of citrus juices and their soluble chemical components. *Proc. Int. Citriculture* **2**: 791-795.
Schirra, M. 1997. Nuove prospettive nella difesa postraccolta degli agrumi mediante termoterapia. *In:* Proc. of the congress Acireale, 13 October, pp. 67-84.
Smagula, J.M. and W.J. Bramlage. 1977. Acetaldehyde accumulation: Is it a course of physiological deterioration of fruits? *Hort. Sci.* **12**: 200-203.
Spurr, H.V. (Jr). 1994. The microbial ecology of fruit and vegetable surfaces: its relationship to postharvest biocontrol. *In*: Biological Control of Postharvest Diseases—Theory and Practice. C.L. Wilson and M.E Wisniewski (Eds). CRC Press, Boca Raton, Florida, pp. 11-23.
Tomkins, R.G. 1938. The effect of ventilation on the wastage of orange in storage. Gt. Brit. Dept. Sci. Ind. Res. Dept. Food Inv. Board.
Wells, A.A.W. 1962. Effects of storage temperature and humidity on loss of weight by fruit. US Dept. Agr. Market Res. Rpt. 539.
Wilson, C.L., Wisniewski, M.E., Droby, S. and E. Chalutz. 1993. A selection strategy for microbial antagonist to control postharvest diseases of fruits and vegetables. *Scientia Horticulturae* **53**: 183-189.

Postharvest Biological Changes and Technology of Citrus Fruit   257

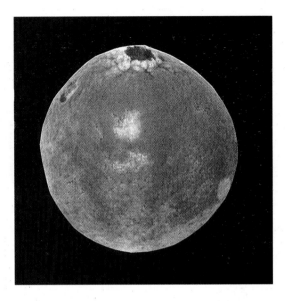

Colour Plate 11.1. Orange fruit affected by green mould rot (*Penicillium digitatum*).

Colour Plate 11.2. Blue mould rot in a grapefruit (*Penicillium italicum*).

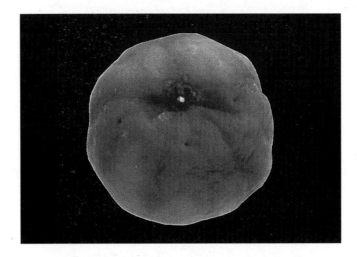

Colour Plate 11.3. Lemon fruit with grey mould (*Botrytis cinerea*).

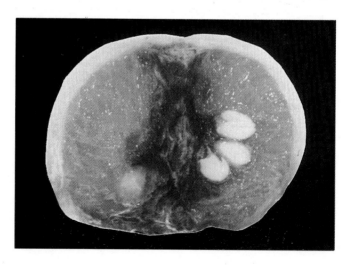

Colour Plate 11.4. Black rot of the carpellary axis in a lemon fruit (*Alternaria alternata*).

Postharvest Biological Changes and Technology of Citrus Fruit **259**

Colour Plate 11.5. Development of sour rot in a lemon fruit (*Geotrichum candidum*).

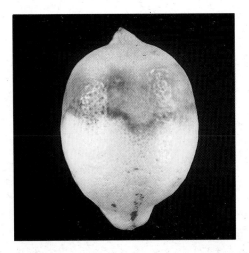

Colour Plate 11.6. Lemon fruit affected by brown rot (*Phytophthora citrophthora*).

Colour Plate 11.7. Epicarp of orange fruit completely colonized by cells of strain A1 *Pichia guilliermondii* (a). A stoma can be seen at the centre (b).

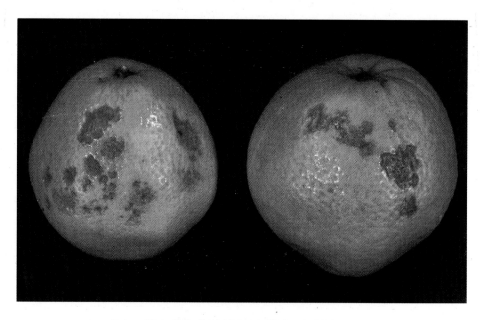

Colour Plate 11.8. Cold pitting in 'Valencia late' orange.

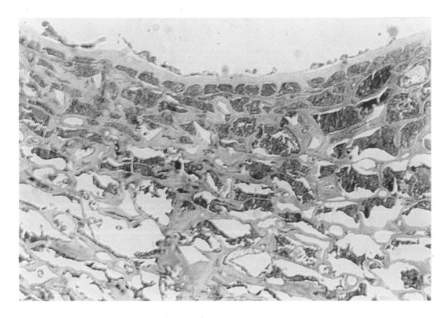

Colour Plate 11.9. 'Valencia late' orange histological and cytological changes from cold pitting.

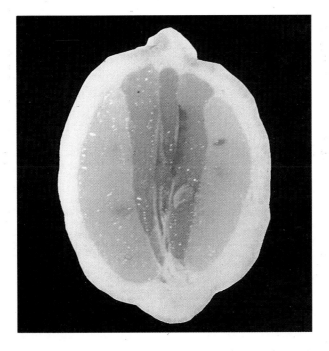

Colour Plate 11.10. Lemon fruit affected by membranous stain.

12

Ethylene Biosynthesis
Role in Ripening and Quality of 'Hayward' Kiwifruit after Harvest, During Storage and Shelf-life

E. Sfakiotakis[1], M.D. Antunes[2], G, Stavroulakis[3] and N. Niklis[4]

[1]*Laboratory of Pomology, School of Agriculture, Aristotle University, Thessaloniki 54006, Greece.*
[2]*Universdade do Algarve, U.C.T.A., 8000 Faro, Portugal.*
[3]*Higher Technical Educational Center, Heraklion, Greece.*
[4]*Ministry of Agriculture, Thermi, Greece*

Abbreviations

ACC   1-aminocyclopropene-1-carboxylic acid
CA   controlled atmosphere
IEC   internal ethylene concentration
kgf   kilo gram force
SSC   soluble solids content
ULO   ultra-low-oxygen

1. Introduction

The kiwifruit known as the Chinese gooseberry belongs to the genus *Actinidia* of the Actinidiacea family. Previously known as *Actinidia chinensis*, it is now classified as *Actinidia deliciosa* (Liang and Ferguson, 1984). Athough the genus contains more than 60 species which produce edible fruits, *Actinidia deliciosa* is the only member which has been developed commercially on a large-scale (Ferguson, 1990). The cultivar 'Hayward' is grown commercially and today comprises 98% of all commercial plantings (Sale, 1983).

Botanically the kiwifruit is a berry with numerous locules filled with many small black seeds (Ferguson, 1984). The thin brown skin including a periderm and hypodermal cells on their surface have small and large hairs (trichomes). No stomates are present, but other openings provide adequate gas exchange. The fruit has three distinctly different tissues: the outer pericarp, the inner pericarp and the white core or collumela (Beevers and Horkirk, 1990). Both the outer and the inner pericarp cells have plastids containing chlorophyll which gives the characteristic green color. Embedded in the inner pericarp is a ring of 20-40 seed locules (carpels) running longitudinally and contain a mucilaginous matrix which supports the small black seeds.

A number of maturity measures have been evaluated for maturity indices. Among them the soluble solids content and the flesh firmness showed a positive relationship with storability and marketing or eating quality to be used as maturity indices. The soluble solids content (SSC) has been the most widely used maturity index. New Zealand uses 6.2 per cent SSC and California 6.5 per cent as minimum acceptable maturity index (Harman, 1981). For long-term storage, the fruit should have an SSC level of 7-9 per cent when harvested (Sale, 1983; Sfakiotakis et al. 1991).

Ripening of kiwifruit is associated with changes in respiration, texture, carbohydrate, organic acids, C_2H_4 production and volatile components which give the characteristic aroma.

Kiwifruit are generally considered to be climacteric fruit with a respiratory pattern somewhat atypical (Pratt and Reid, 1974). The respiratory activity of kiwifruit is low, in the same range as that of apples and grapes, and because of this kiwifruit can be stored for a fairly long time. The respiratory rate measured immediately after harvest at 20° C is less than 20 mg CO_2/kg/h (Pratt and Reid, 1974; Antunes, 1999), whereas mature fruit shows increased rates over 40 mg CO_2/kg/h. During storage, respiration rate is low (3 to 7 mg CO_2/kg/h) at 0° C and increases with increase in temperature. It has a temperature coefficient (Q10) near 3.0. Higher rates of respiration were measured at elevated temperatures (30-40° C) in the range of 35-65 mg CO_2/kg/h (Antunes, 1999). The respiration rate is also sensitive to C_2H_4 (Pratt and Reid, 1974; Given, 1993) or to C_3H_6 (Antunes, 1999).

Kiwifruit texture changes dramatically during ripening. During normal ripening, the flesh of mature kiwifruit changes from being hard and acid into soft and sweet. At harvest, the fruit typically has a firmness of 8 kgf which falls to about 0.5-0.8 kgf at eating ripeness (Beevers and Horkirk, 1990). Softening of kiwifruit at 0° C occurs in two or three phases, depending on maturity at harvest (Crisosto et al. 1984; Sfakiotakis et al. 1991). Texture of kiwifruit shows a rapid drop in firmness after harvest until the fruit reaches a firmness of 2.5 kgf. The rate of softening then slows considerably. If the fruits are exposed to C_2H_4, the rate of fruit softening increases. The outer pericarp of the fruit softens more rapidly than the core of the fruit (Antunes, 1999).

Softening in kiwifruit is one of the most significant quality alterations consistently associated with ripening, and firmness can decline by as much as 95%, from about 6-9 kgf at harvest to 0.5 to 0.6 kgf when ripe (Beevers and Horkirk, 1990). High consumer satisfaction occurs when kiwifruit are purchased and eaten ripen or 'ready to eat' (0.5 N-0.6 kgf). When consumers eat the mature but unripe kiwifruit, they are not satisfied with them and may not eat kiwifruit again. This softening occurs slowly during normal cool storage or may be accelerated with C_2H_4.

Kiwifruits soften rapidly after harvest and ripen completely with time in the storage temperature (0° C). Premature softening during storage is a serious and expensive problem for the kiwifruit industry and slowing the rate of flesh softening is a major goal of any long-term storage program.

Fruits are not acceptable for export if firmness is below 1-1.5 kgf. During storage, fruit softening is accelerated in the presence of C_2H_4 and removing C_2H_4 with scrubbers is the main concern for long storage. Rate of softening in the storage room is slowed but not stopped at low temperatures. It is reduced with a controlled atmosphere (Arpaia et al. 1985, 1986; McDonald and Harman, 1982) or ultra low oxygen (ULO) storage (Thomai and Sfakiotakis, 1996, 1997; Antunes and Sfakiotakis, 1997a). However, the controlled atmosphere (CA) or ULO storage fruit does not ripen even during the shelf-life (Antunes and Sfakiotakis, 1997a).

Flesh firmness declines sharply during the first 1 to 2 months in storage and this is attributed to the conversion of starch to sugars.

Early harvest does not help in maintaining flesh firmness during storage and the immature fruit attains an unacceptable flavor when it is ripe (Sfakiotakis et al. 1991). Leaving the fruit on the vine for late harvest is the best method for long storage.

Starch can constitute around 5-7% of the fresh weight of kiwifruit at harvest (MacRae et al. 1989), then begins to decline coupled with an increase in soluble solids content (SSC). Starch breakdown continues after harvest, even in stored fruit at 0° C. By the time the fruit is ready to be eaten, starch is completely degraded (Mac Rae et al. 1989a, b, 1992).

The main organic acids in kiwifruit are citric, quinic and malic acids (Okuse and Ryugo, 1981). During early development, the fruit is highest in quinic acid, with a lesser amount of malic acid. By harvest the fruit is highest in citric acid, followed by quinic, with a lesser amount of malic acid. Acid content declines as fruit ripens (Mac Rae et al. 1989a).

The characteristic aroma of a ripe kiwifruit is due to the combination of 26 volatile compounds, of which C_2H_4 butanoate, hexanal and trans-hexanal are important components of aroma (Young and Paterson, 1983).

During the postharvest life of kiwifruit great economical losses occur (Horpirk and Clark, 1991). The main causes of deterioration are softening, C_2H_4, water loss, physiological disorders (Arpaia et al. 1986; Lallu, 1997) and pathological breakdown (Sommer et al. 1994). Among them fruit softening and the presence of C_2H_4 in the postharvest environment are serious and costly problems for the kiwifruit industry. Pathological breakdown caused by *Botrytis* infections has caused great losses.

Ethylene is found to be associated with ripening and this growth regulator is being considered as 'the ripening hormone' in many fruits (Abeles et al. 1992). Ethylene plays a major role in the ripening of kiwifruit and monitoring C_2H_4 production, as well as C_2H_4 action which is of great importance in maintaining the quality of the fruit.

Several review articles have been published on postharvest physiology (Cheah and Irving, 1997; Arpaia et al. 1994), technology (McDonald, 1990) and biochemistry (Given, 1993) of the kiwifruit. This chapter reviews the results of several studies undertaken to elucidate the mechanism of the regulation of C_2H_4 biosynthesis in connection with ripening and quality control of 'Hayward' kiwifruit in the above mentioned postharvest stages and discuss published work in this area.

2. Internal C_2H_4 concentration of kiwifruit attached to or detached from the vine

Kiwifruit produces very small amounts of C_2H_4 on the vine and there is no climacteric of C_2H_4 production on the tree. Unlike the apple fruit which shows an increase of internal C_2H_4 concentration (IEC) on the tree (Sfakiotakis and Dilley, 1973a), kiwifruit attached to the tree showed no increase of C_2H_4 production during the maturation period (Fig. 1A). This suggests that kiwivine produces a 'ripening inhibitor' that is translocated to the attached fruits and inhibits C_2H_4 production. It has been suggested with other fruits that this factor not only inhibited the accumulation of 1-aminocyclopropene-1-carboxylic acid (ACC) but also suppressed the

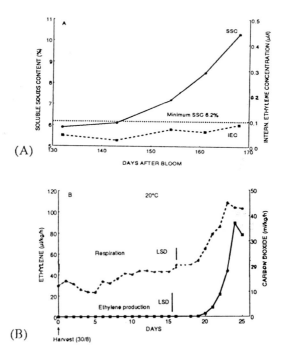

Fig. 1. The effect of harvest (days after bloom) on changes of soluble solids content and internal C_2H_4 concentration of kiwifruit harvested in different stages (A). Internal C_2H_4 concentration determined by the method with needles inserted in fruits (Sfakiotakis and Dilley, 1973a). Climacteric of respiration and C_2H_4 production after harvest kept in a continuous humidifier air stream at 20° C (B) (Antunes, et al., 2000).

development of ACC oxidase (Yang et al. 1986). Therefore, the IEC in kiwifruit cannot be used as a maturity index to determine the harvest period. However, the fruit showed a tendency to soften with an increase of SSC (Fig. 1A) on the tree and this was attributed to the conversion of starch to sugars than to the increase of IEC (MacRae et al. 1989a; Matsumoto et al. 1983).

After harvest, kiwifruit produces negligible amount of C_2H_4 and only after a relatively long shelf-life showed a climacteric pattern of C_2H_4 production and respiration (Pratt and Reid, 1974; Hyodo and Fukazawa, 1985; Yano and Hasegawa, 1993, Arpaia et al. 1994). In our studies fruit harvested in an early stage of maturity (SSC 5.9%) showed a clear climacteric peak of C_2H_4 production and respiration after 20 days shelf-life at 20° C (Fig. 1B). Autogenous autocatalysis of C_2H_4 production started after the fruit reached a threshold level of 0.2 $\mu l/kg/h$ in 19 days. The climacteric of C_2H_4 production followed a typical pattern reaching a peak in 24 days with almost a 20.000 fold increase of the rate of C_2H_4 production. After harvest the respiration measurements showed moderate changes with a 2.5 fold increase in the rate of CO_2 production and this was closely associated with the increase of C_2H_4 production (Fig. 1B).

3. Autocatalysis of C_2H_4 Production in Connection with Ripening

Kiwifruit is very sensitive to exogenous application of C_2H_4 (McDonald and Harman, 1982; Arpaia et al. 1986). Treating kiwifruit with exogenous C_2H_4 induces autocatalysis of C_2H_4 production and ripening. Most of the studies on the effect of C_2H_4 on the ripening of climacteric fruits have been conducted by using different concentrations of exogenous C_2H_4 and following the softening changes as the measure of ripening. The role of C_2H_4 in the ripening of kiwifruit has proved difficult to establish from the literature (Given, 1993). It is unclear whether other physiological processes are equally affected as a direct effect of the C_2H_4 treatment or is it rather the result of the increase of the endogenous C_2H_4. When autocatalysis is studied in climacteric fruits, it is preferable to use the C_2H_4 analog-propylene (C_3H_6) to separate the exogenous gas applied from the endogenous C_2H_4 produced with autocatalysis (Burg and Burg, 1965; McMurchie et al. 1972; Sfakiotakis and Dilley, 1973b). Propylene offers the advantage to study the effect of several parameters (temperature, concentration, time of exposure) on ripening, by measuring either softening or respiration rate without interfering with the effect of the endogenous C_2H_4 produced from autocatalysis.

Treating kiwifruit with C_3H_6 is expected to induce a climacteric of respiration and autocatalysis of C_2H_4 production in a similar manner to what has been found in other fruits like apples (Sfakiotakis and Dilley, 1973b) and avocado (Metzidakis and Sfakiotakis, 1989). Propylene in kiwifruit at 20° C induces C_2H_4 biosynthesis in 75-80 h after exposure, whereas fruit ripening initiates in a few hours without delay (Stavroulakis and Sfakiotakis, 1995). Propylene at ambient temperature (20°C) and low temperature (10°C) induces in kiwifruit a respiration burst (Fig. 2A) like the other climacteric fruit (Antunes, et al., 2000). The fruit is characterized by a burst of C_2H_4 production (Fig. 2B), which follows the respiratory climacteric induced by C_3H_6 at 20°C. The biosynthesis of C_2H_4 was found to be autocatalytic through the methionine and ACC pathway (Stavroulakis and Sfakiotakis, 1993). Autocatalysis involves the stimulation of ACC oxidase, as well as the synthesis of ACC (Antunes, 1999). On the other hand it was found that kiwifruit at low temperatures (10°C) acts as a non-climacteric fruit. It lacks the ability for autocatalysis of C_2H_4 biosynthesis (Stavroulakis, 1991; Antunes, 1999; Antunes et al. 2000).

3.1. Thermoregulation of C_2H_4 Biosynthesis

Ethylene production is highly regulated by temperature in kiwifruit. Stavroulakis (1991) showed that kiwifruit below a critical range of low temperature (>11-14.5° C) lacks the ability of C_2H_4 biosynthesis. The rate limiting factor is the availability of ACC rather than the activity of ACC oxidase. This was confirmed by Antunes et al., (2000), who showed that

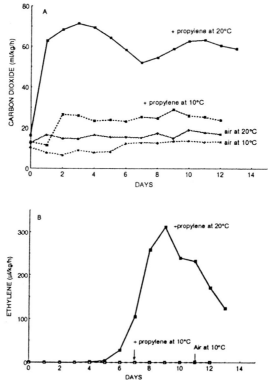

Fig. 2. Respiration (A) and C_2H_4 production (B) of kiwifruit after harvest treated in a continuous humidified air stream with 130 μl/l C_3H_6 and air free of C_3H_6 at 20° and 10° C (Antunes, et al., 2000).

the main reason for the inhibition of the C_3H_6 induced autocatalytic C_2H_4 production in kiwifruit at low temperature was primarily the suppression of the C_3H_6-induced ACC synthase gene expression (Antunes et al. 2000).

The maximum rates of the C_3H_6-induced C_2H_4 occurs in the temperature range from 20° to 35° C (Antunes, and Sfakiotakis, 2000) (Fig. 3A). Below or above this temperature range a reduced rate of C_2H_4 production was found (Antunes and Sfakiotakis, 1997b). Ethylene was drastically reduced below 10° and above 38° C and these changes were correlated with ACC content and the activities of the enzymes ACC synthase and ACC oxidase (Fig. 3B). Northern blot hybridization techniques showed that the main reasons for the inhibition of the C_3H_6-induced autocatalytic C_2H_4 production at low temperature (10° C) were primarily the suppression of the C_3H_6-induced ACC synthase gene expression and the possible-transcriptional modification of ACC oxidase (Antunes, et al., 2000).

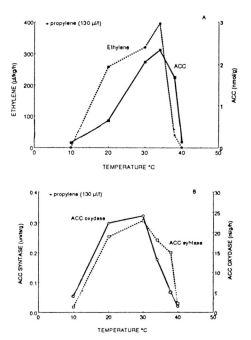

Fig. 3. The effect of temperature on C_2H_4 production and ACC content (A) and activities of ACC synthase and ACC oxidase (B) of 'Hayward' kiwifruit kept in atmosphere containing $130\mu l/l$ C_3H_6. Values were taken four days after the initiation of C_2H_4 autocatalysis. 1 unit/mg = 1pmole ACC/mg protein/2hours (Antunes, 1999).

3.2. Effects of C_3H_6 Concentration and Temperature in Climacteric of Respiration, Ripening and Ethylene Production

3.2.1. Effect on Respiration in Connection with C_2H_4 Production

The application of external C_2H_4 to a climacteric fruit will advance the onset of the climacteric and ripening, this effect being proportional to the concentration of applied C_2H_4 (Tucker and Grierson, 1987).

Kiwifruit treated continuously with C_3H_6 at concentrations 0, 100, 400 and 1000 µl/l at 20° and 10° C for 260 h induced respiration with climacteric peaks (Fig. 4). Respiration at 20° C was increased after the C_3H_6 exposure and reached the peak 45 h after the C_3H_6 exposure. However, 160 h after C_3H_6 exposure the magnitude of the climacteric is relatively independent of the concentration of applied C_3H_6 (Fig. 4A). Treating kiwifruit with different C_3H_6 concentrations (0, 100, 400 and 1000 µl/l) at 10° C advanced ripening similar to non-climacteric fruit (Tucker and Grierson, 1987). Respiration at 10° C was increased after the C_3H_6 exposure and reached the peak 100-120 h after the C_3H_6 exposure, the magnitude of the peak being dependent on the concentration of C_3H_6 (Fig. 4B). A striking difference is noted in the respiratory pattern of fruit held at 20° and 10° C. The magnitude of the respiration of the fruit held at 20° C, 160 h after the C_3H_6 exposure showed to be independent of the C_3H_6 concentration. On the contrary, the magnitude of the respiration of the fruit held at 10° C proved to be dependent on the C_3H_6 concentration.

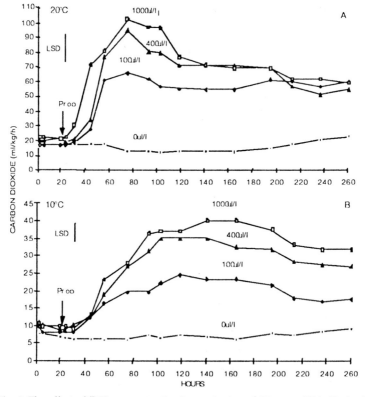

Fig. 4. The effect of C_3H_6 concentration in respiration of 'Hayward' kiwifruit after harvest at 20° C (A) and 10° C in a continuous humidified air stream air + 100 µl/l, air + 400 µl/l and air + 1000 µl/l C_3H_6. LSD at a = 0.05 (Antunes, et al., 2000).

A switching experiment was conducted at low temperatures (12°C) in order to study the dependency of climacteric of respiration on the presence of C_3H_6. Respiration of kiwifruit started to increase significantly in 10 h exposure to C_3H_6 reaching the peak height after 56 h and remained almost constant thereafter (Fig. 5). The removal of C_3H_6 from the air stream caused a reduction in the rate of respiration. On transfer to air, a residual effect persisted for almost one day, then respiration declined. Adding C_3H_6 back again induced an increase of respiration and the removal of C_3H_6 again induced a reduction of respiration with a residual effect for a second time. It was concluded that at low temperature the respiration of the kiwifruit is dependent on the presence of C_3H_6. This response is very similar to what previous workers reported with non-climacteric fruits (Biale, 1964).

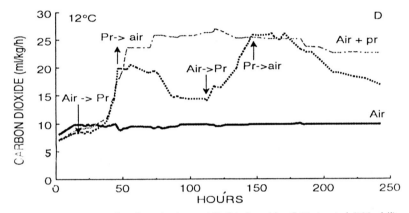

Fig. 5. Time course study of respiration at 12° C induced by C_3H_6-treated (130 µl/l) kiwifruit: a) in continuous humidified air stream, b) in air + C_3H_6, c) treated with C_3H_6 then after 24 h transferred to air, then after 48 h transferred to air + C_3H_6 and finally after 48 h transferred to air (Antunes,1999).

Our data on C_2H_4 production and respiration supports the hypothesis that the kiwifruit stored at room temperatures behaves as a typical climacteric fruit with reference to respiration and C_2H_4 production, while at low temperature like a non-climacteric fruit with reference to the C_2H_4 production. This behavior of the kiwifruit with respect to respiration and C_2H_4 production offers certain advantages in handling operations. After harvest, by keeping the fruit at low temperatures, there is no accumulation of C_2H_4 production in storage rooms. Thus the postharvest life of the fruit is prolonged (Antanes et al., 2000).

3.2.2. Effects on Fruit Softening in Connection with C_2H_4 Production

Kiwifruit treated with propylene at concentrations 0, 5, 10, 50, 100 and 500 µl/l (equivalent to 0, 0.077, 0.380, 0.770 and 3.850 µl/l concentrations of C_2H_4 respectively) for one week at 20° C induced the ripening of the fruit and C_2H_4 production increased with C_3H_6 concentration (Figs. 6A,

B). It was noticed that the ripening induced by C_3H_6 was initiated before the onset of C_2H_4 synthesis (Sfakiotakis et al. 1989). The threshold concentration of C_3H_6 for the induction of C_2H_4 production and ripening was higher than 10 μl/l which was considered to be equivalent to 0.077 μl/l C_2H_4. The concentration of 50 μl/l of C_3H_6 (equivalent of 0.380 μl/l C_2H_4) induced the half-maximum response of fruit softening 6 days after the C_3H_6 exposure at 20° C. With 100 μl/l C_3H_6, the fruit reached the value of 1 kgf in 6 days, while less time (4 days) was required at a higher C_3H_6 concentration (500 μl/l) to reach the eating stage of softening.

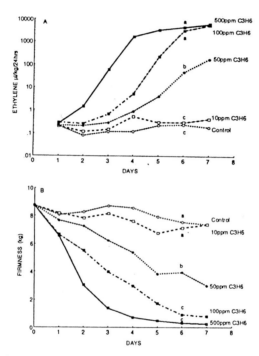

Fig. 6. Effect of C_3H_6 concentration (0, 10, 50, 500 μl/l) on firmness (A) and C_2H_4 production (B).

By storing kiwifruit in chambers with 130 μl/l C_3H_6 at 0°, 10° and 20° C ripening and C_2H_4 production were evaluated during the storage period of 138 days. Propylene at 20° C induced a typical softening of the fruit and autocatalytic C_2H_4 production. Softening of the fruit induced by C_3H_6 was influenced by temperature (Fig. 7A). If we consider the firmness of 1 kgf as the threshold value for consumption, we may estimate the maximum storage life of kiwifruit under 130 μl/l C_3H_6 (equivalent of 1 μl/l C_2H_4) as follows: 4-5 days at 20° C, 18 days at 10° C and 30 days at 0° C (Fig. 7A). A striking difference is noted in the induction of C_2H_4 production by C_3H_6 of fruit held at 20° C and low temperatures (0°, 10° C) (Fig. 7B). Whereas C_3H_6 at 20° C induced almost a 4000-fold increase of C_2H_4

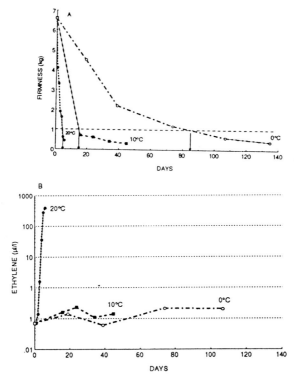

Fig. 7. The effect of C_3H_6 (130μl/l) on firmness (A) and C_2H_4 production (B) of 'Hayward' kiwifruit at 0°, 10° and 20° C.

production in 5 days, the gas was unable to induce autocatalysis of C_2H_4 production at low temperatures, even when the fruit was exposed to C_3H_6 for 138 days.

The above results suggest that, to achieve the best eating quality when artificial ripening is used, it is important to consider the temperature factor and C_2H_4 concentration to establish the ripening protocols.

3.2.3. Effect of Propylene on Ripening at High Temperature

Exposing kiwifruit to ethylene or propylene at unfavorably high temperatures may cause abnormal ripening with adverse effects on fruit quality. Antunes and Sfakiotakis (2000) found that C_3H_6 at 30°-34° C stimulated ripening and C_2H_4 production in an accelerated rate mainly through the induction of ACC synthase and ACC oxidase activities (Figs. 3A, B). However, the fruit failed to ripen normally at 38° C showing a core hard when the flesh was soft, and above 40° C ripening was drastically inhibited. Ethylene production was drastically reduced at 38° C and was almost null at 40° C. This was attributed to the reduced activity of ACC oxidase rather than the reduced activity of ACC synthase. Exposing the fruit to such unusual high temperatures (> 38° C) causes uneven ripening and

3.3. The Effect of Elevated CO_2 and low Oxygen Concentration in Autocatalytic C_2H_4 Production and Ripening

Elevated CO_2 concentrations inhibited the effectiveness of the C_3H_6-induced ripening and autocatalysis of C_2H_4 production (Stavroulakis, 1991). Treatments with high CO_2 (> 15%) drastically delayed and inhibited C_2H_4 production. However, high CO_2 caused moderate reduction in fruit ripening as was evaluated by measuring softening and the increase of soluble solids content.

Low levels of O_2 concentrations suppressed and delayed the C_3H_6-induced fruit ripening and autocatalysis of C_2H_4 biosynthesis (Stavroulakis and Sfakiotakis, 1997). Reduction of oxygen concentration (< 10%) inhibited the effectiveness of C_3H_6 on autocatalytic C_2H_4 production. Exposure of kiwifruit to low oxygen concentration (< 5%) delayed the C_3H_6-induced fruit ripening. Previous studies in other fruits showed that the conversion of ACC to C_2H_4 is O_2 dependent (Yang et al. 1986). Stavroulakis and Sfakiotakis (1997) showed in a similar study that the accumulation of ACC is rather the limiting factor than the activity of ACC oxidase.

4. Induction of Ethylene Biosynthesis by Various Stresses

Stress ethylene is one of the general phenomena observed in plant tissues subjected to various unfavorable conditions (Hyodo, 1981). Various forms of stress in higher plants including temperature extremes, drought, mechanical wounding, hypoxia, and exposure to different gases induce ethylene production (Abeles et al. 1992). Biological stress can result from disease and insect damage. Stress ethylene has been shown to be synthesized via the methionine and ACC pathway (Hyodo, 1981). Besides C_3H_6, we have studied the effect of chilling temperatures in connection with storage period in conventional storage, controlled atmosphere storage (CA) or ultra low oxygen (ULO) storage, wounding and *Botrytis* infections.

4.1. Low Temperature Stress

Low temperature induces stress in horticultural commodities preharvest or postharvest with enhanced ethylene production occuring in chilled pears (Sfakiotakis and Dilley, 1974; Gerasopoulos and Richardson, 1997) or kiwifruit (Hyodo and Fukazawa, 1985). Stress ethylene is produced when plant tissue is returned to warmer non-chilling temperatures. Very often cold treatment is used as a method to synchronize ripening in kiwifruit (Puig et al. 1996).

4.1.1. Chilling Temperatures

Exposing the kiwifruit to chilling temperatures (0°, 5°, 10°, 15° C) for 12 days advanced the onset of C_2H_4 production when transferred to ambient temperatures (20° C) and the rates of ethylene production were correlated with the activities of ACC synthase and ACC oxydase (Fig. 8A) (Antunes, 1999). Fruit chilled at 0° C showed highest rate of ethylene production and higher activities of ACC synthase and ACC oxidase during the shelf-life at 20° C than pretreated at higher temperatures (5°, 10°, 15° C).

4.1.2. Duration of Chilling

Storage of kiwifruit at chilling temperature for five days was not enough to induce autocatalytic C_2H_4 production upon transferring to 20° C (Fig. 8B). Fruits stored for 12 days at 0° C were most-efficient at producing C_2H_4 upon rewarming post-storage with no lag period and gave the highest rate of C_2H_4 production (140 µl/kg/h at the peak). Upon rewarming, C_2H_4 production was accompanied by softening of the fruit.

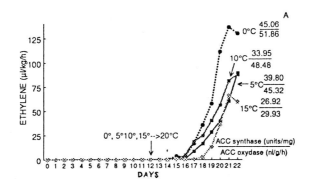

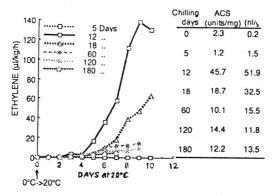

Fig. 8. The effect of chilling in C_2H_4 production and the activities of the enzymes ACC synthase and ACC oxidase of kiwifruit. A. Ethylene production at 20° C pretreated at 0°, 5°, 10° and 15° C for 12 days. B. Ethylene production by kiwifruit at 20° C pretreated with low temperature (0°C) for 5, 12, 18, 60, 120 and 180 days. The table indicates the activities of ACC synthase (ACS) and ACC oxidase (ACO) measured at the end of the experiment. (ACC synthase unit = pmoles of ACC/mg protein/2hrs) (Antunes, 1999).

Longer storage periods (18, 60, 120, 180 days) at 0° C resulted in a decreased capacity to produce C_2H_4 upon rewarming. The lowest value of C_2H_4 production during the shelf-life (5 µl/kg/h) was found after 180 days storage at 0° C. The decrease in the capacity of kiwifruit to produce C_2H_4 with storage time at 0° C was correlated with reduced ACC levels, and reduced activities of ACC synthase and ACC oxidase, as well.

The use of CA (2% O_2 + 5% CO_2) did not affect the chilling induction of C_2H_4 production. Fruits after long storage (180 days) in CA showed also reduced rates of C_2H_4 production upon transfer to 20° C. By using ULO treatments during prolonged storage, the fruit lost the capacity to produce C_2H_4. Fruits removed from ULO storage showed a drastically reduced capacity to produce C_2H_4 mainly due to low ACC oxidase activity and needed to be treated with C_3H_6 or external C_2H_4 to reach the eaten-ripe stage for consumption (Antunes and Sfakiotakis, 1997c).

During the handling and storage operations, C_2H_4 production is suppressed in order to prolong the storage life of kiwifruit. However, during the shelf-life, the fruit needs to restore its ability to produce C_2H_4 and to be able to ripen and obtain the best eating quality. Our data shows the risk, when long storage in conventional storage (CS), CA or ULO, in restoring the capacity of the fruit for C_2H_4 biosynthesis and action in the poststorage period. Fruit needed C_2H_4 aplication in order to ripen.

4.2. Wounding and *Botrytis* Infections

Mechanical wounding refers to wounding caused by bruising or intruding which may cause injury to the surface of the inner tissues of the fruit. In the host-parasite relationship, host cells may be considered mechanically injured as a consequence of invasion or intrusion of pathogens (Hyodo, 1981).

Picking by separating the fruit from the stem leaves a scar inducing some wounding. The stem scar is an obvious possible locus for botrytis infection and a wounded tissue capable of producing C_2H_4. Fruits after harvest, evolved a small amount of C_2H_4 shortly after picking, followed by a marked increase as it was observed in citrus (Aharoni, 1968).

4.2.1. Wounding

We have studied the effect of wounding on C_2H_4 production at low temperatures in kiwifruit which is not known to produce C_2H_4 below the critical range of 11-14.5° C. Wounding the tissue, by peeling the skin, induced C_2H_4 production at 5° C (Sfakiotakis et. al. 1997). The wounded fruit produces considerable amount of C_2H_4 up to 0.35 µl/kg/h at 5° C which is enough to induce ripening of the healthy unwounded fruit at storage temperatures.

In kiwifruit, mechanical injuries may be associated with rough handling during picking, filling and loading of bins or transportation. Injuries to kiwifruit are typically from impacts, cuts and punctures, and abrasion. Bruising resulting from dropping impact causes fruits to produce C_2H_4 at a great rate. The influence of fruit-flesh-firmness level on injury susceptibility has been studied with an evaluation of fruit response to such injuries (Arpaia et al. 1994). When firm fruit equal to or greater than 6 kgf were impacted, the fruit did not show C_2H_4 production. However, below 6 kgf, impact bruising and vibration bruising stimulated respiration and C_2H_4 production. When the fruit softened to about 2.5 kgf, severe injury occurred with a sharp increase of C_2H_4 production that persisted for 1 week.

4.2.2. Botrytis Infections

The most threatening disease of kiwifruit in refrigerated storage has been gray mold, caused by the fungus *Botrytis cinerea* Pers, ex Fr (Sommer et al. 1994). *Botrytis* causes directed losses in the infected fruit, but more severely induces the fruit to produce C_2H_4 which accumulated in the storage rooms subsequently causes excessive softening of the healthy fruit.

Many reports have shown that plants infected with fungi produce stress C_2H_4 (Boller, 1981). Stress C_2H_4 production was associated with rots in many fruits infected with *Botrytis cinerea*.

We have studied in kiwifruit the effect of *Botrytis* infections on C_2H_4 production in storage temperatures. Kiwifruit inoculated with *botrytis* and exposed at different temperatures (30°, 20°, 10°, 5°, 0°, –1°, and –2° C) showed fungus growth at a rate which was influenced significantly by temperature (Fig. 9A) (Niklis et al. 1992). Increased growth rate was found at a higher temperature.

The infected fruit showed a decrease in C_2H_4 production with dropping temperature. Ethylene production at high temperatures (30° and 20° C) measured high rates of C_2H_4 production (1095, 413 $\mu l/kg/h$ respectively) due to autocatalysis and the synergistic interaction of fungus-fruit tissue. Although it is known from previous studies that kiwifruit at low temperatures below 10° C does not produce C_2H_4, the infected kiwifruit at temperatures 10°, 5°, 0°, –1° and –2° C produced a significant amount of C_2H_4 (77, 31, 32, 12, 5 $\mu l/kg/h$ respectively) as a result of the fungus-tissue interaction. It was noticed that the fungus is able to induce C_2H_4 production even at –1° C (Niklis et al. 1992, 1997).

Measurements of the ACC showed accumulation of the ACC at the front area of fungus in the tissue plungers taken from the infected fruit kept at low temperatures (0° and 10°C) (Niklis et al. 1993). This finding gives evidence that the fungus is inducing C_2H_4 biosynthesis in the infected kiwifruit through the methionine pathway.

The *Botrytis* induced ethylene production may have serious consequences in the storage of the healthy fruit. If we consider an example in a storage room of 1000 m³ with 100 kg of infected kiwifruit with *botrytis*, we may calculate the total amount of C_2H_4 produced for 50 days from the infected fruit as follows:

$$C_2H_4 = 5 \; \mu l/kg/h \times 24 \text{ h} \times 50 \text{ days} \times 100 \text{ kg} = 600 \text{ ml}$$

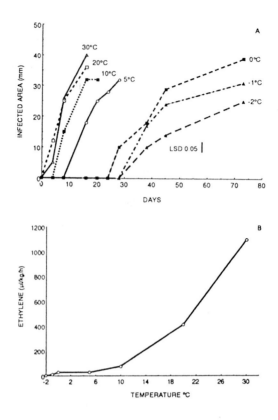

Fig. 9. Growth of *Botrytis cinerea* in inoculated kiwifruit kept at various temperatures (30°, 20°, 10°, 5°, 0°, −1°, and −2° C) (A) and C_2H_4 production measured at the above temperatures (B) in the peakheight (Niklis, 1994).

For a storage room (1000 m^3) without adequate aeration this amount of C_2H_4 it is estimated to give an ambient C_2H_4 concentration of 0.6 µl/l, high enough to induce premature ripening of the healthy fruit in the same storage room.

5. Postharvest Quality Control of Kiwifruit by Monitoring C_2H_4 Concentration in the Postharvest Environment

Ethylene plays a significant role in postharvest quality of kiwifruit—often deleterious, accelerating senescence and reducing shelf-life, and sometimes beneficial, improving the edible quality by promoting faster, more uniform ripening before retail distribution.

In the postharvest handling operations, three stages may be distinguished that may influence the postharvest quality of kiwifruit, the harvest period, the storage and the poststorage period of shelf-life until the product reaches the consumer. Because of the dominating role of C_2H_4 in the ripening of kiwifruit, monitoring C_2H_4 production, as well as C_2H_4

action, is of great importance in maintaining quality at harvest period, during storage and at poststorage shelf-life until the fruit reaches the consumer. The presence of C_2H_4 can have a substantial detrimental effect on storage potential, and much effort is devoted to avoid C_2H_4 exposure of kiwifruit in prolonging storage-life. However, in order to obtain the ultimate optimum edible quality, it is important for the fruit, during the shelf-life, not to lose the potential to produce endogenous C_2H_4 and respond to added ethylene.

There are four levels of manipulation for regulating C_2H_4 responses: (a) by addition or removal of C_2H_4 using C_2H_4 scrubbers, (b) by stimulating or inhibiting C_2H_4 biosynthesis, (c) by modifying C_2H_4 action by changing characteristics of C_2H_4 receptors or the number of receptors using antagonistic of C_2H_4 action such as silver (Ag+), norbornadiene and 1-methylcyclopropane (Sisler, 1991; Sisler and Serek, 1997) and (d) by manipulating C_2H_4 gene expression (Theologis, 1994). Among them manipulation of C_2H_4 biosynthesis and C_2H_4 action offers several advantages in controlling quality of kiwifruit.

Many factors may control autocatalysis of C_2H_4 biosynthesis such as temperature, the concentration of ethylene, CO_2 and O_2, stress by chilling, duration of storage, wounding and *botrytis* infections. Monitoring these factors may influence the rate of ripening and keeping the edible quality of the fruit. Among these factors, temperature was found to have the strongest influence and thermoregulation of C_2H_4 production is the prevailing factor in monitoring the hormone gas internally and in the fruit environment (Fig. 10).

Kiwifruit is a unique climacteric fruit which lacks the ability for autocatalysis of C_2H_4 production at a low temperature. On the other hand, chilling and other stresses induce the fruit to produce ethylene. Therefore, C_2H_4 production is highly regulated by temperature and the control of temperature during the handling operations is the key factor to regulate the C_2H_4 concentration in the postharvest life of kiwifruit.

Our results on the autocatalytic C_2H_4 production in climacteric fruits are compatible with the concept that two systems of C_2H_4 production are involved in the ripening process of kiwifruit. It has been suggested that system I is the low level of C_2H_4 present in fruits before the onset of ripening, while system II, represents the autocatalytic burst of C_2H_4 production which accompanies ripening (McMurchie et al. 1972; Sfakiotakis and Dilley, 1973b). Furthermore, for system I, Yang et al. (1986) suggested that in the preclimacteric fruit the resistance to ripening or the resistance to C_2H_4 action is mediated through a 'ripening inhibitor'. Our results support the view that in preclimacteric kiwifruit, the system I is responsible for low C_2H_4 production and the application of C_3H_6 or C_2H_4 changes the status of fruit from system I to system II resulting in the autocatalytic burst of C_2H_4 production. Evidence has been shown that in kiwifruit, temperature plays an essential role in the autocatalytic C_2H_4 production probably suggesting that low temperature strongly inhibits

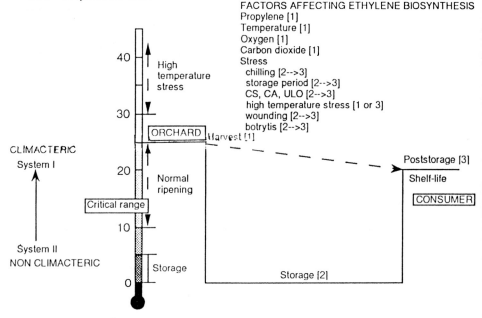

Fig. 10. Factors affecting C_2H_4 biosynthesis within the temperature range from 0° C to 45° C and the postulated change from system I to system II by increasing temperature in the critical range 11°-14.5° C.

the conversion from system I to system II (Fig. 10). System II operates in the temperature range of 14.5° to 38° C and can be induced by C_3H_6 (or C_2H_4) or stresses (chilling, wounding).

5.1. Avoiding Ethylene

Ethylene has detrimental effects on the postharvest quality of kiwifruit by accelerating softening, inducing autocatalysis of C_2H_4 production and indirectly by enhancing the growth of postharvest diseases (*Botrytis*). Therefore avoiding C_2H_4 throughout the postharvest life, during storage, transportation and distribution is of great importance.

After harvest, kiwifruit at ambient temperatures produces very small amount of C_2H_4 and in an open place, the internal C_2H_4 concentration remains very low. It takes a long time (17-20 days) to initiate C_2H_4 production during the shelf-life unless the fruit has been exposed to external ethylene. Therefore, there is no risk during curing 2-5 days at 15° C to stimulate C_2H_4 production in a C_2H_4-free environment.

Exposing the fruit after harvest to external C_2H_4 accelerates softening and induces autocatalysis of C_2H_4 production. Exposure of a few hours (> 6-12 h) to C_2H_4 before cooling initiates rapid softening that continues during subsequent cold storage (Mitchell et al. 1994). Ethylene exposure at ambient temperatures induces softening and autocatalysis of C_2H_4 production in the kiwifruit.

After precooling, the high quality fruit stored at low temperatures produce negligent amounts of C_2H_4 that is very low to accelerate ripening. However, the wounded or infected fruit produces significant amount of C_2H_4 at low temperature and accumulation of the C_2H_4 can cause excessive softening of the healthy kiwifruit stored in the same room. The fruit infected with *Botrytis* may produce enough C_2H_4 (1.5 µl/kg/h at 0° C and 9 µl/kg/h at 10° C) to induce ripening of the healthy fruit (Niklis, 1994). Other sources of C_2H_4 contamination must be avoided. Kiwifruit cannot be stored with or near other climacteric fruit (e.g. apples, pears) which produce significant amounts of C_2H_4 even under CA storage.

Premature softening during storage is a serious and expensive problem for the kiwifruit industry and slowing the rate of flesh softening is a major goal for any long-term storage program. Fruits are not acceptable for export if the firmness value is below 1-1.5 kgf.

Kiwifruit is extremely sensitive to ethylene in ambient air (Retamales and Campos, 1997). During storage, fruit softening is accelerated in the presence of C_2H_4 and removing C_2H_4 with scrubbers is the main concern for long storage. During storage, at low temperature, the presence of C_2H_4 stimulates only ripening. The magnitude of flesh softening depends largely on the C_2H_4 concentration, the fruit temperature and composition of the storage atmosphere. When C_2H_4 is present above the 20 ppb (ideally less than 10 ppb) in the storage room, it is recommended to provide fresh-air ventilation to remove accumulating C_2H_4 from the storage room (Kader, 1990). More effectively C_2H_4 scrubbers (with $KMnO_4$ or catalytic combustion) in commercial storage room with conventional or controlled atmosphere of kiwifruit are used commercially to reduce the C_2H_4 concentration in the gas atmosphere (Kader, 1990).

High CO_2 or low O_2 inhibits C_2H_4 production and fruit ripening. The combination of high CO_2 and low O_2 in controlled atmosphere storage drastically reduced fruit softening (Arpaia et al. 1985; McDonald and Harman, 1982; Harman and McDonald, 1983) or ULO storage (Thomai and Sfakiotakis, 1996, 1997; Antunes and Sfakiotakis, 1997c). However, the CA or ULO storage fruit does not ripen even during the shelf-life (Antunes and Sfakiotakis, 1997c).

Removing kiwifruit from storage temperatures produces C_2H_4 enough to induce ripening during the shelf-life and indirectly enhances the growth of postharvest diseases (*Botrytis*). However, the CA or ULO stored fruit may have a problem in ripening evenly. The fruit may not be able to produce C_2H_4 and the application of external C_2H_4 is needed to ripen the fruit.

5.2. Pre-conditioning with C_2H_4 for Ripening

Although avoiding C_2H_4 exposure of kiwifruit is important in prolonging storage life, very often C_2H_4 treatment is desirable to improve the eating quality. Most consumers prefer to buy kiwifruit which are near their full ripeness (0.5-0.8 kgf). When consumers buy mature, but not ripe kiwifruit, they usually do not know how to properly ripen them. Ethylene treatment may be desirable during harvest and early storage or when the fruit is removed from CA storage in order to benefit from ripening for immediate marketing or consumption.

For fresh harvest or early storage ripening can be achieved by applying ethylene at concentrations between 10 and 100 ppm at least for 24 h to kiwifruit that are warmed (20° to 25° C). For artificial ripening of kiwifruit similar procedure as in banana ripening is used (Arpaia et al. 1994). Treating the freshly harvested fruit with C_2H_4 for 24 hours induces ripening and subsequently autocatalysis of C_2H_4 production results and ripening is complete in 5 or 6 days. Ripening changes are slowed down by decreasing fruit temperature. For fruit packed with plastic wraps, it is preferable to apply 100 ppm C_2H_4 for 48 h. For artificial ripening, avoid using elevated temperatures. Experiments with C_3H_6 showed abnormal ripening with hard core when the flesh was soft at high temperatures (> 38° C) (Antunes, 1999). Exposing the fruit to such unusually high temperatures causes uneven ripening and has a detrimental effect on fruit quality parameters such as firmness, SSC and flesh color.

Removing the fruit from the storage after the fruit has been exposed to the chilling temperature induces the fruit to produce C_2H_4 and there is no need for conditioning the fruit with an external application of ethylene. For fruit which has been exposed to storage temperature for more than 12 days, after fruit removal from the storage there is no need for C_2H_4 treatment since after fruit removal, chilling induces C_2H_4 production in 3 days and subsequently (in 6-7 days) fruit ripening (Sfakiotakis et al. 1997; Antunes, 1999).

Fruit removed from CA or ULO storage may show abnormal ripening during shelf-life. Prolonged storage in CA and mainly in ULO may damage the C_2H_4 producing system (Antunes and Sfakiotakis, 1997c; Antunes, 1999) and this may necessitate the use of C_2H_4 for artificial ripening.

A pre-conditioning protocol at the shipping point and a second protocol dealing with handling preconditioned and ripening for kiwifruit receivers are recommended for the California Kiwifruit growers (Crisosto, 1997). For long distance shipping (2-3 weeks), it is recommended to apply 100 ppm C_2H_4 for 12 h at 0°-10° C to induce uniform softening. After venting, cold C_2H_4 treated kiwifruit can be stored back in cold storage (0° C) for 3-6 weeks. For short distance shipping (4-7 days) the cold C_2H_4 treated fruit should be placed at temperatures adjusted according to the anticipated consumption schedule.

6. Conclusions

Considerable information is now available on the physiology and regulation of ethylene biosynthesis and ripening in kiwifruits which differ from other climacteric fruits in several ways. For example, there is an increase of ethylene production neither on the tree nor of the fruits during the cold storage. After harvest, the fruits rarely produce ethylene for a long time without exposure to exogenous ethylene. During storage there is no ethylene production. The fruits show a typical climacteric respiration rate and ethylene production at moderate temperature (16-36° C), while at chilling temperatures (> 11-14.5° C) behave as non-climacteric fruits.

Several factors can initiate autocatalysis and other factors may also affect the rate of ethylene production in the harvested fruits and the control of these factors can be of significant importance in prolonging the storage life and maintaining the quality of kiwifruits. Ethylene production can be induced autocatalytically by exogenous ethylene or propylene, exposure to chilling temperature, mechanical wounding and *Botrytis* infection.

Exogenous ethylene or propylene, which was shown to be more appropriate for these studies, stimulates ethylene and ACC production at warm temperatures by inducing the activities of ACC synthase and ACC oxidase. Although the propylene-treated fruit do not produce ethylene at cold storage temperatures, ripening proceeds independently of autocatalytic ethylene production. There is an inhibition of propylene-induced ethylene production in kiwifruit below a critical temperature range of 11-14.5° C. This inhibition of ethylene production at low temperatures is attributed primarily to the inhibition of the expression of the propylene-induced ACC synthase gene and to the possible post-transcriptional modification of ACC oxidase.

Increased CO_2 concentrations (5-10%) of the storage atmosphere, drastically reduce the rate of propylene-induced ethylene production of kiwi, while reduced O_2 concentrations (1-5%) inhibits the propylene-induced ethylene production at the ACC synthesis level.

Exposure of the fruit to chilling temperatures for at least 12 days, stimulates ethylene production upon rewarming by inducing ACC synthase and ACC oxidase activity. However, the fruit removed from prolonged storage at 0° C in conventional or CA showed reduced capacity to produce ethylene, due to reduced ACC production and ACC synthase and ACC oxidase activities. The ULO storage drastically inhibited the induction of ethylene production by chilling which destroys the system that converts ACC into ethylene.

Tissue wounding induces ethylene production even at cold temperatures and this may cause early softening of the unwounded fruit at storage temperature.

Botrytis cinerea plays a crucial role in the postharvest life of kiwifruit. Fruits infected by *Botrytis* produces considerable ethylene through the methionine pathways and causes early softening of the healthy fruit in the storage rooms.

The above findings can be used to draw guidelines for an integrated handling system to prolong the storage life and improve the edible quality of the kiwifruit. Ethylene has detrimental and beneficial effects in the postharvest quality of kiwifruit, during the harvest period, the storage and the poststorage period of shelf-life until the product reaches the consumer. Monitoring ethylene production, as well as ethylene action, is of great importance in maintaining quality. Avoiding exposure of kiwifruit to ethylene can be accomplished by removal of ethylene from the postharvest environment. However, the best way to avoid the detrimental effects of ethylene is to control ethylene production by monitoring temperature and avoiding the stress ethylene (mechanical wounding and *botrytis* infections). CA and MA with high CO_2 and low O_2 may also have an effect on both ethylene action and ethylene production and may indirectly influence the postharvest quality. However, when such treatments are used, it is important to understand that during the shelf-life, the fruit needs to restore its ability to produce C_2H_4 and be able to ripen and obtain the best eating quality.

Although much progress has been made in understanding how various factors and storage conditions directly or indirectly influence the cause of postharvest quality losses of kiwifruit, many challenges and opportunities remain and must be addressed by postharvest physiologists and biotechnologists. We are beginning to discover how ethylene biosynthesis is controlled at a genetic and biochemical level in such fruits. The kiwifruit is a unique fruit with a mechanism of thermoregulation of ethylene biosynthesis. Therefore, this mechanism at molecular level offers potentialities to transfer it by genetic manipulation to other fruits to prolong the storage life and improve the postharvest quality.

References

Abeles, F.B., Morgan, P.W. and M. E. Salveit. 1992. *Ethylene in Plant Biology*. Academic Press.

Aharoni, Y. 1968. Respiration of oranges and grapefruits harvested at different stages of development. *Plant Physiology* **43**: 99-101.

Antunes, M.D.C. 1999. Biochemical basis of thermoregulation and autocatalytic ethylene production in ripening 'Hayward' kiwifruit prestorage and, subjected to chilling and under controlled atmosphere storage conditions. Ph.D. Thesis, submitted to Universdade do Algarve, Portugal

Antunes, M.D.C. and E. Sfakiotakis. 1997a. The effect of controlled atmosphere and ultra low oxygen on storage ability and quality of 'Hayward' kiwifruit, *Acta Horticulturae* **444**: 613-618.

Antunes, M.D.C. and E. Sfakiotakis. 1997b. Biochemical basis of thermoregulation of ethylene production and ripening of 'Hayward' kiwifruit, *Acta Horticulturae* **444**: 541-546.

Antunes, M.D.C. and E. Sfakiotakis. 1997c. Ethylene production of 'Hayward' kiwifruit after ultra low oxygen and controlled atmosphere. *Acta Horticulturae* **444**: 535-540.

Antunes, M.D.C., Pateraki, I. Ververidis, P., Kanellis, A.K. and E. Sfakiotakis. 1999. Differential effects of low temperature inhibition on kiwifruit ripening and ethylene production. In: *Biology & Biotechnology of the Plant Hormone Ethylene II*. Kluwer Academic Publishers, Dordrecht, pp. 433-436.

Antunes, M.D.C. and E.M. Sfakiotakis, 2000. Effect of high temperature stress on ethylene biosynthesis respiration and ripening of "Hayward" Kiwifruit *Postharvest Biology and Technology*: in press

Antunes, M.D.C., Pateraki, I., Kanellis, A. and E. Sfakiotakis, 2000. Differential effects of low temperature inhibition on the propylene induced autocatalysis of ethylene production and respiration on ripening of "Hayward" Kiwifruit. *Journal of Horticultural Science and Biotechnology* **75**(4) in press

Arpaia, M, Mitchell, F.G. and A.A. Kader. 1994. Postharvest physiology and causes of deterioration. In: *Kiwifruit Growing and Handling*. J.K. Hasey et al. (Eds). Publ. No. 3344, Univ. Calif. DANR Publ., Oakland, CA, pp. 88-93.

Arpaia, M.L., Mitchell, F.G., Kader, A.A. and G. Mayer. 1985. Effects of 2% O_2 and varying concentrations of CO_2 with or without ethylene on the storage performance of kiwifruit. *Journal of the Americ Society for Horticultural Science* **110**: 200-203.

Arpaia, M.L., Mitchell, F.G., Kader, A.A. and G. Mayer. 1986. Ethylene and temperature effects on softening and white inclusions of kiwifruit stored in air or controlled atmosphere. *Journal of the American Society for Horticultural Science* **109**: 768-770.

Beevers, D.J. and G. Horkirk. 1990. Fruit development and fruit physiology, In: *Kiwifruit: Science and Management*. I.J. Warrington and G.C. Weston (Eds), N.Z. Soc. Hort. Sci., Auckland, pp. 97-126.

Biale, J.B. 1964. Growth maturation and senescence in fruits. *Science* **146**: 880-888.

Boller, T. 1981. Ethylene in pathogenesis and disease resistance. In: *The Plant Hormone Ethylene*. A.K. Mattoo, and J.C. Suttle (Eds), CRC Press, pp. 293-314.

Burg, S.P. and E.A. Burg. 1965. Ethylene action and the ripening of fruits. *Science* **148**: 1190-1196.

Cheah, L.H. and D.E. Irving. 1997. Kiwifruit. In: *Postharvest Physiology and Storage of Tropical and Subtropical Fruits*, S.K. Mitra (Ed.), pp. 209-227.

Crisosto, C.H. 1997. Final preconditioning guidelines for kiwifruit shippers. *Central Valley Postharvest Newsletter* **6**: 1-27. (www.uck.ac.edu).

Crisosto, G.T., Mitchell, F.G., Arpaia, M.L. and G. Mayer. 1984. The effect of growing location and harvest maturity on the storage performance and quality of 'Hayward' kiwifruit. *Journal of the American Society for Horticultural Science* **109**: 584-587.

Ferguson, A.R. 1984. Kiwifruit—a botanical review. *Horticultural Reviews* **6**: 1-64.

Ferguson, A.R. 1990. The genus Actinidia. In: *Kiwifruit Science and Management*, Ray Richards, in association with New Zealand Society for Horticultural Science, Auckland, pp. 15-35.

Gerasopoulos, D. and D.G. Richardson. 1997. Differential propylene induced ethylene production in 'Anjou' pears during storage at chilling and non-chilling temperatures. *Journal of Horticultural Science* **72**: 571-575.

Given, N.K. 1993. Kiwifruit. In: *Biochemistry of Fruit Ripening*, G.B. Seymor, J.E. Taylor and A. Tucker (Eds), Chapman & Hall, London, pp. 235-254.

Harman, J.H. 1981. Kiwifruit maturity. *Orchardist of New Zealand.* **54**: 126-127, 130.

Harman, J.E. and B. McDonald. 1983. Controlled atmosphere storage of kiwifruit: effects on storage life and fruit quality. *Acta Horticulturae* **138**: 195-201.

Horpirk, G. and C. Clark. 1991. Postharvest fruit losses in the New Zealand kiwifruit Industry. *Acta Horticulturae* **297**: 611-616.

Hyodo, H. 1991. Stress/wound ethylene. In: *The Plant Hormone Ethylene*. A.K. Mattoo and J.C. Suttle. (Eds), CRC Press, pp. 43-63.

Hyodo, H. and R. Fukazawa. 1985. Ethylene production in kiwifruit. *J. Japan. Soc. Hort. Sci.* **54**: 209-215.

Kader, A. 1990. The effect of storage methods on the quality of fresh horticultural crops. In: *Qualyta dei Prodoti Ortofrutticoli Postraccolta*. pp. 11-26.

Lallu, N. 1997. Low temperature breakdown in kiwifruit. *Acta Horticulturae* **444**: 579-585.

Liang, C.F. and A.R. Ferguson. 1984. Emendation of the Latin name of *Actinidia chinensis* P. var. Hispida C.F. Lang. *Guihata* 4. 181-182.

MacRae, E., Lallu, N., Searle, A.N. and J.H. Bowen. 1989a. Changes in the softening and composition of kiwifruit (*Actinidia deliciosa*) affected by maturity at harvest and postharvest treatments. *Journal of the Science of Food and Agriculture* **49**: 413-430.

MacRae, E.A., Bowen, J.H. and M.G.H. Stec. 1989b. Maturation of kiwifruit (*Actinidia deliciosa* cv. Hayward) from two orchards: differences in composition of the tissue zones. *Journal of the Science of Food and Agriculture* **47**: 401-416.

MacRae, E., Quick, W.P., Benker, C. and M. Stitt. 1992. Carbohydrate metabolism during postharvest ripening in kiwifruit. *Planta* **188**: 314-323.

Matsumoto, S., Obara, T. and B.S. Luh. 1983. Changes in chemical constituents of kiwifruit during preharvest ripening. *Journal of Food Science.* **48**: 607-611.

McDonald, B. 1990. Precooling, storage and transport of kiwifruit. In: *Kiwifruit, Science and Management*, J. Warrington and G.C. Weston (Eds), N.Z. Soc. Hort. Sci., Auckland, pp. 429-459.

McDonald, B. and J. E. Harman. 1982. Controlled-atmosphere storage of kiwifruit. 1. Effect on fruit firmness and storage life. *Scientia Horticulturae* **17**: 113-123.

McMurchie, E.J., McGlasson, W.B. and I.L. Eaks. 1972. Treatment of fruit with propylene gas gives information about the biogenesis of ethylene. *Nature* **237**: 235-236.

Metzidakis, J. and E. Sfakiotakis. 1989. The control of auytocatalytic ethylene production and ripening in avocado fruit by temperature, high carbon dioxide and low oxygen. *In: Biochemical and Physiological Aspects of Ethylene Production in Lower and Higher Plants*, J.C. Pech, A. Lathche and C. Balaque (Eds), Kluwer Academic Publishers, pp. 201-210.

Mitchell, F.G., Arpaia, M. and G. Mayer. 1994. Harvesting and preparation for market. *In: Kiwifruit Growing and Handling* J.K. Hasey, R.S. Johnson, J.A. Grant and W.O. Reil (Eds). ANR Publications, University of California, USA, pp. 99-107.

Niklis, N. 1994. The role of *Botrytis cinerea* Pers fungus in ethylene production and biosynthesis of 'Hayward' kiwifruit during storage under low temperature and low oxygen. Ph.D. submitted to Aristotle Univesrity of Thessaloniki (in Greek).

Niklis, N., Thanassoulopoulos, C.C. and E. Sfakiotakis. 1992. Ethylene production and growth of *Botrytis cinerea* in kiwifruit as influenced by temperature and low oxygen storage. *In: Recent Advances in Botrytis Research*, L.K. Verhoeff, N.E. Malathrakis and B. Williamson (Eds), Pudoc Scientific Publishers, Wageningen.

Niklis, N., Sfakiotakis, E. and C.C. Thanassoulopoulos. 1993. Ethylene biosynthesis in 'Hayward' kiwifruit infected by *Botrytis cinerea*. *In: Biochemical and Physiological Aspects of Ethylene Production in Lower and Higher Plants*, J.C. Pech, A. Lathche and C. Balaque (Eds), Kluwer Academic Publishers, pp.130-135.

Niklis, N., Sfakiotakis, E. and C.C. Thanassoulopoulos. 1997. Ethylene production by *Botrytis cinerea*, kiwifruit and *Botrytis* rotted kiwifruit under several storage temperatures. *Acta Horticulturae* **444**: 733-737.

Okuse, I and K. Ryugo. 1981. Compositional changes in the developing 'Hayward' kiwifruit in California. *Journal of the American Society for Horticultural Science* **106**: 73-76.

Pratt, H.K. and M.S. Reid. 1974. Chinese gooseberry: Seasonal patterns in fruit growth and maturation, ripening, respiration and the role of ethylene. *Journal of the Science of Food and Agriculture* **25**: 747-757.

Puig, L. Varga, D.M., Chen, P.M. and E.A. Miekle. 1996. Synchronizing ripening in individual 'Bartlett' pears with ethylene. *Hort Technology* **6**: 24-27.

Retamoles and Campos, 1997. Extremely low ethylene levels in ambient all still critical for kiwifruit storage. *Acta Horticulturae* **444**: 573-578.

Sale, P.R. 1983. *In: Kiwifruit Culture*, D.A. Williams, (Ed.), New Zealand Government Printer, Wellington, p. 95.

Sfakiotakis, E.M. and D.R. Dilley. 1973a. Internal ethylene concentrations in apple fruits attached or detached from the tree. *Journal of American Horticultural Science* **98**: 501-503.

Sfakiotakis, E.M. and D.R. Dilley. 1973b. Induction of autocatalytic ethylene production in apple fruits by propylene in relation to maturity and oxygen. *Journal of American Horticultural Science* **98**: 504-508.

Sfakiotakis, E. and D.R. Dilley. 1974. Induction of ethylene production in 'Bosc' pears by postharvest cold stress. *Hort Science* **9**: 336-338.

Sfakiotakis, E., Stavroulakis, G., Ververides Ph. and D. Gerasopoulos. 1989. The control of autocatalytic ethylene production and ripening by propylene in 'Hayward' kiwifruit, *In: Biochemical and Physiological Aspects of Ethylene Production in Lower and Higher plants*, H. Clijsters, M. De Proft, R. Marcelle and M. Van Poucke (Eds), Kluwer Academic Publishers, pp. 173-178.

Sfakiotakis, E., Stavroulakis, G., Ververides Ph., Gerasopoulos, D. and Lionakis. 1991. Storage performance of 'Hayward' kiwifruit harvested at different maturity stages from two growing locations of Pieria and Rethymno. *Agricultural Research* **15**: 463-482 (in Greek).

Sfakiotakis, E., Antunes, D., Stavroulakis, G., Niklis. N., Ververidis, P. and D. Gerasopoulos. 1997. Ethylene biosynthesis and its regulation in ripening 'Hayward' kiwifruit. *In: Biology and Biotechnology of the Plant Hormone Ethylene*. A.K. Kanellis, C. Chang, H. Kende and D. Grierson (Eds), Kluwer Academic Publishers, pp. 47-56.

Sisler, E.C. 1991. Ethylene-binding components in plants. *In: The Plant Hormone Ethylene*. A.K. Matoo and J.C. Suttle (Eds), CRC Press, pp. 81-99.

Sisler, E.C. and M. Serek. 1997. Inhibitors of ethylene responses in plants at the receptor level: recent developments. *Physiologia Plantarum* **100**: 577-582.

Sommer, N.F, Suadi, J.E and R.J. Fortlage. 1994. Postharvest storage diseases. *In: Kiwifruit Growing and Handling*. J.K. Hasey et al. (Eds), Publ. No. 3344. Univ. Calif. DANR Publ., Oakland, CA, pp. 116-121.

Stavroulakis, G. 1991. The effect of temperature, oxygen, carbon dioxide in the propylene induced ripening and biosynthesis of ethylene of 'Hayward' kiwifruit. Ph.D. submitted to Aristotle University of Thessaloniki (in Greek).

Stavroulakis, G. and E.M. Sfakiotakis. 1993. Regulation by temperature of the propylene induced ethylene biosynthesis and ripening in 'Hayward' kiwifruit. *In: Biochemical and Physiological Aspects of Ethylene Production in Lower and Higher plants*, J.C. Pech, A. Lathche and C. Balaque (Eds), Kluwer Academic Publishers, pp. 142-143.

Stavroulakis, G. and E. Sfakiotakis. 1995. Time course study of thermoregulation in ethylene biosynthesis and ripening of 'Hayward' kiwifruit induced by propylene. *Acta Horticulturae* **397**: 429-436.

Stavroulakis, G. and E.M. Sfakiotakis. 1997. Regulation of the propylene induced ripening and ethylene biosynthesis by oxygen in 'Hayward' kiwifruit. *Postharvest Biology and Technology* **10**: 189-194.

Theologis, A. 1994. Control of ripening. *Current Biology* **3**: 369-371.

Thomai, T.K. and E.M. Sfakiotakis. 1996. Effect of low-oxygen atmosphere on storage behavior of kiwifruit. *Advances Horticultural Science* **11**: 137-141.

Thomai, T. and E. Sfakiotakis. 1997. Effect of low-oxygen early on quality changes, acetaldehyde and ethanol accumulation and late harvest of 'Hayward' kiwifruit. *Acta Horticulturae* **444**: 593-597.

Tucker, G.A. and D. Grierson. 1987. Fruit Biochemistry. *In: Biochemistry of Plants*. Academic Press Inc. **12**: 265-317.

Yang, S.F., Liu, Y. and O.L. Lau. 1986. Regulation of ethylene biosynthesis in ripening apple fruits. *Acta Horticulturae* **179**: 711-720.

Yano, N. and Y. Hasegawa. 1993. Ethylene production in harvested kiwifruit with special references to ripe rot. *Journal of Japan Society of Horticultural Science* **62**: 443-449.

Young, H.V.J. and D.J.W. Paterson. 1983. Volatile aroma constituents of kiwifruit. *Journal of the Science of Food and Agriculture* **34**: 81-85.

13

Postharvest Pests and Physiological Disorders of Cassava

Weston Msikita[1], Henry T. Wilkinson[2] and Robert M. Skirvin[2]

[1]*Ohio State University, Department of Plant Biology, 1735 Neil Avenue, Columbus, OH 43210 USA*
[2]*University of Illinois, Department of Natural Resources and Environmental Sciences, Urbana, IL 61801, USA*

Abbreviations

CIAT	Centro Internacional de Agricultura Tropical
FAO	Food and Agriculture Organization
ha	hectare
IITA	International Institite of Tropical Agriculture
K	Potassium
N	Nitrogen
P	Phosphorus
PPD	Postharvest

1. Introduction

Cassava (*Manihot esculenta* Crantz), also known as yuca, manioc, and tapioca is a herbaceous plant largely cultivated for its starchy roots. Cassava is an important food crop for an estimated 400 million people in Africa, Asia, and Latin America (Cock, 1985; El-Sharkawy, 1993), the majority being resource-poor. Cassava is the fourth most extensively cultivated food crop in the tropics (FAO, 1989). Africa produces an estimated 80 million tons of cassava annually, largely by small land (average farm of 0.6 ha) holders (Nweke and Lynam, 1997). Virtually every part of the plant is usable: storage roots are consumed raw (for low cyanide containing varieties), cooked or fried or processed into food, beverage or industrial products. Stems are used as planting material, firewood or chopped up and fed to animals. Leaves are cooked and consumed like a vegetable in parts of Central, East and West Africa or chopped up together with stems and fed to animals (Cock, 1985). Cassava roots are a good source of carbohydrates, and is therefore a good source of energy, and its leaves are high in calcium and protein (Nweke and Lynam, 1997). Because of its many uses, cassava production in all three continents is on the increase.

About 40% of global cassava output comes from Africa (Nestel, 1993). The crop provides food security to millions of people in Africa, Asia, and Latin America at times of drought and famine. In parts of Central and West Africa, cassava is also assuming an important economic role because it is planted for sale, and provides sole income for the households (Nweke, 1998).

In many parts of Africa, cassava production is on the increase largely due to the factors given below:

(a) Intrinsic crop characteristics endear cassava to resource-poor farmers. These intrinsic characteristics include the ability of the crop to tolerate drought, and produce a good yield under low management. Most cassava growers in Africa have little or no knowledge of modern crop production practices. For instance inorganic fertilizers are rarely used for cassava production in Africa (Nweke and Lynam, 1997).

(b) Cassava is also endearing to resource-poor farmers because it is amenable to partial harvest (also known as piecemeal harvesting), permitting the grower to harvest the quantity needed, leaving the remainder of the plant standing in the ground for later harvest. Piecemeal harvest permits the farmer to maintain a supply of roots in the ground. After maturation some varieties may be left unharvested for up two years.

(c) Civil wars have resulted in large displacement of populations, making it difficult for the affected populations to plant crops such as maize which require more intense management than cassava. Displaced populations have spread it to new areas or have adopted cassava in the new areas they have moved into. In addition to civil wars, parts of Africa constantly face droughts, making it increasingly difficult to grow food crops such as maize and forcing farmers to turn their attention to drought tolerant crops such as cassava and sorghum.

(d) Availability of improved cultivars and information on production practices. International agricultural research centers such as the International Institute of Tropical Agriculture (IITA) in Africa, and the Centro Internacional de Agricultura Tropical (CIAT) in South America in conjunction with the national programs have developed early maturing, high yielding cultivars that are more resistant to problem pests and diseases. For instance the African cassava mosaic geminivirus transmitted by the whitefly (*Bemisia tabaci* Gennadius), and cassava bacterial blight, incited by *Xanthomonas campestris* pv. *campestris* (Berthet et Bondar) Dye have been significantly reduced in parts of Africa where new cultivars have been adopted (Mignouna and Dixon, 1997). Agronomic packages have been developed and tested and adapted on farms in many countries, further making it easier to grow the crop. Increased cassava production has in turn stimulated cassava agro-industries such as starch and beverage production, and processing for animal feed (Nweke and Lynam, 1997).

A major impediment to cassava assuming a leading economic role in most countries is that it is highly perishable once harvested, and its roots are bulky. Cassava growers lack capital to purchase mechanical processing equipment, resulting in their processing the roots manually, a process which is slow and cumbersome. Additionally, most farmers lack storage facilities to store the crop once harvested, resulting in an estimated loss of 40-50% (FAO, 1989). Pests, diseases and physiological disorders further hamper increased cassava production in all three continents. In this chapter, we review pre- and postharvest practices and their impact on postharvest diseases, and physiological disorders of cassava.

2. Planting and Field Management

2.1. Planting Material

Cassava is almost exclusively planted from stem cuttings. True botanical seeds are mostly used for breeding purpose. The planting material (called 'cuttings' or 'stakes') is obtained from the lignified lower stems. Stem cuttings 10-20 cm long (with 6-10 nodes) are obtained using a sharp knife, and planted on flat beds, mounds or ridges. Cassava is also planted from 'ministem cuttings' (small stem pieces, 5-10, with one to six nodes depending on the degree of lignification of the stem). Ministem cuttings are planted in nursery beds (at about 10 cm × 10 cm) or in polythene bags containing sterilized soil. Ministem cuttings are left in the nursery bed or polythene bags for 4 to 6 weeks, and planted in the field at the spacing mentioned above. A modification of the ministem cutting technique has been developed by Cock (1985). By this method, lignified cuttings with two nodes are cut from a healthy mature plant and grown in sterilized soil. When sprouts reach 5-10 cm, the upper part is cut off and rooted using an auxin, after which they are transferred to the field. The axillary buds of the lower part of the sprout of the stem will form a new sprout, and the procedure can be repeated several times.

Cassava plants may also be propagated using tissue culture. Various types of explants, including meristems, leaf lamina, petiole, internodes, and seed pieces (Raemakers et al. 1997) have been successfully used on various media. Micropropagation of cassava has mainly been used to eliminate pathogens from planting stock. Meristem cultures has been successfully used to eliminate viruses from cassava (Anonymous, 1990), and the virus-free plants have been used to start a new stock.

Cassava plants started using these different propagation practices (e.g. ministem stem cuttings, micropropagation) are more sensitive to biotic and abiotic stresses, and when planted out in the field, may require more intense management, and may take longer to mature than cassava propagated by conventional methods, although its root yield is not significantly affected (Anonymous, 1990). There is paucity of information elucidating the various planting methodologies with postharvest pests, diseases and physiological disorders.

2.2. Field Planting

In Africa it is common to plant more than one cultivar in the same field. Depending on the cultivar, cuttings are spaced at 25-30 cm within row, and 90-100 cm between rows. Stem cuttings are planted on a slant, or completely buried in 5-10 cm deep furrows. It is also common to intercrop cassava with legumes (e.g. groundnuts, beans, cowpeas), cereals (maize, sorghum), and vegetables. Cassava intercropping system may be simple (consisting of only two crop species) or complex mixture (comprising three or more crops). Usually cassava is planted one to two months before the intercrop. Advantages of intercropping cassava with other crops include: efficient use of field space, control of some pests and diseases, and some of the intercrops, particularly leguminous crops, provide nutrients to the cassava plants (Ikeorgu et al. 1998). When incorrectly used, intercropping cassava with other crops has its disadvantages, including competition for nutrients and moisture, shading, and introduction of pests and diseases. Because some intercrops have the potential to affect cassava growth, and even the availability of nutrients while in the field, they could potentially affect cassava postharvest quality and yield of roots. Unfortunately there is little information elucidating intercroppings with postharvest ameliorations, deteriorations, pests and/or disease problems.

2.3. Field Maintenance

Like many crops, weed management is particularly important during the first four weeks after planting. Later cassava forms a canopy inhibiting weed growth. Weeding is usually done mechanically using a hoe. The number of times a cassava field has to be weeded depends on the type of weeds, precipitation in the area, and the rate at which cassava forms a canopy. Apart from competing with cassava for water and nutrients, weeds harbor pests and diseases. Unfortunately little information exists associating weeding frequency, and/or type of weeds, with pests and diseases in the field to postharvest related problems of cassava.

3. Nutrient Requirements

Cassava has no specific soil requirement, and is therefore grown on many types of soil, being limited only by salinity, alkalinity and poor drainage (Lozano et al. 1980). Salt inhibited growth is observed in soils with high salt (particularly sodium) concentration. Cassava does not tolerate pH > 7.8, and a conductivity of more than 0.5 mmhos/cm, and sodium saturation of more than 2.5% (Lozano et al. 1981). Cassava has an extensive root system making it suitable to grow in nutrient-deficient soils by extending its roots to areas less accessible to other crops. Although most traditional, small-scale growers do not apply organic or inorganic fertilizer, Lozano et al. (1981) have found that cassava responds positively to nutrient (macro- and micro) applications. Among the major elements, nitrogen and phosphorus are more important in producing a good tuber yield than a large supply of potassium (Anonymous, 1990). Actual amounts of NPK needed varies with cultivar and quantity available in the soil. Nitrogen deficiency leads to stunted growth, and pale green, narrow leaves whereas deficiency of phosphorus in soil produces stunted growth, accompanied by a violet discoloration of leaves (Anonymous, 1990). Potassium deficiency also results in stunted growth, while the color of leaves is often dark, and necrosis occurs at the margins. Deficiency of potassium is also reported to lower starch content in the roots and results in general diminished resistance to pests and diseases (Anonymous, 1990). Whereas the response of cassava to the availability of nutrients has been well documented, there is paucity of information correlating cassava postharvest related problems: either nutritional supplies or nutritional status of the crop in the field. There are no specific nutrient application recommendations to minimize post-harvest pests, diseases or disorders.

4. Harvesting and Transport

The duration from planting to harvesting varies depending on cultivar, crop management, and climate. Some cultivars of cassava are ready for harvest 9 to 12 months after planting while others require up to 24 months after planting. For some varieties, the longer the crop is left unharvested after maturation, the greater the fiber content in the roots, thereby reducing the quality of the roots. There are no technical guidelines for pre-harvest management practices to reduce postharvest physiological disorders, pest, and disease problems. However, some of the pre-harvest management practices the crop is subjected to are known to affect postharvest quality of the roots. For instance, some farmers will harvest part of the plant, leaving the remainder standing in the ground (known as 'partial harvesting'). The stage and type of harvest (piecemeal or complete) has an effect on the quality of the roots. Piecemeal harvest inflicts wounds on the plant, and may open an entry for pest and disease attack. Research needs to determine the optimum developmental stage, and the time of harvest for each cultivar, for each climatic region, to elucidate the effect of partial harvest on postharvest related diseases and physiological disorders.

Harvesting is done by hand, using various types of implements. In parts of West Africa, one to six months prior to harvest, cassava farmers cut down (a practice called 'pre-harvest pruning') the above-ground part of the plant at the end of the growing season, and store the stems under a tree for later use. Although the object is to save stems for later use, pre-harvest pruning has been reported to significantly reduce postharvest physiological deterioration (PPD) (O'Brien et al. 1997). In addition to reducing PPD, pre-harvest pruning could reduce the possibility of infection of roots due to stem infecting pests and diseases. Research is yet to determine the optimum crop age and season to impose pre-harvest pruning to derive maximum benefits resulting from a reduction in postharvest diseases and disorders, including PPD.

A lot of the losses incurred on fresh cassava roots emanate from poor harvesting and handling practices which cause damage to the harvested roots (Nduguru et al. 1998). After harvest, roots are transported by carrying on the head ('headloads'), by bicycle, carts drawn by animals or by motor vehicles to markets or to factories.

In many parts of Africa, cassava leaves are harvested from about six months until the roots are ready for harvesting. Intensity and duration of leaf harvest vary, and significantly affect damage by pests and diseases (Lahai et al. 1997). Unfortunately the effect of leaf removal on post-harvest quality of cassava remains unknown.

5. Long-term Storage Products and Their Utilization

Improved storage methods for fresh cassava are based on techniques involving freezing, irradiation, waxing (Ashagbley et al. 1994), control of storage environment, moisture, temperature (Annan et al. 1994), and the use of a fungicide treatment such as thiabendazole (Nduguru et al. 1998). These methods and their applications are being studied and tested in parts of Africa, Latin America, and Asia (Anonymous, 1990; Nduguru et al. 1998). The simplest method involves washing the freshly harvested roots, dipping them in a fungicide, and placing them in a polythene bag which is then sealed. Root respiration within the bags causes the roots to cure, and can extend shelf-life up to 10 days. There are three methods by which cassava is stored long-term: storage underground, drying the roots, and processing the roots into various food products.

5.1. Ground Storage

Some farmers in Africa, store cassava 'in ground' (also known as 'in situ' storage) i.e. cassava is unharvested for 12-36 months after maturation or partially harvest individual plants (remove large roots from individual plants, leaving small roots to enlarge further). Cassava cultivars respond differently to 'in ground' storage. Some cultivars (particularly early maturing ones) develop a fibrous consistency in the central cortical region the longer they remain unharvested or are subjected to piecemeal harvesting. Some cultivars take long to heal from wounds inflicted during piecemeal harvest, and thus an avenue for insect, pest and disease attack is created.

'Pit' or 'trench' storage is another form of ground storage for cassava (Anonymous, 1990). By this method cassava roots are harvested, stacked in a shallow trench (2 m long, 1.5 m wide and 1m deep), covered with dried leaves and mounded with soil. Sites for trenches should be well-drained, adequately shaded (natural vegetation or a constructed thatched roof over the trench) and slightly sloping with the trenches running parallel to the slopes to minimize waterlogging. Under arid conditions, fresh cassava roots can be stored for up to four months by this method.

5.2. Drying Roots

After harvesting, roots may be peeled, sliced, grated immediately after peeling or dried whole to a form a product called 'chips'. Chips are dried for two to seven days on a raised platform, on roof tops, or on the ground. Usually the smaller the chip size, the shorter the drying time, and less storage rot (Anon, 1994). Chips are the main form of stored cassava in Africa.

Chips are stored in various locally home-made structures, baskets, polythene or jute bags. Home-made structures are made of straw, smeared with mud, and built on stilts (to keep the roots dry and away from rodents). Dried cassava roots are stored for six to eight months prior to use.

Chips are milled, and fed to animals or poultry or the flour is prepared into various food products (e.g. bread) or after mixing with flour from cereals such as maize. Cassava flour is also made by peeling roots, grating, and mechanically pressing it (to remove water), drying the grates, and milling it into flour. The flour is then dried in the sun, and may be roasted into granules referred to by various names, including 'gari' in West Africa, 'farinha' in Brazil.

Alternatively, fresh roots may be peeled and boiled, dried in the sun, and stored in baskets or jute bags. Parboiled roots may be stored for up to three months. Pillai and Rajamma (1987) have shown that parboiled cassava roots were less subjected to insect infestation in storage than the non-parboiled roots.

5.3. Processing into Food Products

Fermenting whole roots, and storing them for later use has been shown to be successful in reducing insect infestation in storage (Rajamma et al. 1996). Fermentation is done by peeling the roots, and soaking roots in water for 3-5 days or by heaping the roots for several days on a raised platform. After soaking in water, roots are pounded, and the mash kneaded into balls (5-10 cm in diameter). The balls are dried in the sun or smoked, and stored for later use. Balls coated with smoke are believed to repel insects and mold fungi and can be stored for up to six months.

Starch, a product of fermentation is an important industrial product (Balagopalan and Sundar, 1997) used in laundering for sizing paper, in making glue, and as a food ingredient or additive (Pedroarias et al. 1997). Alcohol is another product made from fermented cassava roots (Essers and Nout, 1997). To make starch, roots are crushed following soaking in water, and starch separated from the crushed materials by filtration, sedimentation, and separating the waste product as supernatant.

6. Postharvest Deterioration, Diseases and Insects

A major problem with cassava is the poor storability of roots once harvested. The condition known as postharvest physiological deterioration (PPD) sets in within twenty hours after harvesting, and limits marketability of cassava, and necessitates either prompt consumption or processing of harvested roots (Beeching et al. 1994). PPD occurs in two phases: primary and secondary deterioration (Raemakers et al. 1997). Symptoms of primary PPD are vascular streaking in the root, a ring of blue-brownish coloration in the intervening region between the outermost and innermost parts of the parenchymatous tissue. In the secondary phase of deterioration, roots become infested with microorganisms resulting in rot (Plumbley and Rickard, 1991). To reduce PPD, roots are rapidly cooled after harvest or coated with wax. Unfortunately most cassava growers cannot afford facilities to rapidly cool the roots or to buy the wax.

In storage, cassava roots are affected by a plethora of microorganisms, insects and rodents. In Table 1, we summarize the microorganisms associated with fresh and dried cassava products in storage.

Symptoms of microorganism infestation include black, green, yellow or red discoloration of the product. Because some of these pathogens excrete carcinogenic products, there is public concern to manage them better. Fumonisins, for example, are the most prominent class of mycotoxins produced by *Fusarium moniliforme* (Jardine and Leslie, 1999). *F. moniliforme* was found in association with stored cassava products (Msikita et al. 1996). Though the ability to produce fumonisin on cassava is yet to be elucidated, fumonisin is known to be toxic to livestock, particularly horses, and carcinogenic in test animals (Schumacher et al. 1995).

In surveys carried out in three countries (Benin, Ghana and Nigeria) of West Africa, we found that *Rhizopus* (47.5% of total samples) and *Aspergillus* spp (29.6% of total fungi isolated) were the main fungi affecting cassava chips in storage (Msikita, unpublished data). Both genera of fungi are saprophytic and capable of causing unsightly discolorations of chips. Also present on cassava chips were fungal genera known to be pathogenic (e.g. *Botryodiplodia theobromae* (Pat.), *Fusarium*, and *Pythium* spp). Their presence may suggest that infection of cassava occurred in the field, during harvest as well as in storage. We determined the presence of *Fusarium* spp infecting fresh and dried cassava roots and showed that infection varied by season and moisture in the roots (Table 2).

Table 1. Microorganisms and insects associated with rot of fresh and dried cassava products in storage

Product and associated /microorganism/insect	Reference
Fresh root	
(a) Bacteria	
Bacillus sp.	Majumder, 1955
(b) Fungi	
Armillaria heimii Sacc.	Makambila and Loubacky, 1998
Armillaria mellea (Vahl:Fr.) P. Kumm.	Lozano et al. 1981; Makambila and Loubacky, 1998
Aspergillus flavus Link:Fr	Noon and Booth, 1977; Ekundayo and Daniel, 1973
Aspergillus niger van Tiegh.	Ekundayo and Daniel, 1973
Aspergillus spp	Msikita et al. 1997
Corticium rolfsii Sacc.	Ranomenjanahary et al. 1994
Cylindrocarpon candidum (Link) Wollen	Ekundayo and Daniel, 1973
Diplodia manihotis Sacc.	Anonymous 1992
Fomes lignosus (Klot.) Bres. Syn.	Affran, 1976; Doku, 1969; Lozano et al. 1981
Rigidoporus lignosus (Klotzsch) Imazeki	
Fusarium oxysporum Schlechtend.:Fr.	Anonymous 1992
F. moniliforme J. Sheld	Msikita et al. 1996
F. solani (H. Mart. Sacc.)	Anonymous, 1992; Noon and Booth, 1977; Msikita et al. 1996; Ranomenjanahary
Lasiodiplodia theobromae (Pat.) Griffon & Maubl. (syn. *Botryodiplodia theobromae* Pat.)	et. al. 1994; Sanusi and Ikotun, 1998
Macrophomina phaseolina (Tassi) Goidanich	Ekundayo and Daniel, 1973; Msikita et al. 1997; Noon and Booth, 1977;
Nattrassia mangiferae (Syd. & P. Syd.) B. Sutton & Dyko (syn. *Hendersonula toruloidea* Nattrass)	Ranomenjanahary et al. 1994
	Msikita et. al. 1998a; Msikita et al. 1997
Penicillium spp	Msikita et al. 1997
Phytophthora cryptogea Pethybr & Lafferty	Msikita et al. 1997
P. drechsleri Tucker	Anonymous, 1992
P. erythroseptica Pethybr.	Anonymous, 1992; Lozano et al. 1981
P. nicotianae pv. *nicotianae* G.M. Waterhouse	Anonymous, 1992

(Contd.)

Table 1 (*Contd.*)

Product and associated /microorganism/insect	Reference
Pythium spp	Anonymous, 1992
Rhizopus spp	Lozano et al. 1981; Msikita et al. 1997
Rosellinia necatrix Prill.	Msikita et al. 1997
Sclerotium rolfsii Sacc.	Lozano et al. 1981
Scytalidium sp.	Lozano and Booth, 1974; Sanusi and Ikotun, 1998
Sphaerostilbe repens Berk. & Broome	Anonymous, 1992
Trichoderma harzianum Rifia	Sanusi and Ikotun, 1998
Trichoderma spp	Noon and Booth, 1977; Ekundayo and Daniel, 1973
Verticillium dahliae Kleb.	Msikita et al. 1997
(c) Nematodes	Anonymous, 1992
Helicotylenchus erythinae Sher	Ikotun and Osiru, 1990
Helicotylenchus spp.	Daudi and Saka, 1998
Meloidogyne incognita (Kofoid & White) Chitwood	Ikotun and Osiru, 1990
Meloidogyne javanica (Treub) Chitwood	Daudi and Saka, 1998
Pratylenchus branchyurus (Godfrey) Filipjev & Schuurmans-Stekhoven	Ikotun and Osiru, 1990
Pratylenchus spp	
Rotylenchus reniformis Linford & Oliveira	Daudi and Saka, 1998
Rotylenchus spp	Ikotun and Osiru, 1990
(d) Virus and virus-like	Daudi and Saka, 1998
Cassava brown streak virus	Frison and Feliu, 1991; Thresh et al. 1998
Frogskin disease	Frison and Feliu, 1991
(e) Insects	
Cydnidae sp	Lozano et al. 1981
Cyrtomenus bergi Froeschner	Lozano et al. 1981
Stictococcus vayssierei Richard	Ngeve, 1998; Lutete et al. 1997
Dried root	

(*Contd.*)

Table 1 (Contd.)

Product and associated /microorganism/insect	Reference
(a) Bacteria	
Agrobacterium spp	Noon and Booth, 1977
Bacillus spp	Noon and Booth, 1977
Erwinia spp	Noon and Booth, 1977
Xanthomonas spp	Noon and Booth, 1977
(b) Fungi	
Aspergillus spp	Msikita (unpublished)
B. theobromae	Msikita (unpublished)
Diplodia manihotis Sacc.	Burton, 1970
Fusarium spp	Burton, 1970
F. avenaceum (Fr.:Fr.) Sacc.	Msikita et al. 1998b
F. decemcellulare C. Brick	Msikita et al. 1998b
F. moniliforme	Msikita et al. 1996; Msikita et al. 1998b
F. oxysporum Schlechtend.:Fr.	Msikita et al. 1998b
F. proliferatum (Matsush.) Nirenberg	Msikita (unpublished)
F. semitectum Berk. & Ravenel	Msikita et al. 1998b
F. solani	Msikita et al. 1996; Msikita et al. 1998b
F. subglutinas (Wollenweb. & Reinking)	Msikita (unpublished); P.E. Nelson et al.
Pythium spp	Msikita (unpublished)
Penicillium spp	Msikita (unpublished)
Rhizopus spp	Burton, 1970; Msikita (unpublished)
Syncephalastrum sp	Burton, 1970; Msikita (unpublished)
Trichoderma spp	
(c) Insects	
Araecerus fasciculatus DeGeer	Pillai and Rajamma, 1987; Rajamma et al. 1996
Dinoderus bifoveolatus Linné	Pillai and Rajamma, 1987
Dinoderus minutus Fab.	Pillai and Rajamma, 1987; Haines, 1988; Rajamma et al. 1996
Heterobostrychus brunneus De Geer	

(Contd.)

Table 1 (*Contd.*)

Product and associated /microorganism/insect	Reference
Lasioderma serricorne Fab.	Anon, 1994
Palorus subdepressus (Wollaston)	Pillai and Rajamma, 1987; Rajamma et al. 1996
Prostephanus truncatus Horn	Anon, 1994
Rhyzopertha dominica Fab.	Anon, 1994
Sitophilus oryzae Linné	Anon, 1994
Sitophilus zeamais Linné	Rajamma et al. 1996
Stictococcus vayssierei	Anon, 1994
	Ngeve, 1998
Tribolium castaneum Herbst	Pillai and Rajamma, 1987
Tribolium confusum Duval	Anon, 1994

Table 2. Identity and prevalence of *Fusarium* spp on fresh and dried, stored cassava roots. Values are proportion (%) of the total *Fusarium* spp (n = 56) isolated in the dry and rainy seasons

Fusarium specie	Dry season		Rainy season	
	Fresh roots	Dried roots	Fresh roots	Dried roots
F. avenaceum	0.0^a	0.0^a	0.0^a	5.0^a
F. decemcellulare	2.1^a	0.0^a	0.0^a	0.0^a
F. moniliforme	3.1^a	9.5^c	1.0^a	30.0^c
F. oxysporum	6.3^b	5.0^b	17.4^c	1.0^a
F. proliferatum	0.0^a	0.0^a	5.1^b	0.0^a
F. semitectum	1.0^a	0.0^a	6.2^b	25.0^c
F. solani	2.1^a	0.0^a	55.6^d	10.0^b
F. subglutinans	0.0^a	0.0^a	0.0	5.0^a

Interestingly, *F. moniliforme* known to be carcinogenic to humans and animals (Schumacher et al. 1995) was found in large quantities on dried cassava products.

Some of the microorganisms associated with cassava rot in storage such as *T. harzianum*, *Pseudomonas fluorescens* Migula and *P. putida* have been shown to be antagonistic agents of some of the pathogenic agents such as *D. manihotis*, *F. solani*, *F. oxysporum*, *P. drechsleri* (Anonymous, 1992), *S. rolfsii* and *S. repens* (Anonymous, 1992; Sanusi and Ikotun, 1998); their role in storage rot remains unclear. Also fresh cassava roots tends to be infested with endophytic fungi e.g. *Septoria nodorum* Berk., *Alternaria tenuissima* (Kunze:Fr.) Wiltshire, *Torula* sp., *Nigrospora* sp. (Anonymous, 1992, Lozano and Laberry, 1993) whose role in postharvest deteriorations of cassava remains poorly understood.

Many species of nematodes have been found associated with fresh cassava roots. Their enzymes result in galls, their feeding results in wounds through which root rotting fungi may gain entry. The role of nematodes in subsequent postharvest deterioration of cassava is yet to be elucidated. Similarly cassava brown streak virus, and frog skin disease are known to harden the cassava roots, and render the roots unusable. Farmers sort and discard roots with visible symptoms of malformations, and/or discolorations prior to processing the roots for storage.

Dried cassava products are attractive to a wide range of insects. Common insects found on cassava include members of the Bostrychidae, particularly the larger grain borer (*Prostephanus truncatus* Horn). Large quantities of starch as found in cassava, appear to attract some of the insects (El-Sebay et al. 1995). Insects eat stored products, drill holes in the roots, and can turn the stored roots into worthless powder. It has been suggested that some may mechanically transmit fungi and other pathogens to stored cassava products.

7. Markets

In most of Africa, fresh and dried cassava (chips) are sold at local markets or transported from producing areas to neighboring communities.

Fresh and dried cassava roots are sold throughout the year with peak sales occurring when staples are in short supply. In Zambia for instance, trade in cassava is at the peak from December to February when locally grown maize is in short supply. Fresh and dried cassava products fetch premium prices, and are exported out of farmholds to major cities.

Some millers buy fresh cassava and process it into flour. Distillers process cassava into alcohol or starch. The market for cassava starch comprises a number of end users, including textiles, pharmaceuticals, paper, food and adhesive industries.

From West Africa, dried cassava products (mostly chips) are exported to Europe and the United States. In parts of West Africa, stems of improved cultivars are sold as planting materials just prior to the planting season (April to May). Fresh stems are sold in bundles, directly at farm gate or in open air markets.

8. Conclusions and Recommendations

Although cassava is an important food crop for millions of people in Africa, Asia and Latin America, it appears that little research emphasis has been placed on post-harvest systems for managing pest and diseases of stored products (Anonymous, 1994). Loss of stored cassava due to a combination of pest and diseases has been estimated at 20-40% after six months (Anonymous, 1994). Due to lack of knowledge on proper chemical use, and the cost of chemicals, cassava farmers very rarely use chemicals to protect their crop: only 3.3% of farmers use chemicals to protect their stored cassava (Anonymous, 1994). Postharvest rot could be avoided if basic information on minimizing physical damage to roots during harvesting, transportation, and trading was available (Nduguru et al. 1998).

Cassava farmers use wide ranging pre-harvest management practices and their effects on stored products remain unknown. For instance, it is well documented that small-scale farmers often intercrop cassava with legumes, cereals, or even tuber crops (Nweke and Lynam, 1997) which ameliorate the cassava crop, and affect the quality of the roots harvested (Anonymous, 1990; Ikeorgu et al. 1998). Unfortunately little information is available on a range of these pre-harvest management practices on postharvest related problems (e.g. PPD, pests and diseases) encountered in cassava. Other pre-harvest management practices whose effects on post-harvest related problems need elucidation include mulching (live and dead mulches), weeding, fertilizing practices (rate, type of fertilizer, and timing), tillage or no tillage, cultivars, spacing, leaf harvesting practices, and pruning the stems before harvest.

Similarly, for long-term storage, farmers use different types of storage structures. The overall objective is to keep the roots dry and protected from insect pests, diseases and rodents. It is quite common to find fermented cassava kneaded into balls of various sizes, which are later smoked on fire. Coating the balls with smoke repels insects, mold and rodents.

There are a lot of unknowns to this practice (e.g. optimum size of the balls, duration of smoking the balls, cultivar responses). Similarly, in parts of West Africa it is not uncommon to find farmers adding cow dung to the mud smeared on the storage structure to repel insects and micro-organisms affecting stored cassava products. The full benefits of such practices remain to be elucidated. There is therefore, need to investigate such indigenous knowledge, and even improve on it by incorporating locally available but proven biopesticides such as neem (*Azadirachta indica* L.) into the storage structure.

A plethora of micro-organisms and insects are associated with cassava in storage. Some of the micro-organisms infect cassava roots while in the field, and some are attracted to the roots during storage. Conditions of storage, and moisture content are some of the factors that attract some of the micro-organisms in storage (Msikita et al. 1998b). Optimum conditions that repel the micro-organisms and insects need to be determined. Toxin producing fungi such as *Fusarium moniliforme*, and some fungi known to cause diseases on humans and animals such as *Nattrassia mangiferae* (syn. *Hendersonula toruloidea*) (Frankel and Rippon, 1989; Jones et al. 1985; Knudtson and Kirkbride, 1992) have been isolated from fresh and stored cassava products. Effects on humans of such potentially harmful fungi isolated from cassava need to be determined. Encouraging is that cultural practices, when understood and implemented, could improve cassava production and storage without requiring increase in pesticide use.

References

Affran, D.K. 1976. Cassava and its economic importance. *Ghana Farmer* **12(4)**: 172-178.
Annan, K.M., Oldham, J.H., Amoko-Attah, C.K., Ashagbley, J.A., Ellis, W.O., Dzogbefia, V.P. and B.KK. Simpson. 1994. An innovative method of preserving cassava using a passive evaporative cooler. *In:* 10th Symposium of the International Society for Tropical Root Crops (ISTRC). Salvador, Bahia, Brazil, 13-19 November 1994 (Abstract).
Anonymous, 1990. Cassava in tropical Africa. A reference manual. International Institute of Tropical Agriculture, Ibadan, Nigeria, 176 p.
Anonymous, 1992. Cassava program 1987-1991. Working document No 116. Centro Internacional de Agricultura Tropical, Cali, Colombia, 473 p.
Anonymous, 1994. Survey on post-harvest systems of yams and cassava. Post-Harvest Project/Ministry of Food and Agriculture/GTZ, Accra, Ghana, 29 p.
Ashagbley, J.A., Amoako-Attah, C.K., Oldham, J.H., Annan, K.M., Ellis, W.O., Dzogbefia, V.P. and B.K. Simpson. 1994. A novel method for preserving cassava using coating of gums. *In:* 10th Symposium of the International Society for Tropical Root Crops (ISTRC). Salvador, Bahia, Brazil, 13-19 November 1994 (Abstract).
Balagopalan, C. and P. Sundar. 1997. Integrated technology for the treatment and management of cyanoglucosides in cassava starch and sago factory waste waters. *African Journal of Root and Tuber Crops* **2(1,2)**: 69-73.
Beeching, J.R., Dodge, A.D., Moore, K.G., Phillips, H.M. and J.E. Wenham. 1994. Physiological deterioration in cassava: possibilities for control. *Tropical Science*. **34**: 335-343.
Burton, C.L. 1970. Diseases of tropical vegetables on the Chicago Market. *Tropical Agriculture (Trinidad)* **47(4)**: 303-313.
Cock, J.H. 1985. *Cassava, a New Potential for a Neglected Crop*. Westview Press, Boulder, CO, 191 p.

Daudi, A.T. and V.W. Saka. 1998. Diseases and nematodes of cassava and sweet potato in Malawi. *In:* Root Crops and Poverty Alleviation. *In:* Proceedings of the Sixth Triennial Symposium of the International Society for Tropical Root Crops—Africa Branch. M.O. Akoroda and I.J. Ekanayake (Eds), IITA, Ibadan, Nigeria, pp. 44-47.

Doku, E.V. 1969. Cassava in Ghana. Ghana University Press. Accra, Ghana.

Ekundayo, J.A. and T.M. Daniel. 1973. Cassava rot and its control. *Transactions of the British Mycological Society* **61**: 27-32.

El-Sebay, Y, Helal, H. and H. Hoda. 1995. Wood-borers and wooden trees relationship. *Egyptian Journal of Agricultural Research* **73 (1)**: 135-153.

El-Sharkawy, M.A. 1993. Drought-tolerant cassava for Africa, Asia, and Latin America. *BioScience* **43(7)**: 441-451.

Essers, A.J.A. and M.J.R. Nout. 1997. Application of microbial starter cultures for new and traditional cassava products. *African Journal of Root and Tuber Crops* **2(1,2)**: 110-113.

Frankel, D.H. and J.W. Rippon. 1989. *Hendersonula toruloidea* infection in man. Index cases in the non-endemic North American host, and a review of the literature. *Mycopathologia* **105(3)**: 175-186.

FAO (Food and Agriculture Organization). 1989. Utilization of tropical foods: Roots and tubers. Rome, Italy.

Frison, E.A. and E. Feliu. 1991. FAO/IBPGR Technical guidelines for the safe movement of cassava germplasm. Food and Agriculture Organization of the United Nations/International Board for Plant Genetic Resources, Rome, Italy, 47 p.

Haines, C.P. 1988. A new species of predatory mite (Acarina: Cheyletidae) associated with bostrichid beetles on dried cassava. *Acarologia* **29 (4)**: 361-375.

Ikeorgu, J.E.G., Eke-Okoro, O.N. and S.O. Odurukwe. 1998. Evaluation of cowpea varieties for intercropping with cassava in southeast Nigeria. *In:* Root Crops and Poverty Alleviation. Proceedings of the Sixth Triennial Symposium of the International Society for Tropical Root Crops—Africa Branch. M.O. Akoroda and I.J. Ekanayake (Eds), IITA, Ibadan, Nigeria, pp. 393-395.

Ikotun, T. and D.S.O. Osiru. 1990. Manioc en Afrique. Un manuel de référence. IITA/ UNICEF, Ibadan, Nigeria, 190 p.

Jardine, D.J. and J.F. Leslie. 1999. Aggressiveness to mature maize plants of *Fusarium* strains differing in ability to produce fumonisin. *Plant Disease* **83**: 690-693.

Jones, S.K., White, J.E., Jacobs, P.H. and C.R.C. Porter. 1985. *Hendersonula toruloidea* infections of nails in Caucasians. *Clinical and Experimental Dermatology* **10(5)**: 444-447.

Knudtson, W.U. and C.A. Kirkbride. 1992. Fungi associated with bovine abortion in the northern plain states (USA). *Journal of Veterinary Diagnostic Investigation* **4(2)**: 181-185.

Lahai, M.T., Ekanayake, I.J. and J.B. George. 1997. Reaction of leaf retention and pest and disease incidence of cassava (*Manihot esculenta* Crantz) genotypes to leaf harvesting. *In:* 10th Symposium of the International Society for Tropical Root Crops (ISTRC). Salvador, Bahia, Brazil, 13-19 November 1994 (Abstract).

Lozano, J.C. and R.H. Booth. 1974. Diseases of cassava (*Manihot esculenta*). *Pans* **20(1)**: 30-51.

Lozano, C. and R. Laberry. 1993. Endophytic fungi are also found in cassava. *Cassava Newsletter* **17(2)**: 4-6.

Lozano, J.C., Byrne, D. and A. Bellotti. 1980. Cassava ecosystem relationships and their influence on breeding strategy. *Tropical Pest Management* **26**: 180-187.

Lozano, J.C., Bellotti, A. Reyes, J.A., Howeler, R., Leihner, D. and J. Doll. 1981. Field problems in cassava. Centro Internacional de Agricultura Tropical, Cali, Colombia, 205 p.

Lutete, D., Tata-Hangy, K. and T. Kasu. 1997. Occurrence in Zaire of *Stictococcus vayssierei*, a pest on cassava (*Manihot esculenta*). *Journal of African Zoology* **111 (1)**: 71-73.

Majumder, S.K. 1955. Some studies on microbial rot of Tapioca. *Bulletin of the Central Food Technology Research Institute (Mysore)* **4**: 164.

Makambila, C. and V.U. Loubacky. 1998. Description d'une methode d'evaluation du pouvoir pathogene des isolats de *Armillaria heimii*, agent pathogene du pourridie du manioc. *In:* Root crops and poverty alleviation. Proceedings of the Sixth Triennial Symposium of the International Society for Tropical Root Crops—Africa Branch. M.O. Akoroda and I.J. Ekanayake (Eds), IITA, Ibadan, Nigeria, pp. 144-145.

Mignouna, H.D. and A.G.O. Dixon. 1997. Genetic relationships among cassava clones with varying levels of resistance to African mosaic disease using RAPD makers. *African Journal of Root and Tuber Crops* **2(1&2)**: 28-32.

Msikita, W., Nelson, P.E., Yaninek, J.S., Ahounou, M. and R. Fagbemissi. 1996. First report of *Fusarium moniliforme* causing cassava root, stem, and storage rot. *Plant Disease* **80(7)**: 823.

Msikita, W., Yaninek, J.S., Ahounou, M., Baïmey, H. and C.O. Fagbemissi. 1997. First report of *Nattrassia mangiferae* root and stem rot of cassava in West Africa. *Plant Disease* **81(11)**: 1332.

Msikita, W., James, B., Wilkinson, H.T., and J.H. Juba. 1998a. First report of *Macrophomina phaseoling* Causing pre-harvest cassava root rot in Bénin and Nigeria. *Plant Disease* **82(12)**: 1402.

Msikita, W., Yaninek, J.S., Ahounou, M., Fagbemissi, C.O., Hountondji, F. and K. Green. 1998b. Identification of *Fusarium* spp associated with cassava chips. *In:* Root Crops and Poverty Alleviation. Proceedings of the Sixth Triennial Symposium of the International Society for Tropical Root Crops—Africa Branch. M.O. Akoroda and I.J. Ekanayake (Eds), IITA, Ibadan, Nigeria, pp. 136-139.

Nduguru, G.T., Modaha, F., Bancroft, R.D., Digges, P.D., Kleih, U., Westby, A. and F. Mashamba. 1998. The use of needs assessment methodologies to focus technical interventions in root and tuber crop post-harvest systems: A case-study to improve the marketing and post-harvest handling of cassava entering Dar-es-Salam, Tanzania. *In:* Root Crops and Poverty Alleviation. Proceedings of the Sixth Triennial Symposium of the International Society for Tropical Root Crops—Africa Branch. M.O. Akoroda and I.J. Ekanayake (Eds), IITA, Ibadan, Nigeria, pp. 76-82.

Nestel, B. 1993. A new outlook for an ancient crop. *Biotechnology* **1**: 53-59.

Noon, R.J.A. and R.H. Booth. 1977. Nature of postharvest deterioration of cassava roots. *Transactions of the British Mycological Society* **69**: 287-290.

Ngeve, J.M. 1998. Outbreak of a new tuberous root mealybug (*Stictococcus vayssierei*) [Homoptera: Stictococcidae] of cassava (*Manihot esculenta* Crantz) in Cameroon. *In:* Root Crops and Poverty Alleviation. Proceedings of the Sixth Triennial Symposium of the International Society for Tropical Root Crops—Africa Branch. M.O. Akoroda and I.J. Ekanayake (Eds), IITA, Ibadan, Nigeria, pp. 153-156.

Nweke, F.I. 1998. The role of cassava production in poverty alleviation. *In:* Root Crops and Poverty Alleviation. Proceedings of the Sixth Triennial Symposium of the International Society for Tropical Root Crops—Africa Branch. M.O. Akoroda and I.J. Ekanayake (Eds), IITA, Ibadan, Nigeria, pp. 102-110.

Nweke, F.I. and J.K. Lynam. 1997. Cassava in Africa. *African Journal of Root and Tuber Crops* **2(1&2)**: 10-13.

O'Brien, G.M., van Oirschot, Q., Orozco, O., Chaves, A.L. and J. Mayer. 1997. The effects of pre-harvest pruning of cassava upon post-harvest deterioration potential, scopoletin and dry matter contents. *African Journal of Root and Tuber Crops* **2(1,2)**: 120-123.

Pedroarias, V., Kafka, A., Rovedo, C., de Fabrizio, S.V., Suarez, C. and G. Chuzel. 1997. Behavior of starch suspensions from different cassava varieties subjected to thermal treatment and lactic fermentation. *African Journal of Root and Tuber Crops* **2(1,2)**: 128 (Abstract).

Pillai, K.S. and P. Rajamma. 1987. Storage pests of tuber crops. In Annual Progress Report, India Central Tuber Crops Research Institute, Kerala, India, pp. 89-90.

Plumbley, R.A. and J.E. Rickard. 1991. Post-havest deterioration of cassava. *Tropical Science* **31**: 295-303.

Raemakers, C.J.J.M., Jacobsen, E. and R.G.F. Visser. 1997. Micropropagation of *Manihot esculenta* Crantz (cassava). *In:* Biotechnology in Agriculture and Forestry, Vol. 39. High-Tech and Micropropagation V. Y.P.S. Bajaj (Ed.), Springer-Verlag, Berlin, p. 77-102.

Rajamma, P., George, M. and K.R. Lakshmi. 1996. A comparative study on insect infestation of fermented and non-fermented cassava chips. *Journal of Root Crops* **22(2)**: 82-87.

Ranomenjanahary, S., Andrianaivo, A., Andrianiahanana, M. and R. Rasolofo. 1994. Epidemologie des maladies du manioc sur les Hauts Plateux de Madagascar. *In:* 10th Symposium of the International Society for Tropical Root Crops (ISTRC), Salvador, Bahia, Brazil, 13-19 November (Abstract).

Sanusi, A.F. and T. Ikotun. 1998. Rhizosphere microorganisms of cassava and their effects on plant health. *In:* Root Crops and Poverty Alleviation. Proceedings of the Sixth Triennial Symposium of the International Society for Tropical Root Crops—Africa Branch. M.O. Akoroda and I.J. Ekanayake (Eds), IITA, Ibadan, Nigeria, pp. 183-189.

Schumacher, J., Mullen, J., Shelby, R., Lenz, S., Ruffin, R.D. and B.W. Kemppainen. 1995. An investigation of the role of *Fusarium moniliforme* in dudenitis/proximal jejunitis of horses. *Veterinary and Human Toxicology* **37**: 39-45.

Thresh, J.M., Fargette, D. and G.W. Otim-Nape. 1998. The viruses and virus diseases of cassava in Africa: what is the magnitude of the problem? *In:* Root Crops and Poverty Alleviation. Proceedings of the Sixth Triennial Symposium of the International Society for Tropical Root Crops—Africa Branch. M.O. Akoroda and I.J. Ekanayake (Eds), IITA, Ibadan, Nigeria, pp. 126-130.

14

Environmental Control for Storage of Rooted Propagation Material

Nihal C. Rajapakse
Department of Horticulture, Clemson University, Clemson, SC 29634, USA

Abbreviations

ABA	Abscisic acid
ACC	1-aminocyclopropane-1-carboxylic acid
CA	controlled atmosphere
CO_2	carbon dioxide
F_o	minimum flurescence
F_m	maximum fluorescence
F_v	variable fluorescence $F_m - F_o$
GA	gibberellins
IAA	indoleacetic acid
LP	low pressure
PS	photosystem

1. Introduction

Most horticultural crops, especially bedding plants, are produced as rooted plug transplants from seeds, vegetative cuttings or tissue culture plantlets. Under ideal conditions, storage of plugs is not necessary because they should be transplanted in the field when they reach the correct size. However, ideal conditions rarely exist in the 'real world' and storage of transplants is necessary due to a variety of reasons.

Short-term storage (holding) of several days up to about two weeks is often necessary because of labor shortages and unfavorable weather conditions prevailing during field planting. In such conditions, short-term storage in the greenhouse with chemical growth regulator applications or withholding water and nutrients has been traditionally used to slow down the growth of plug transplants. Greenhouse holding is disadvantageous to propagators because it reduces the production capacity. In addition, the use of chemical growth regulators has been banned on vegetable transplants due to perceived risks to humans thus making greenhouse holding difficult. Short-term storage of transplants in separate rooms with environmental control provides a better alternative to greenhouse holding because this can free the valuable greenhouse space, increase production capacity, facilitate crop scheduling, and minimize the stress to stored plants, if proper storage conditions are used.

Plant propagators in temperate regions are often faced with difficulties in meeting the demand for plant material because of the high demand within a narrow market window. Long-term storage methods of four weeks or more would allow plant propagators in these regions to produce plants early in the season and release when the demand is high thus avoiding shortage of planting material during peak demand.

Considerable amounts of plant material are now produced in tropical regions and are shipped to temperate regions because of the favorable environmental conditions and low operating and labor costs prevailing in tropical countries. These materials mostly include tropical plants that are sensitive to low temperature. Plant propagators in tropical regions can benefit from both short- and long-term storage methods because this would allow them to use relatively inexpensive but lengthy truck or ship transportation in place of costly but fast air transportation for movement of plant material.

Many perennial plants with temperate origin require exposure to low temperatures (5-9° C) for breaking dormancy and floral induction. Most commercial nurseries achieve this low temperature requirement by 'overwintering' rooted liners in greenhouses. However, in mild winter regions, greenhouse 'over-wintering' may not provide sufficient chilling to break the dormancy. As a result, shoot emergence and flowering may be delayed and plants may be weak. In these regions, temperature controlled storage rooms can be used not only to hold the plants but also to satisfy the chilling requirement for breaking bud dormancy and flower induction (Wood and Cameron, 1989).

Low temperature storage is widely used to delay the postharvest senescence of horticultural commodities. However, storage requirements for rooted plant material are different from non-rooted material or other horticultural produce because of the active growth of rooted material. The main objectives of rooted plant material storage are to minimize growth and development during storage, sustain photosynthetic potential, and maintain visual quality so that the consumer acceptance or the growth after field planting will not be adversely affected. In this chapter, environmental controls for preserving quality of transplants and physiological changes taking place during storage of these plant materials will be discussed.

2. Requirements for Successful Storage

Survival of plants during storage and recovery after removal from storage depends on plant factors and the environmental conditions under which they are stored. For successful storage, high quality plants that can withstand storage stress should be selected. Storage environment must be chosen to minimize the damage during storage and to ensure the growth after transplanting in the field.

2.1. Plant Factors

2.1.1. Genotype and Cultivar Differences

Storage can be beneficial or detrimental depending on the sensitivity of a crop to low temperature. The genetic makeup of species and cultivars influences their sensitivity to low temperatures. Generally, plants from temperate regions and high altitudes are less prone to chilling injury whereas plants of tropical and sub-tropical origin are sensitive to chilling injury when stored below a critical temperature. The differences in cultivar sensitivity to chilling injury have been reported with many crops including tomato [*Lycopersicon esculentum* Mill.] (Patterson and Payne, 1983), bean [*Phaseolus vulgaris* L.] (Dickson and Boettger, 1984), and *Aglaonema* (Hummel and Henny, 1986).

Chilling injury results in loss of visual quality as damaged plants have symptoms such as wilting due to excessive water loss and/or reduced water uptake (Rikin and Richmond, 1976; Wright, 1974; Wilson, 1976). When sensitive plants are exposed to temperatures below the critical chilling temperature, lipid structure of cell membranes changes thus, disrupting the physical and chemical organization (Murata, 1990; Murata and Nishida, 1990). As a result, membrane permeability and solute leakage can be increased in injured plants. If the damage is not severe, plants can recover from chilling injury once transferred to non-chilling temperatures. However, prolonged exposure to chilling temperatures or the exposure to very low temperatures below the critical chilling temperature can cause irreversible damage leading to death of transplants (Creencia and Bramlage, 1971; Wright and Simon, 1973). Although mildly damaged plants can recover from chilling injury symptoms, these plants have been reported to have reduced photosynthesis and therefore, growth after field planting can be retarded (Lasley et al. 1979; Yakir et al. 1986).

2.1.2. Quality of Plants

The initial quality of plant material plays an important role in determining the length of storage without significant quality loss. Plants that are intended for storage should be free of diseases and physical damages because physical injuries can induce ethylene production, increase transpiration, increase respiration rates, and allow the spread of diseases, all of which can deteriorate the quality of plants during storage and shorten the acceptable storage duration.

Water stress, carbohydrate reserves, and fertilization also contribute to the initial quality of plant material. Plants should be adequately watered, as both excess water and water stress can shorten storage life. If plants are intended for long-term dark storage, treatments to improve carbohydrate reserves may be beneficial before storage. It has been shown that rooted chrysanthemum (*Dendranthema grandiflorum* Ramat. Kitamura) cuttings with 'good' storage potential (4 to 6 weeks at 0°C or 3°C) contained higher levels of soluble sugars and fructans and had a lower respiration rate than the cultivars that had 'poor' storage potential (Rajapakse et al. 1996).

Feeding with high nitrogen should be avoided because excessive nitrogen promotes shoot growth and increases the susceptibility to diseases. Heins et al. (1994a) reported that fertilization of *Petunia* plug transplants with 200 to 400 mg L^{-1} of nitrogen and potassium for one week prior to storage increased the incidence of *Botrytis cinerea* during low temperature (7.5° C) storage. Nitrogen fertilization prior to storage has been also reported to increase the mortality of geranium (*Pelargonium* x *hortorum* Baily) seedlings due to infection by *Pythium* (Gladstone and Moorman, 1989; Kaczperski et al. 1996). Application of fungicides prior to storage can reduce the risk of developing fungal diseases during storage (Heins et al. 1994b). However, foliage should be allowed to dry after the fungicide treatment before placing in storage room.

2.1.3. Pre-conditioning

Transplants are commercially produced under optimal environmental conditions for growth and development. Sudden changes in environmental conditions to sub-optimal levels during storage can result in loss of plant quality because of chilling injury, increase in leaf yellowing, excessive stem elongation, and abscission of leaves and buds. Preconditioning or hardening of plants by gradual reduction of temperature (Hatton, 1990; Wheaton and Morris, 1967; Wilson, 1976) and light (Krizek et al. 1985), withholding water and nutrients (Wilson, 1976), or chemical growth regulator application (Jaffe and Isenberg, 1965; Rikin and Richmond, 1976; Sasson and Bramlage, 1981) can enhance the ability of most plants to withstand low temperature stress without detrimental effects.

Low temperature preconditioning can enhance the chilling tolerance of sensitive plants by changing the fatty acid composition of membrane lipids. Wilson and Crawford (1974) reported that preconditioning of bean seedlings at 12° C for 4 days prior to exposure to 5° C increased the degree of unsaturation of fatty acids in the phospholipid fraction but did not affect the fatty acid composition of the glycolipid fraction. They also reported that the degree of unsaturation and the phospholipid content decreased as the physiological age increased thus, requiring longer exposure of old plants to attain the maximum protection against chilling injury.

Carbohydrate level of plants has been shown to play a role in improving low temperature tolerance of sensitive plants. King et al. (1988) reported that tomato seedlings were most sensitive to chilling injury if plants were exposed to low temperature at the end of the dark period. Brief exposure to light (as short as 10 minutes for 1200 μmol m^{-2} s^{-1}) prior to storage improved the low temperature tolerance. They also reported that low intensity light (120 μmol m^{-2} s^{-1}) was less effective than high intensity (1200 μmol m^{-2} s^{-1}) in reducing chilling sensitivity. In contrast, Krizek et al. (1985) reported that coleus (*Coleus blumei* Benth) plants exposed to low light (8 μmol m^{-2} s^{-1}) prior to exposure to low temperatures (5° C) showed increased protection against chilling injury compared to plants exposed to high light (320 μmol m^{-2} s^{-1}).

Chemical growth regulators have been shown to improve the low temperature tolerance of seedlings. Jaffe and Isenberg (1965) reported that B-nine [Butanedioic acid mono (2,2-dimethylhydrazide)] treated cucumber (*Cucumis sativus* L.) seedlings were less sensitive to chilling injury. Abscisic acid (10^{-4}M ABA) has been shown to reduce chilling injury symptoms in cucumber seedlings by reducing water loss (Rikin and Richmond, 1976; Sasson and Bramlage, 1981). Tantau and Dorffling (1991) reported that poinsettia (*Euphorbia pulcherrima* Willd.) cultivars with higher ABA content were more tolerant to low temperatures. Several triazole plant growth regulators, paclobutrazol [($\pm$)-(R*, R*)-β- ((4-chlorophenyl)methyl)-α-(1,1-dimethylethyl)-1H-1,2,4-triazole-1-ethanol)], uniconazole [E-1-(p-chlorophenyl)-4,4-dimethyl-2-(1,2,4-triazol-1-yl)-1-penten-3-ol], and triadimefon [1-(4-chlorophenoxy)-3,3-dimethyl-1-(1,2,4-triazol-1-yl)-2-butanone] have been also shown to reduce the chilling sensitivity of plants possibly by protecting the cell membrane (Asare-Boamah and Fletcher, 1986; Lee et al. 1985; Senaratna et al. 1988). Triazole compounds are known to inhibit gibberellin (GA) biosynthesis and the increased tolerance of triazole treated plants to low temperatures has been attributed to the reduction in GA content and/or the increase in ABA/GA ratio (Davis et al. 1988; Reid et al. 1974; Waldman et al. 1975). Although growth regulators are shown to improve low temperature tolerance, these should be used carefully because of the recent restrictions on food crops.

2.2. Environmental Factors

Temperature, light, gas composition and relative humidity are among the major environmental factors that affect the quality of stored plant material. Environmental conditions under which plants are stored or shipped need to maintain plants in a slow-growth stage, minimize water loss, respiration rate, spread of fungal diseases, and loss of chlorophyll while maintaining photosynthetic potential and visual quality.

2.2.1. Temperature

Lowering temperature is the most effective method in reducing metabolic processes that lead to deterioration of horticultural crops. Low temperature can also suppress the development and spread of diseases. However, as discussed earlier, the extent to which temperature can be reduced depends on the sensitivity of plants to low temperature. In most low temperature sensitive crops, chilling injury occurs when stored below 10° C to 12° C (Couey, 1982). In general, severity of chilling injury increases as the temperature to which the plants are exposed decreases and/or the duration of exposure to temperature below critical temperature increases. Therefore, the quality of chilling sensitive crops can be better maintained by storing those plants above the critical temperature that chilling damage occurs and/or by storing for a shorter duration. Heins et al. (1994a) evaluated the chilling sensitivity and storage potential of several widely used ornamental bedding plant species at various low temperatures and their results and optimum temperature recommendations are shown in Table 1.

Table 1. Maximum storage durations, in weeks, for selected bedding plant species stored at various temperatures under 1 μmol m^{-2} s^{-1} of photosynthetic photon flux from fluorescent bulbs (L) or in dark (D). Adapted from Heins et al. 1994

Species	0° C		2.5° C		5° C		7.5° C		10° C		12.5° C	
	L	D	L	D	L	D	L	D	L	D	L	D
Ageratum (*Ageratum houstonianum* Mill.)	1	1	2	1	6	5	6	6	6	3	6	3
Alyssum (*Lobularia maritima*	6	5	6	4	6	4	6	2	3	1	3	1
Begoina Fibrous (*Begonia* Semperflorens-Cultorum)	0	0	6	4	6	6	6	5	6	5	4	0
Begonia Tuberous(*Begonia x tuberosa*)	3	3	4	3	6	3	6	4	4	3	4	3
Celosia (*Celosia argenta*)	0	0	1	1	1	1	1	1	3	2	3	2
Cyclamen (*Cyclamen persicum*)	6	6	6	6	6	6	6	5	6	5	6	4
Dehlia (*Dahlia x hybriba*)	1	1	4	2	5	2	6	2	6	2	5	2
Geranium (*Pelargonium x hortorum*)	4	4	4	4	4	4	4	4	4	4	4	2
Impatients (*Impatiens walleriana*)	0	0	2	2	3	3	6	6	6	5	6	4
Lobelia (*Lobelia erinus*)	0	0	5	5	6	6	5	5	4	4	4	0
Marigold French (*Tagetes patula*)	0	0	3	1	6	3	6	3	5	2	3	2
New Guniea Impatients (*Impatiens hawkeri*)	0	0	0	0	0	0	0	0	0	0	3	2
Pansy (*Viola x wittrockiana*)	6	6	6	6	6	6	6	6	6	6	6	6
Petunia (*Petunia x hybrida*)	6	6	6	6	6	6	6	5	6	5	6	4
Portulaca (*Portulaca grandiflora*)	1	1	3	3	5	5	5	5	6	6	4	6
Salvia (*Salvia splendens*)	0	0	0	0	6	6	6	6	6	4	6	4
Tomato (*Lycopersicon esculentum*)	0	0	1	1	2	2	3	3	1	1	1	1
Verbena (*Verbena x hybrida*)	3	1	2	2	1	1	1	1	4	1	4	1
Vinca (*Catharanthus roseus*)	0	0	2	2	3	3	6	4	6	5	6	5

Temperate plants can generally tolerate near-freezing temperatures, as they are less sensitive to chilling injury. These crops can be stored for longer durations at or near freezing temperatures. The freezing temperature varies among species but, in general, freezing occurs below the freezing point of water due to the freezing point depression by solutes. Rudnicki et al. (1991) reported that rooted chrysanthemum cuttings could be stored at –0.5° C to –1.6° C for 3 to 6 weeks without freeze damage. Rajapakse et al. (1996) reported that rooted chrysanthemum cuttings could be stored between 0° C and 3° C but the quality upon removal from storage varied among cultivars. For example, their results show that the visual quality of garden chrysanthemum cultivars 'Anna' and 'Debonair' did not deteriorate during 4 weeks of storage at 0° C or 3° C, but 'Emily' and 'Naomi' cultivars were of unacceptable quality after 1 week in storage due to leaf necrosis. Wilson et al. (1998a, 1999b) reported that the quality and the recovery of in vitro broccoli (*Brassica oleracea* L.) seedlings stored at 1° C was better than those stored at 5° C or 10° C.

2.2.2. Relative Humidity

Relative humidity in the storage room is an important environmental factor that determines the water relations of stored plant material. Water stress has been shown to induce production of ethylene and abscisic acid, both of which can promote leaf and bud abscission, promote chlorophyll degradation and accelerate senescence (Spikman, 1986). Since herbaceous plants contain a considerable amount of water (over 90%) in their tissues, a small change in water content (2% to 5%) can create a water stress and adversely affect the developmental processes. Because of difficulties in irrigation during storage, plant water content during storage must be maintained mainly by reducing the water loss from the plants and soil. Plug transplants with large soil volume can hold more water and therefore, plants grown in large containers can generally maintain a better water status than the small plugs during storage.

Water loss from a plant primarily depends on the water vapor pressure deficit between leaves and the storage atmosphere and the magnitude of resistance to water vapor. At a given temperature, a decrease in relative humidity increases the water vapor pressure deficit between the tissue and the surrounding environment. For example, a decrease in relative humidity from 80% to 60% nearly doubles the water vapor pressure deficit (Table 2). Therefore, low relative humidity in storage increases the risk of desiccation and water stress because of the water loss through transpiration and direct evaporation from the medium.

Table 2. Effect of temperature and relative humidity on water vapor pressure deficit (VPD)

Temperature (° C)	Relative humidity (%)	VPD (mm Hg)
25	80	0.63
	60	1.27
20	80	0.47
	60	0.94
15	80	0.34
	60	0.68
10	80	0.25
	60	0.49
5	80	0.17
	60	0.35
0	80	0.12
	60	0.24

Saturated water vapor pressure at 25, 20, 15, 10, 5, and 0° C are 3.16, 2.33, 1.70, 1.22, 0.87, and 0.61 kPa, respectively.
Source: Handbook of chemistry and physics. CRC Press, Boca Raton, FL.

Maintaining high humidity (95-100%) is optimal for low temperature storage of herbaceous plants because it reduces water loss and possibly prevents chilling injury to sensitive crops. Wright and Simon (1973) reported that 100% relative humidity prevented the water loss and the development of chilling injury symptoms in cucumber seedlings exposed to 5° C whereas 85% relative humidity increased the chilling injury symptoms. Although the respiration rate and photosynthesis were reduced in plants exposed to 5° C at 100% humidity, they reported that both respiration and photosynthesis returned to initial levels upon return to warm conditions. However, in plants exposed to 5° C at 85% humidity, respiration and photosynthesis did not recover indicating that those plants were permanently damaged.

At a given temperature, misting in the storage room can increase relative humidity. However, it is difficult to maintain high humidity by misting at low temperatures, especially near freezing point. Alternatively, relative humidity can be increased by covering plants with plastic films that are impermeable to water vapor. However, if light is provided in storage, covering with plastic films can affect photosynthesis due to rapid depletion of CO_2 caused by limited ventilation. A rapid depletion of CO_2 to very low levels (< 130 $\mu L\ L^{-1}$) inside chrysanthemum plug trays wrapped with plastic film and stored under light (23 or 34 $\mu mol\ m^{-2}\ s^{-1}$ of photosynthetic photon flux from cool white fluorescent bulbs) has been observed (Rajapakse et al. 1996). Heins et al. (1994a) reported that exposure to 1 $\mu mol\ m^{-2}\ s^{-1}$ of light during storage greatly improved the overall quality of many bedding plants. Under such low light levels, CO_2 depletion may not be a problem.

At a given relative humidity, lowering temperature decreases the water vapor pressure deficit between plant tissue and the storage environment because saturated water vapor pressure decreases as the air temperature decreases. For example, at 60% relative humidity, water vapor pressure deficit at 25° C is over 3 times greater than at 5° C (Table 2). Therefore, lowering temperature provides an effective mean of reducing water loss at a given humidity.

Water loss from plants mainly takes place through stomata (90 to 95%) and to a lesser extent through the cuticle (5 to 10%). Therefore, water loss can be reduced by increasing the resistance of leaf surface to water vapor by the use of antitranspirants (Martin and Link, 1973; Nitzsche et al. 1991). However, Whitcomb et al. (1974) found that several antitranspirants were beneficial only under high humidity conditions because the antitranspirant coating tends to peel off under low humidity. High air velocity in storage rooms can decrease the boundary layer resistance to water vapor thus increasing the water loss from the plant.

2.2.3. Light

Storage in dark is the common practice but, dark storage often results in reduction of plant quality and poor field establishment when transplanted in the field (Koranski et al. 1989). Dark storage induces loss of chlorophyll, leaf abscission, loss of carbohydrate reserves, and susceptibility to pathogens (Behrens, 1988; Conover, 1976; Curtis and Rodney, 1952; Hansen et al. 1978; Smith, 1982). Paton and Schwabe (1987) reported that low temperature (4° C) dark storage reduced rooting ability of *Pelargonium* cuttings while pretreatment of cuttings with 2% to 5% sucrose for 24 h or providing light in storage improved rooting ability.

Chlorophyll containing tissues, such as plug transplants, have the potential to photosynthesize during storage. Therefore, by providing light in storage, photosynthesis can be maintained and loss of carbohydrates can be minimized. Heins et al. (1994a) reported that storage duration of several bedding plant species was increased when light as low as 1 μmol m^{-2} s^{-1} was provided during storage (Table 1). Kubota and Kozai (1994) reported that the plant quality, dry weight, and photosynthetic potential of in vitro broccoli seedlings were best maintained when stored at low temperatures (5° C) near the light compensation points (2 μmol m^{-2} s^{-1}). Light in storage (10 μmol m^{-2} s^{-1}) has been shown to greatly improve the overall appearance and reduce the development of leaf necrosis in chrysanthemum cultivars stored at 0° C or 3° C (Rajapakse et al. 1996). Marino et al. (1985) reported that *Prunus* rootstocks were better-stored in darkness than in light at –3° C whereas they stored better in light than in darkness at 4° C or 8° C.

Quality of light can also influence the plant quality during storage. Kubota et al. (1996) reported that 2 μmol m^{-2} s^{-1} of white light maintained the visual quality of in vitro broccoli plantlets better than red or blue light during 6 weeks storage at 5° C. Red and blue light caused stem elongation and reduction in chlorophyll content when compared with white light. Similar stem elongation was observed in chrysanthemum plants stored under low irradiance red and blue light (Unpublished data, Rajapakse). Wilson et al. (1998a), however, reported that red light was better than white or blue light in maintaining photosynthetic ability and carbohydrate pools of in vitro broccoli seedlings stored up to 12 weeks at 5° C. Kubota et al. (1997) also reported that red and blue light increased hexose sugar accumulation compared to white light.

Although providing light is beneficial in storage, high irradiance can adversely affect some crops that are sensitive to chilling injury. It has been reported that the chilling injury is much more pronounced when strong light accompanies the chilling period than when plants were exposed to low temperature in dark (Powles et al. 1983; Taylor and Rowley, 1971; Van Hasselt and Van Berlo, 1980). Van Hasselt (1972) reported that exposure of cucumber leaves to light (1300 μmol m^{-2} s^{-1}) during exposure to low temperature (1° C) resulted in bleeding of chloroplast pigments and decreased the apparent photosynthesis and chlorophyll florescence. Powles et al. (1983) reported that exposure of bean plants to low temperature (6° C) under high irradiance (2000 μmol m^{-2} s^{-1}) resulted in substantial inhibition of the quantum yield of CO_2 uptake indicating a reduction of photochemical efficiency of photosynthesis. They also reported that the extent of inhibition was dependent on the photon flux density during low temperature exposure and that no inhibition occurred under low irradiance (70 μmol m^{-2} s^{-1}). Their results also show that no inhibition of photosynthesis occurred, regardless of irradiance, when temperatures were above 11.5° C. Krizek et al. (1985) reported that exposure to light prior to the exposure to low temperature may be as important or even more important than light during low temperature exposure. Their results show that *Coleus* plants exposed to 320 μmol m^{-2} s^{-1} for 48 h prior to exposure to 5° C resulted in severe chilling injury while those plants exposed to 8 μmol m^{-2} s^{-1} had increased protection against chilling injury. These findings indicate that high irradiance should be avoided in low temperature storage and that exposure to low irradiance prior to low temperature storage may be as effective as low irradiance during low temperature storage.

2.2.4. Gas Composition

Modification of gas composition surrounding the produce by controlled atmosphere (CA) or low pressure (LP) methods has been used widely to extend the life of horticultural commodities. In CA storage, atmospheric modification is achieved by reducing the O_2 concentration and/or increasing the CO_2 concentration in the storage room. In LP storage, atmospheric pressure is reduced thereby reducing the partial pressure of all gases. Low O_2 and/or high CO_2 concentrations can extend the storage life by reducing the rate of respiration, reducing the production and the activity of ethylene, and reducing the spread of pathogens (Bridgen and Staby, 1981; Burg and Burg, 1967; Kader et al. 1989; Lougheed et al. 1978; Mathooko, 1996; Zagory and Kader, 1988). Significant advances in commercial CA and LP storage of fruits, vegetables, and cut flowers have been made during past three decades but very little work has been done on CA/LP storage of propagation material possibly because of the high cost. Eisenberg et al. (1978) evaluated the possibility of using LP storage to extend life of ornamental cuttings. Their results show that low pressure (4.5 kPa) in combination with refrigeration (5° C) extended the life of geranium and poinsettia cuttings by 2 weeks more than the refrigerated storage alone. Low pressure in combination with low temperature reduced loss of chlorophyll and the development of diseases compared to low temperature alone. Andersen (1986) reported that LP storage at 2 kPa reduced leaf yellowing and senescence of *Epiprenum* (L.) but did not affect the respiration rate or the chlorophyll degradation of *Hibiscus* (L.) cuttings. Kirk et al. (1986) reported that the LP storage at 2 kPa caused stomata to remain open compared to those stored under normal pressure causing a greater water loss from *Hibiscus* cuttings after removal from storage.

Another benefit of atmospheric modification may be its potential to reduce deterioration of plants that are sensitive to chilling. Lyons (1973) proposed that accumulation of toxic volatile substances such as acetaldehyde and ethanol might be involved in the development of chilling injury. Low pressure can reduce the boiling point of these substances thereby reducing the accumulation of toxic substances that cause the development of chilling injury symptoms. Several studies have focused on the development of chilling injury symptoms in fruits and vegetables in response to atmospheric modification. In fruits and vegetables, low O_2 and high CO_2 levels have been shown to be beneficial (e.g. okra, peach, zucchini, grapefruit), detrimental (e.g. cucumbers, peppers) or have no effect (e.g. papaya, tomato, lemon) on the development of chilling damage. (For details see Wang, 1993).

3. Physiological Changes During Storage

The reduction in the rate of metabolic processes that leads to deterioration is the key to maintaining quality and regrowth following storage. Following sections briefly discuss major physiological changes taking place during storage.

3.1. Rate of Respiration and Loss of Carbohydrate Reserves

Respiration is a key process in all living organisms that provides energy through the oxidative breakdown of organic compounds. Part of the energy produced in respiration is utilized for metabolic processes and the rest is released as heat. High respiration rate is a characteristic of 'poor' storing cultivars. Higher rates of respiration during storage deplete the respiratory substrates faster and shortens storage life. Rajapakse et al. (1996) reported that the respiration rate of rooted cuttings of chrysanthemum cultivars with 'good' storage potential was lower than the cultivars with 'poor' storage potential.

The primary benefit of low temperature is to reduce the rate of respiration, which in turn reduces the rate of substrate depletion and prolongs storage life. The rate change of biochemical reactions in response to temperature is characterized by Q_{10}, which is the ratio of a reaction at one temperature (T_1) versus the rate at $T_1+10°$ C (rate at $T_1+10°$ C/rate at T_1). For most plants, the Q_{10} for respiration is near 2 to 2.5 in 5° C to 25° C range indicating that respiration at least doubles with every 10° C rise in temperature. As the temperature increases above 25° C or decreases below 5° C, Q_{10} for most crops decreases. At very high or freezing temperatures, respiration can be depressed due to the inactivation of enzymes.

Although respiration is reduced down at low temperatures, the depletion of carbohydrate reserves continues during dark storage. Loss of carbohydrate pools has been implicated in poor survival during storage, poor root growth, and delayed field establishment after transplanting. Lieten et al. (1995) reported that starch content of strawberry crown tissue decreased about 50% during 7-week dark storage at −1.5° C and continued to decrease further as the storage duration increased. They reported that the plant vigor and the productivity decreased as the storage duration increased. Rajapakse et al. (1996) reported that soluble sugar content of chrysanthemum cultivars decreased considerably during dark storage at 0° C or 3° C and that loss of soluble sugars was greater in chrysanthemum cultivars that had a 'poor' storage potential than the cultivars that had 'good' storage potential (Figs. 1 and 2). Light in storage can prevent the depletion of carbohydrates and the loss of dry weight of stored plant material. Kubota et al. (1997) reported that the respiration rate and the carbohydrate levels of in vitro broccoli seedlings decreased when stored for 6 weeks in dark at 5° C but the carbohydrate levels were maintained at the pre-storage level when stored in light. Wilson et al. (1998b) reported that the soluble sugar levels of photoautotrophic (grown in sugar free culture medium) broccoli seed-

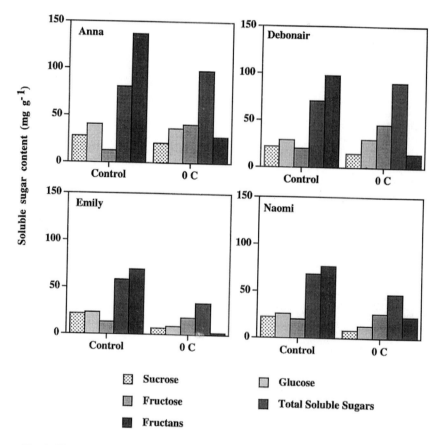

Fig. 1. Change in leaf soluble sugar content of four chrysanthemum cultivars after four week storage at 0° C. Control is the sugar content before storage. Modified from Rajapakse et al. (1996). 'Anna' and 'Debonair' are cultivars with 'good' storage potential whereas 'Emily' and 'Naomi' are cultivars with 'poor' storage potential.

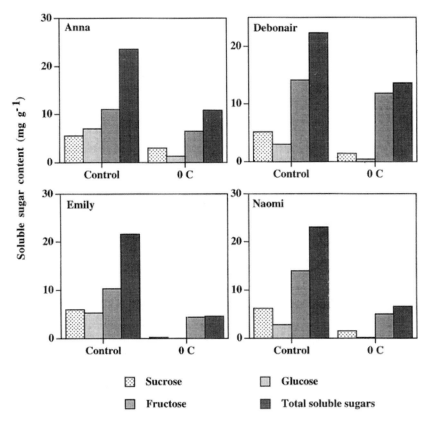

Fig. 2. Change in stem soluble sugar content and fructans of four chrysanthemum cultivars after four week storage at 0° C. Control is the sugar content before storage. Modified from Rajapakse et al. (1996). 'Anna' and 'Debonair' are cultivars with 'good' storage potential whereas 'Emily' and 'Naomi' are cultivars with 'poor' storage potential.

lings rapidly decreased during 4 week storage in dark at 5° C and that lack of respiratory substrate caused death of plantlets after 4 weeks in storage. However, photoautotrophic seedlings stored in light (5 μmol m^{-2} s^{-1}) and photomixotrophic (with 2% sucrose in the culture medium) seedlings stored in dark survived 8 weeks at 5° C suggesting that carbohydrate supply either through the medium or through photosynthesis is essential to maintain plant quality.

3.2. Loss of Chlorophyll and Photosynthetic Ability

Most visible evidence of quality loss of plant material is the loss of chlorophyll and the subsequent leaf yellowing and/or the development of necrotic areas. Both chlorotic and necrotic leaves reduce the consumer acceptability and increase the susceptibility to pathogens. Dark storage induces leaf chlorosis because light is required for chlorophyll synthesis. However, low temperature can delay the induction of leaf chlorosis by slowing down the breakdown of chlorophyll during storage. Arteca et al.

(1996) reported that storage at 4° C reduced the loss of chlorophyll of *Pelargonium* cuttings compared to those stored at 20° C or 25° C. Dark storage of rooted chrysanthemum cuttings at 0° C or 3° C has been shown to induce loss of chlorophyll and the development of necrotic areas, especially in poor storing cultivars (Rajapakse et al. 1996). However, light (as low as 10 μmol m^{-2} s^{-1} of PPF from fluorescent bulbs) in storage reduced the development of leaf necrosis and loss of chlorophyll in chrysanthemum cuttings. Kubota and Kozai (1994) reported that in vitro broccoli seedlings stored at 5° C in darkness retained more chlorophyll than those stored at 15° C after 3 or 6 weeks of storage. Wilson et al. (1999b) reported that the leaf yellowing of both photoautotrophic and photomixotrophic hosta plantlets increased as the storage duration and the temperature increased. However, their results indicated that leaf yellowing was greater in photoautotrophic plantlets than in photomixotrophic plantlets suggesting that carbohydrate supply can affect the loss of chlorophyll.

Low temperature dark storage can also affect chlorophyll fluorescence, a process that reflects the photosynthetic processes taking place in the chloroplast. Chlorophyll fluorescence characteristics have been used as an indicator of the integrity of photosystems and photosynthetic efficiency. Within physiological temperatures, chlorophyll fluorescence is predominantly emitted by chlorophyll *a* in photosystem II (PSII) with a band maximum at 685 nm. At low temperatures, fluorescence can be emitted from chlorophyll *a* of both photosystems (PS II and PS I) with band maxima at 684-697 nm for PS II and at 720-740 nm for PS I (Papageorgiou, 1975). Kubota and Kozai (1994) reported that in vitro broccoli seedlings stored for 3 and 6 weeks at 5° C in darkness or 2 μmol m^{-2} s^{-1} of PPF had similar chlorophyll content in their leaves. However, chlorophyll fluorescence ratios (F_m/F_0 at 695 nm and 730 nm) of dark stored plantlets were reduced considerably compared to light stored plantlets. Their results show that chlorophyll fluorescence of plantlets stored in light was maintained at the pre-storage level. Wilson et al. (1998b) reported that chlorophyll fluorescence (F_v/F_m) of photoautotrophic broccoli seedlings stored in dark at 1° C decreased over 90% after 4 weeks into storage. However, in photomixotrophic seedlings, chlorophyll fluorescence was only slightly decreased during the first 8 weeks followed by a substantial reduction after 12 weeks in storage. Lichtenthaler (1988) reported that when the reaction center of PSII is damaged due to environmental stress the chlorophyll fluorescence could be reduced. Smillie and Nott (1979) also reported that low temperature injury decreased chlorophyll fluorescence.

3.3. Changes in Growth Substances and Shoot Emergence/Flowering

Ethylene is one of the major hormones that affects quality of plants in storage. Many forms of stress stimulate ethylene production resulting in accelerated senescence (Abeles et al. 1992). Most herbaceous plants are sensitive to ethylene. Loss of chlorophyll, leaf and flower abscission, reduced bud break, and high mortality after planting have been reported in response to exposure to ethylene in storage (Knee, 1991; Nichols and Frost, 1985). It is generally considered that ethylene exerts little effect at low temperatures in most plants (Blanpied et al. 1972). However, low temperature has been shown to induce ethylene production from chilling sensitive plants (Yang and Hoffman, 1984). Wang and Adams (1980, 1982) reported that ACC (1-aminocyclopropane-1-carboxylic acid) levels, ACC synthase activity, and ethylene production remained low during storage but increased rapidly when transferred to warm temperatures. Arteca et al. (1996) reported that the ACC synthase activity was not detected in *Pelargonium* cuttings at 4° C whereas it was high in cuttings stored at 25° C following 3 to 4 days of storage. They reported that plants stored at 4° C accumulated high ACC levels and produced very little ethylene compared to plants stored at 25° C.

Very little work has been done on the changes in other endogenous growth substances during storage. Cappiello and Kling (1994) studied the changes in growth regulators (indoleacetic acid (IAA), ABA, zeatin, and zeatin riboside) in roots and shoots of *Cornus sericea* L. cuttings during low temperature storage. They reported that the levels of previously mentioned growth regulators in shoots and roots remained relatively unchanged during a 60-day storage period at 4° C. Upon removal from storage, all growth regulator levels in shoots increased three to four fold and bud break was associated with the increase in cytokinin and IAA levels. In roots, zeatin and zeatin riboside levels increased 3 to 4-fold, IAA increased slightly, and ABA levels decreased.

Bud break and flowering of many temperate perennial plants may be delayed without sufficient exposure to low temperature. Low temperature storage can satisfy the chilling requirement of these crops and stimulate bud break and flower induction. Keever et al. (1999) reported that low temperature storage of hosta at 4° C reduced the time to shoot emergence and produced larger plants compared to non-stored plants. They also reported that time to shoot emergence decreased as the storage duration at 4° C increased. Runkle et al. (1999) reported that low temperature storage (5° C) of *Rudbeckia* Ait. 'Goldsturm' for 15 weeks hastened flowering by 3 to 4 weeks and reduced the critical day length requirement for flowering. Huang et al. (1999) reported that the storage of *Thalictrum delavayi* Franch for 6 weeks at 8° C hastened flowering.

4. Conclusion

Both short- and long-term storage methods for rooted cuttings and rooted liners are beneficial for a variety of reasons, such as to store plants during unexpected weather conditions and labor shortages, maintain a continuous plant supply during peak demand period, and fulfill cold requirements of temperate perennials. The required length of storage can vary from several days for short-term storage to several weeks for long-term storage.

Although growers and transplant producers have recognized the benefits of storage techniques, they are typically used only when absolutely necessary. One reason for this under-utilization is that storage cannot be 'pre-scheduled' for most crops due to lack of information on optimum storage conditions. Future research on both practical and physiological aspects can greatly enhance our knowledge for improving environmental control of rooted plant material. Research on temperature limits, storage duration, optimum light levels, and other treatments to improve the quality of a wide range of crops and cultivars will benefit producers. Storage conditions for crops are complicated by the growing conditions and these facts should be considered when developing storage protocols for various crops. Research into improving survival and field establishment of stored transplants will enhance the long-term storage capability of transplants.

Many crops traditionally propagated by cuttings are now micropropagated. Information on environmental conditions for storage of micropropagules is limited, and certainly the micropropagation industry could benefit from such research in the future. Acclimatization of micropropagules following low temperature storage has been a major challenge. More effective acclimatization techniques are needed to assure successful transfer into the greenhouse of viable plantlets stored for prolonged periods in low temperature and low illumination conditions. Identifying better acclimatization methods such as misting, environmental control, and CO_2 enrichment during acclimatization may help improve acclimatization following storage at low temperature.

A better understanding of how storage conditions change physiological responses of plants such as endogenous growth regulator levels, metabolic processes, and food reserves will help develop efficient storage methods.

Acknowledgements

I wish to thank Dr Jeff Adelberg of Clemson University, Dr Royal Heins of Michigan State University, Dr Chieri Kubota of Chiba University, Japan and Dr Sandra Wilson of University of Florida, Indian River Research and Education Center for critically reviewing this paper.

References

Abeles, F.B., Morgan, P.W. and M.E. Saltviet, (Jr). 1992. *Ethylene in Plant Biology*. 2nd Edition. Academic Press, Inc. San Diego, CA.

Andersen, A.S. 1986. Low pressure storage of herbaceous cuttings. *Acta Horticulturae* **181**: 305-312.

Arteca, R.N., Arteca, J.M. Wang, T-W. and C.D. Schlagnhaufer. 1996. Physiological, biochemical, and molecular changes in *Pelargonium* cuttings subjected to short-term storage conditions. *Journal of the American Society for Horticultural Science* **121**: 1063-1068.

Asare-Boamah, N.K. and R.A. Fletcher. 1986. Protection of bean seedlings against heat and chilling injury by triadimefon. *Physiologia Plantarum* **67**: 353-358.

Behrens, V. 1988. Storage of unrooted cuttings. Advances in Plant Science Series. Vol. 2. Dioscorides Press, Portland, OR, 235-247.

Blanpied, G.D., Cadun, O. and T. Tamura. 1972. Ethylene in apple and pear experimental controlled atmosphere chambers. *Journal of the American Society for Horticultural Science* **97**: 207-209.

Bridgen, M.P. and G.L. Staby. 1981. Low pressure and low oxygen storage of *Nicotina tabacum and Chrysanthemu x morifolium* tissue cuttings. *Plant Science Letters* **22**: 177-186.

Burg, S.P. and E.A. Burg. 1967. Molecular requirements for the biological activity of ethylene. *Plant Physiology* **42**: 144-152.

Cappiello, P.E. and G.J. Kling. 1994. Changes in growth regulator and carbohydrate levels in roots and shoot tips of *Cornus sericea* during cold storage and emergence from dormancy. *Journal of the American Society for Horticultural Science* **119**: 785-788.

Conover, C.A. 1976. Postharvest handling of rooted and unrooted cuttings of tropical ornamentals. *HortScience* **11**: 127-128.

Couey, H.M. 1982. Chilling injury of crops of tropical and sub-tropical origin. *HortScience* **17**: 162-165.

Creencia, R.P. and W.J. Bramlage. 1971. Reversibility of chilling injury to corn seedlings. *Plant Physiology* **47**: 389-392.

Curtis, O.F. and D.R. Rodney. 1952. Ethylene injury to nursery trees in cold storage. *Proceedings of the American Society for Horticultural Science* **50**: 104-108.

Davis, T.D., Steffens, G.L. and N. Sankhla. 1988. Triazole plant growth regulators. *Horticultural Reviews* **10**: 63-105.

Dickson, M.H. and M.A.A. Boettger. 1984. Emergence, growth, and blossoming of bean (*Phaseolus vulgaris*) at sub-optimal temperatures. *Journal of the American Society for Horticultural Science* **109**: 257-260.

Eisenberg, B.A., Staby, G.L. and T.A. Fretz. 1978. Low pressure and refrigerated storage of rooted and unrooted ornamental cuttings. *Journal of the American Society for Horticultural Science* **103**: 732-737.

Gladstone, L.A. and G.W. Moorman. 1989. Pythium root rot of seedling geraniums associated with various concentrations of nitrogen, phosphorus, and sodium chloride. *Plant Disease* **73**: 733-736.

Hansen, J., Stromquist, L.H. and A. Ericsson. 1978. Influence of the irradiance on carbohydrate content and rooting of cuttings of pine seedlings (*Pinus sylvestris* L.). *Plant Physiology* **61**: 975-979.

Hatton, T.T. 1990. Reduction of chilling injury with temperature manipulation, In: *Chilling Injury of Horticultural Crops*, C.Y. Wang (Ed.), CRC Press, Boca Raton, FL 269-280.

Heins, R., Lange, N., Wallace (Jr), T.F. and W. Carlson. 1994a. Plug storage A to Z. *Greenhouse Grower*, 1-20.

Heins, R., Si, Y. and M. Hausbeck. 1994b. Control of *Botrytis* during storage of bedding-plant plugs. Research Report No. F-9406. Bedding Plant Foundation, Inc., Lancing, MI, 1-7.

Huang, N., Funnel, K.A. and B.R. MacKay. 1999. Vernalization and growing degree-day requirements for flowering of *Thalictrum delavayi* 'Hewitt's Double'. *HortScience* **34**: 59-61.

Hummel, R.L. and R.J. Henny. 1986. Variation in sensitivity to chilling injury within the Genus *Aglaonema*. *HortScience* **21**: 291-293.

Jaffe, M.J. and F.M. Isenberg. 1965. Some effects of N-dimethyl amino succinamic acid (B-nine) on the development of various plants, with special reference to the cucumber, *Cucumis sativus*. *Proceedings of the American Society for Horticultural Science* **87**: 420-428.

Kaczperski, M.P., Armitage, A.M. and P.M. Lewis. 1996. Performance of plug-grown geranium seedlings preconditioned with nitrogen fertilizer or low-temperature storage. *HortScience* **31**: 361-363.

Kader, A.A., Zagory, D. and E.L. Kerbel. 1989. Modified atmosphere packaging of fruits and vegetables. *Critical Reviews in Food Science and Nutrition* **28**: 1-30.

Keever, G.J., West, M.S. and J.R Kessler (Jr). 1999. Chilling effects on shoot emergence and subsequent growth in hosta. *Journal of Environmental Horticulture* **17**: 84-87.

King, A.I., Joyce, D.C. and M.S. Reid. 1988. Role of carbohydrate in diurnal chilling sensitivity of tomato seedlings. *Plant Physiology* **86**: 764-768.

Kirk, H.G., Andersen, A.S. Veierskov, B. Johansen, E. and Z. Aabrandt. 1986. Low-pressure storage of *Hibiscus* cuttings. Effect on stomatal opening and rooting. *Annals of Botany* **58**: 389-396.

Knee, M. 1991. Role of ethylene in chlorophyll degradation in radish cotyledons. *Journal of Plant Growth Regulation* **10**: 157-162.

Koranski, D., Karlovich, P. and A. Al-Hemaid. 1989. The latest research on holding and shipping plugs. *GrowerTalks* **53(8)**: 72-79.

Krizek, D.T, Semeniuk, P., Moline, H.E., Mirecki, R.M. and J.A. Abbott. 1985. Chilling injury in coleus as influenced by photosynthetically active radiation, temperature and abscissic acid pretreatment. I. Morphological and physiological responses. *Plant, Cell, and Environment* **8**: 135-142.

Kubota, C. and T. Kozai. 1994. Low-temperature storage for quality preservation and growth suppression of broccoli plantlets cultured in vitro. *HortScience* **29**: 1191-1194.

Kubota, C., Rajapakse, N.C. and R.E. Young. 1996. Low-temperature storage of micropropagated seedlings under selected light environments. *HortScience* **31**: 449-452.

Kubota, C., Rajapakse, N.C. and R.E. Young. 1997. Carbohydrate status and transplant quality of micropropagated broccoli plantlets during low temperature storage under different light environments. *Postharvest Biology and Technology* **12**: 167-173.

Lasley, S.E., Garber, M.P. and C.F. Hodges. 1979. After effects of light and chilling temperatures on photosynthesis in excised cucumber cotyledons. *Journal of the American Society for Horticultural Science* **104**: 477-480.

Lee, E.H., Byun, J.K. and G.L. Steffens. 1985. Increased tolerance of plants to SO_2, chilling, and heat stress by a new GA biosynthesis inhibitor, paclobutrazol (PP333). *Plant Physiology* **77**(Suppl.): 135.

Lichtenthaler, H.K. 1988. In vivo chlorophyll fluorescence as a tool for stress detection in plants, In: *Applications of chlorophyll fluorescence in photosynthesis research, stress physiology, hydrobiology and remote sensing*, H.K. Lichtenthaler (Ed.), Kluwer Academic Publishers, The Netherlands, 129-142.

Lieten, F., Kinet, J.M. and G. Bernier. 1995. Effect of prolonged cold storage on the production capacity of strawberry plants. *Scientia Horticulturae* **60**: 213-219.

Lougheed, E.C., Murr, D.P. and L. Berard. 1978. Low pressure storage of horticultural crops. *HortScience* **13**: 21-27.

Lyons, J.M. 1973. Chilling injury in plants. *Annual Review of Plant Physiology* **24**: 445-466.

Mathooko, F.M. 1996. Regulation of respiratory metabolism in fruits and vegetables by carbon dioxide. *Postharvest Biology and Technology* **9**: 247-264.

Marino, G., Posati, P. and F. Sagrati. 1985. Storage of in vitro cultures of *Prunus* rootstocks. *Plant Cell Tissue and Organ Culture* **5**: 73-78.

Martin, J.D. and C.B. Link. 1973. Reducing water loss of potted chrysanthemums with pre-sale application of antitranspirants. *Journal of the American Society for Horticultural Science* **98**: 303-306.

Murata, T. 1990. Relation of chilling stress to membrane permeability. In: *Chilling Injury of Horticultural Crops*, C.Y. Wang (Ed.), CRC Press, Boca Raton, FL, 201-209.

Murata, N. and Nishida, I. 1990. Lipids in relation to chilling sensitivity of plants. In: *Chilling Injury of Horticultural Crops*, C.Y. Wang (Ed.), CRC Press, Boca Raton, FL, 181-199.

Nichols, R. and C.E. Frost. 1985. Post-harvest effects of ethylene on ornamental plants. In: *Ethylene and plant development*, J.A. Roberts and G.A. Tucker (Eds), Butterwoths, London, 343-351.

Nitzsche, P., Berkowitz, G.A. and J. Rabin. 1991. Development of a seedling-applied antitranspirant formulation to enhance water status, growth, and yield of transplanted bell pepper. *Journal of the American Society for Horticultural Science* **116**: 405-411.

Papageorgiou, G. 1975. Chlorophyll fluorescence: an intrinsic probe of photosynthesis. In: *Bioenergetics of Photosynthesis*, W. Govindjee (Ed.), Academic Press, New York, 319-371.
Paton, F. and W.W. Schwabe. 1987. Storage of cuttings of *Pelargonium* x *hortorum* Bailey. *Journal of Horticultural Science* **62**: 79-87.
Patterson, B.D. and L.A. Payne. 1983. Screening for chilling resistance in tomato seedlings. *HortScience* **18**: 340-341.
Powles, S.B., Berry, J.A. and O. Bjorkman. 1983. Interaction between light and chilling temperature on the inhibition of photosynthesis in chilling-sensitive plants. *Plant,Cell, and Environment* **6**: 117-123.
Rajapakse, N.C., Miller, W.B. and W. Kelly. 1996. Low temperature storage of rooted chrysanthemum cuttings; relationship to carbohydrate status of cultivars. *Journal of the American Society for Horticultural Science* **121**: 740-745.
Reid, D.M., Pharis, R.P. and D.W.A. Roberts. 1974. Effect of four temperature regimes on the gibberellin content of winter wheat cv Kharkov. *Physiologia Plantarum* **30**: 53-57.
Rikin, A. and A.E. Richmond. 1976. Amelioration of chilling injuries in cucumber seedlings by abscissic acid. *Physiologia Plantarum* **38**: 95-97.
Rudnicki, R.M., Nowak, J. and D.M. Goszczynska. 1991. Cold storage and transportation conditions for cut flowers, cuttings and potted plants. *Acta Horticulturae* **298**: 225-236.
Runkle, E.S., Heins, R.D., Cameron, A.C. and W.H. Carlson. 1999. Photoperiod and cold treatment regulate flowering of *Rudbeckia fulgida* 'Goldsturm'. *HortScience* **34**: 55-59.
Sasson, N. and W.J. Bramlage. 1981. Effect of chemical protectants against chilling injury of young cucumber seedlings. *Journal of the American Society for Horticultural Science* **106**: 282-284.
Senaratna, T., Mackay, C.E., McKersie, B.D. and R.A. Fletcher. 1988. Uniconazole-induced chilling tolerance in tomato and its relationship to antioxidant content. *Journal of Plant Physiology* **133**: 56-61.
Smillie, R.M. and R. Nott. 1979. Assay of chilling injury in wild and domestic tomatoes based on photosystem activity of the chilled leaves. *Plant Physiology* **63**: 796-801.
Smith, P.M. 1982. Diseases during propagation of woody ornamentals. *Proceeding of the 21st International Horticulture Congress* **2**: 884-893
Spikman, G. 1986. The effect of water stress on ethylene production and ethylene sensitivity of freesia inflorescence. *Acta Horticulturae* **181**: 135-140.
Tantau, H. and K. Dorffling. 1991. Effects of chilling on physiological responses and changes in hormone levels in two *Euphorbia pulcherrima* varieties with different chilling tolerance. *Journal of Plant Physiology* **138**: 734-740.
Taylor, A.O. and J.A. Rowley. 1971. Plants under climatic stress: I. Low temperature, high light effects on photosynthesis. *Plant Physiology* **47**: 713-718.
Van Hasselt, P.R. 1972. Photooxidation of leaf pigments in *Cucumis* leaf discs during chilling. *Acta Botanica Neerlandica* **21**: 539-548.
Van Hasselt, P.R. and H.A.C. Van Berlo. 1980. Photooxidative damage to the photosynthetic apparatus during chilling. *Physiologia Plantarum* **50**: 52-56.
Waldman, M., Rikin, A., Dovrat, A. and A.E. Richmond. 1975. Hormonal regulation of morphogenesis and cold-resistance. *Journal of Experimental Botany* **26**: 853-859.
Wang. C.Y. 1993. Approaches to reduce chilling injury of fruits and vegetables, In: *Horticultural Reviews*, J. Janick (Ed.), John Wiley and Sons, Inc. New York, 63-95.
Wang, C.Y. and D.O. Adams. 1980. Ethylene production by chilled cucumbers (*Cucumis sativus* L.). *Plant Physiology* **66**: 841-843.
Wang, C.Y. and D.O. Adams. 1982. Chilling induced ethylene production in cucumbers (*Cucumis sativus* L.). *Plant Physiology* **69**: 424-427.
Wheaton, T.A. and L.L. Morris. 1967. Modification of chilling sensitivity by temperature conditioning. *Proceedings of the American Society for Horticultural Science* **91**: 529-533.
Whitcomb, C.E., Hall, G.C., Davis, L.T. and G.S. Southwell. 1974. Potentials of antitranspirants in plant propagation. *Proceedings of the International Plant Propagators Society* **24**: 342-348.

Wilson, J.M. 1976. The mechanism of chill- and drought-hardening of *Phaseolus vulgaris* leaves. *New Phytologist* **76**: 257-270.

Wilson, J.M. and R.M.M. Crawford. 1974. The acclimatization of plants to chilling temperatures in relation to the fatty-acid composition in polar lipids. *New Phytologist* **73**: 805-820.

Wilson, S.B., Rajapakse, N.C. and R.E. Young. 2000a. Media composition and light affects storability and post storage recovery of micropropagated host plantlets. *HortScience* (in press).

Wilson, S.B., Rajapakse, N.C. and R.E. Young. 2000b. Use of low temperature and light to improve storage of in vitro broccoli seedlings. *Journal of Vegetable Crop Production* (in press).

Wilson, S.B., Iwabuchi, K., Rajapakse, N.C. and R.E. Young. 1998a. Responses of broccoli seedlings to light quality during low temperature storage in vitro. I. Morphology and survival. *HortScience* **33**: 1253-1257.

Wilson, S.B., Iwabuchi, K., Rajapakse, N.C. and R.E. Young. 1998b. Responses of broccoli seedlings to light quality during low temperature storage in vitro. II. Sugar content and photosynthetic efficiency. *HortScience* **33**: 1258-1261.

Wood, T. and A.C. Camaron. 1989. Cold storage: Over wintering your softwood cuttings in the refrigerator releases valuable greenhouse space. *American Nursery Magazine* **170(7)**: 49-54.

Wright, M. 1974. The effect of chilling on ethylene production, membrane permeability and water loss of leaves of *Phaseolus vulgaris*. *Planta* **120**: 63-69.

Wright, M. and E.W. Simon. 1973. Chilling injury in cucumber leaves. *Journal of Experimental Botany* **24**: 400-411.

Yakir, D., Rudich, J. and B.A. Bravdo. 1986. Adaptation to chilling: photosynthetic characteristics of the cultivated tomato and a high altitude wild species. *Plant Cell and Environment* **9**: 477-484.

Yang, S.F. and N.E. Hoffman. 1984. Ethylene biosynthesis and its regulation in higher plants. *Annual Review of Plant Physiology* **35**: 155-189.

Zagory, D. and A.A. Kader. 1988. Modified atmosphere packaging of fresh produce. *Food Technology* **42**: 70-77.

15

Minimizing Postharvest Rots and Quality Loss in New Zealand Carrots

L-H Cheah and D.W. Brash
New Zealand Institute for Crop & Food Research Ltd, Private Bag 11-600, Palmerston North, New Zealand

1. Introduction

Fresh carrots (*Daucus carota* L.) have become one of the promising new crops for export from New Zealand to Asia. In the last five years exports have risen from NZ$ 2.4 million in 1994 to over NZ$ 12 million in 1999, and further increases in demand are expected. Carrots are transported by sea to Japan, Hong Kong, Singapore, Thailand and Malaysia. Fresh whole carrots obtain premium prices, although some carrots are also processed for juice.

Majority of New Zealand carrots are grown in the central regions of the North Island in volcanic soil (andesitic sand with peat) with an average soil pH of about 6. Two main carrot types are grown: Nantes (e.g. Explorer) and Kuroda (e.g. 'Koyo II'). Nantes types have been grown for many years for the domestic market. They are suited to mechanical harvesting and ground storage over winter. Kuroda types are preferred in Japan. They are less suited to mechanical harvesting and cannot be ground stored over winter. Carrots are sown from October to December and harvested from February to July in the following year.

The carrots are washed, hydro-cooled, packed and cool-stored before sale and export. Exported carrots are packed in 10 kg cartons and placed in refrigerated (0-2° C) sea containers. These carrots may be stored in containers for three to four weeks before reaching overseas markets. The temperature and relative humidity at the destination (e.g. Singapore) are usually higher than ideal for carrots, and chilled storage conditions are often scarce. The temperature inside the closed cartons often increases. Such conditions favour micro-organisms which cause rot.

Storage rots, especially those caused by fungal pathogens, are responsible for substantial postharvest losses (Cheah et al. 1996a, b), reducing both grower returns and market confidence. The major factors causing storage rots and quality loss of carrots are pathogens (e.g. fungal infection), poor postharvest handling (e.g. injuries), and exposure to undesirable storage conditions (e.g. high ambient temperature). All of these factors arise along the handling chain from harvest to the market and involve field practices (e.g. crop protection and harvesting methods), postharvest handling (e.g. washing and cooling) and packaging and storage (e.g. cartons and storage temperature). This chapter presents an overview of some of the storage rot problems and the physiological disorders of carrots. The influence of field diseases and cultural practices on the level of storage rot incidence and quality of carrots are described. Methods for minimizing these problems are also suggested.

2. Storage Rots and Physiological Disorders

Carrots are susceptible to a number of postharvest diseases and disorders. Seven species of fungi, *Cladosporium* sp., *Fusarium* spp., *Penicillium* spp., *Pythium* spp., *Rhizoctonia carotae*, *Sclerotinia sclerotiorum* and *Thielaviopsis basicola*, and one bacterial species, *Erwinia carotovora*, have been isolated from rotter carrots (Cheah et al. 1996a). Pathogenecity tests on carrots have shown that *S. sclerotiorum* and *T. basicola* are highly pathogenic. Some of the major storage rots and disorders are described below.

2.1. Diseases

2.2.1. Sclerotinia Rot

Sclerotinia rot, caused by *Sclerotinia sclerotiorum*, is the most common disease of stored carrots and can also affect crops in the field. It is a common disease in many vegetable crops.

Symptoms of the disease are soft, watery rots on stored carrots (Colour Plate 15.1). The infected tissue is covered with white cottony mycelium and small (1-3 mm) black sclerotia amid the mycelium. These symptoms can be distinguished from lesions caused by *Rhizoctonia* or *Fusarium* spp., fungi which do not produce sclerotia.

Primary infection probably results from colonization of leaf and root tissue by mycelium produced from sclerotia in the soil (Howard et al. 1994). Root infection may take place after the foliage and crown become infected. Disease development in storage usually starts from inflections that have occurred in the field or at harvest (Cheah, 1999). Schwarz (1977) found less rotting by *S. sclerotiorum* at 0° C than at storage temperatures of 2-4° C, and the optimum temperature for disease development is 13-18° C. Mycelium from a single infected carrot can spread to adjacent carrots, producing radiating pockets of infection (nesting) on roots stored in pallet boxes, plastic bags or in bulk storage (Colour Plate 15.1).

2.2.2. Black Root Rot

Black root rot, caused by *Thielaviopsis basicola*, was first recorded in New Zealand in 1995 when it caused severe postharvest rot on cool-stored carrots (Cheah et al. 1996b). Since then there have been sporadic occurrences of the disease on exported carrots from New Zealand. The disease has also been reported to cause severe losses in harvested carrots in British Columbia (Punja et al. 1992).

Initial signs of the disease are very fine growth of mould, which later become dark grey and powdery. In mature lesions *T. basicola* may be distinguished by its deep black colour (Colour Plate 15.2). Rotting is superficial, but the dark patches constitute a serious blemish which can render a consignment worthless. In culture *T. basicola* can be readily distinguished by its production of pale, thin-walled spores (endoconidia) and dark, thick-walled spores (macroconidia) which become resting spores (chlamydospores) (Cheah et al. 1996b).

The fungus survives in plant debris or as chlamydospores in the soil (Tsao and Brecker, 1966). Disease development is associated with freshly harvested carrot roots stored at high temperatures and high relative humidity. The disease is rarely a serious problem if carrots are stored under optimum conditions (0-1° C and 98-100% relative humidity). Wounding appears to be a prerequisite for infection. Disease can also occur at sites where infected debris is attached to the surfaces of the carrot roots.

2.2.3. Black Ring

Black ring is characterized by shallow black lesions at the leaf bases on carrots. The lesions do not usually spread or increase in severity during storage. Black ring is not normally a problem as importers accept small amounts of black stain. However, the problem can become serious if storage temperatures increase due to myceliae growth from the lesions. In such conditions, affected carrots can be rejected at the export destination.

A number of fungi have been isolated; *Fusarium* spp. are the most common microorganisms associated with this problem, although species of *Alternaria*, *Pythium* and *Cladosporium* have been occasionally isolated. If the carrots are not harvested when mature or left in the ground for long periods (e.g. over winter in the ground), saprophytic microorganisms may start to invade further down into the root.

2.2.4. Other Fungal Rots

Other fungi such as *Botrytis cinerea* (grey mould), *Rhizoctonia carotae* (crater rot) and *Alternaria radicina* (black rot), have also been reported to cause postharvest losses in carrots, but they seldom cause major problems.

2.2.5. Bacterial Soft Rot

Bacterial soft rot is caused by the bacteria *Erwinia carotovora* or *Pseudomonas* spp. Bacteria soft rot is not an important disease when carrots are stored at 0° C, but can cause heavy losses at ambient temperature and at high humidity.

Infected flesh is often brownish in colour, and may be extremely soft and slimy. Often the tips of carrots are affected. The presence of secondary bacteria frequently results in the production of a putrid odour, and makes accurate diagnosis difficult.

Bacteria survive on plant debris in the soil and may infect growing crops (Towner and Berah, 1976). Rotting can occur in field-stored roots (Tamietti and Matta, 1981) or during transport and storage. Roots are likely to be surface contaminated at the time of harvest and infection during washing and hydrocooling is likely. Decay is largely determined by postharvest handling practices and the storage environment. Build-up of bacteria in washing tanks can lead to severe rotting of roots, especially those packed in polyethylene and held at ambient temperatures (Segall and Dow, 1973).

2.2.6. Yeast Rot

Yeast rot, caused by species of *Candida*, has become a problem on carrots exported to Asia. The problem was reported in 1998 and again in 1999 on carrots arriving at Singapore and Japan markets. There is little published information about this rot on carrots. Snowdon (1991) found that yeast rot was associated with inadequate cooling and ventilation.

2.3. Physiological Disorders

2.3.1. Skin Browning

Skin or surface browning are characterized by brown spots or patches on the surfaces of carrots. In severe cases, these patches may look like a dark skin covering almost the entire root. The discoloration is usually only superficial. The symptoms develop when carrots are washed and stored in cool rooms for long periods before packing.

Abrasion caused by mechanical washing often removes the epidermal layer of the roots, exposing the inner tissue to oxidation of phenolic compounds, which turn brown or black (Galati and McKay, 1995). Under a microscope the affected cells are brown, crusted and dead. Den Outer (1990) reported that a fatty layer of dead cells is responsible for the discoloration.

2.3.2. Silvering (or white blush)

Washed carrots often develop scaly surfaces on storage. This symptom appears when carrots become slightly dehydrated or dry during storage. Overbrushing or 'polishing' can remove this layer but may make the carrots prone to storage rots. One of the authors (Cheah unpublished) has found that Nantes carrots are more susceptible than Kuroda types to silvering because the skin of Nantes types is harder to remove than that of Kuroda types. As a result, more skin pieces remain on the roots. Silvering is a major criticism of New Zealand-grown carrots in Asia.

2.3.3. Root Splitting (breakage)

Root splitting can occur in the field in association with rapid growth or over-maturing in some varieties. However, the incidence of splitting is much higher under cool, wet conditions. Damage during harvest and in packhouses can cause splitting as well. Kuroda type carrots are more susceptible to splitting than Nantes type carrots.

3. Minimizing Rot Incidence

Most of the storage rots mentioned above originate in the field and affect carrots after harvest (Cheah, 1999). Therefore, these rots should be minimized by adopting appropriate field cultural practices, growing resistant cultivars, applying chemicals and harvesting the crop carefully. Physiological disorders can be minimized by washing carrots gently and keeping them cool and wet at all times (Galati et al. 1995).

3.1. Field practices

3.1.1. Cultural Practices

Site selection plays an important role in crop quality (Wood, 1995). Carrots should be grown in soil with excellent water drainage and at a density that allows good air circulation through the crops. Carrots are best grown on light textured soils to allow both long straight roots to form and to facilitate harvesting. Carrots grown in wet, heavy soils tend to develop irregular shaped roots and are susceptible to infection by soilborne pathogens, including bacterial rot (Snowdon, 1991).

The effects of crop rotation on the suppression of soilborne diseases (Cook, 1988) has been recognized for centuries. In recent years, crop rotation has been largely neglected by vegetable growers because of the availability and cost-effectiveness of chemical fertilizers, pesticides and soil fumigants. However, Howard et al. (1994) recommend that crop rotation with non-host crops such as onion, beet, spinach, grasses, cereals and corn, for three years reduce the level of soilborne inoculum and controls soilborne pathogens.

Increasing air circulation within the leaf canopy by reducing plant density (Wood, 1995) and by controlling weeds (Howard et al. 1994) may also reduce the incidence of foliar infections by these pathogens.

Adopting hygienic crop practices can also reduce disease incidence. Early detection of disease in the field and removal of diseased plants before sclerotia form are recommended (Snowdon, 1991). Infected soil should not be spread by machinery or animals.

3.1.2. Resistant Cultivars

Resistant cultivars can form part of an integrated disease management program along with cultural practices. Some carrot cultivars are more resistant to soilborne pathogens than others, e.g. 'Paramount' and 'Dess

Dan' are more resistant to *S. sclerotiorum* than 'Six Pak II' (Howard et al. 1994) while 'Top Pak', 'Nandor' and 'Crusader' have moderate cavity spot (*Pythium* spp.) resistance (Galati and McKay, 1995).

3.1.3. Chemical Control

Soil fumigation with chemicals (e.g. metham sodium or methyl bromide) can be prohibitively expensive for carrot crops and does not always give effective control of soilborne diseases (Cheah and Page, 1999a). Biofumigation using brassicas which contain high levels of glucosinolates, (e.g. broccoli residues) may offer an alternative method of control (Subbarao and Hubbard, 1996).

Spraying the tops of carrots before harvesting with benomyl, thiophanate-methyl or vinclozoline reduced or prevented decay caused by *S. sclerotiorum* and *B. cinerea* during storage (Tahvonen, 1985). The best time to treat the tops is the day before harvesting. We observed that weekly field application of chlorothalonil and macozeb gave excellent leaf blight (*Alternaria dauci*) control and also reduced the incidence of black ring (*Fusarium* spp.) (unpublished data). Application of copper oxychloride may control bacterial blight caused by *Xanthomonas campestris*, but no chemical was effective against bacterial soft rot caused by *Erwinia carotovora* (Anonymous, 1986).

3.1.4. Harvest Maturity

Carrots are mature when they have reached their full size (appropriate diameter, length, taper and 'tip fill' for the cultivar). Mature carrots store better than immature carrots, most probably because immature carrots are growing rapidly and have high respiration rates. Phan and Hsu (1973) indicated that 'biochemical maturity' was reached when carotene and carbohydrate concentrations reached their maximum.

A harvest maturity indicator has not been identified to help decide whether a line of carrots will store well. Ewaldz (1997) stored carrot samples at 10° C and 100% relative humidity for 6 weeks as a measure of suitability for 6 months storage. Lines with good and poor storage ability were clearly identified but 25% of the samples were unpredictable. These samples had disease incidences of intermediate values.

3.1.5. Crop Nutrition and Irrigation

The main aim of crop fertilizer practices is to meet nutritional requirements for maximum economic yield. Shibario et al. (1998) found no link between potassium nutrition and postharvest moisture loss. Liming to increase soil pH (rather than raise soil calcium levels) reduced the incidence of cavity spot caused by *Pythium cobratum* (El-Tarabily et al. 1997).

Lack of soil water during carrot growth adversely affects carrot yields. Sorenson et al. (1997) found that drought stress at any stage reduced total yield. Drought stress just prior to harvest did, however, increase marketable yields by reducing root splitting. Drought stress during early growth increased the risk of infection by common scab (*Streptomyces scabies*).

3.1.6. Ground Storage

In New Zealand, carrots are stored in the ground over winter in the main growing areas. This option is cheaper than building and running a coolstore.

The practice is used in other countries such as the UK and France where the winter weather is not harsh (Rubatzky et al. 1999). In New Zealand, soil is spread over the carrot crowns to a depth of 20-30 mm to prevent frost damage and pigmentation of the crown tissue. Some growers use black polythene film followed by a layer of soil to provide insulation and maintain low soil moisture levels in carrot fields during storage periods.

Galati et al. (1995) suggested that carrots are more susceptible to decay and rots when 'overmature', presumably when left in the soil for too long. Cheah and Page (1999a) found that ground storage has exposed carrot crops to violet root rot (*Helicobasidium purpureum*) and black ring (*Fusarium* spp.). We recommend early harvest if disease is detected in the field (Cheah and Page, 1999b). The peak export season for New Zealand carrots is late autumn to early winter, which fortunately coincides with a requirement for only a short period of ground storage after the tops have been frosted off. Brash (1996) found that ground-stored carrots harvested early in spring had a higher incidence of root rots in storage than those from winter harvests.

3.1.7. Harvesting Methods

Minimizing bruising and cracking of carrots during harvesting will reduce storage rots caused by many pathogens (Sherf and MacNab, 1986). Compared to many other crops, carrots are a high volume crop well-suited to mechanical harvesting. Mechanical harvesting does, however, lead to damage. Damage may allow rots to develop in storage. Apeland (1974) showed up to 30% broken and 13% split roots following harvesting by machine (four harvesters and hand harvesting were compared). Mechanically harvested carrots had a 30% higher level of storage rots than hand-harvested carrots after 5 to 7 months in storage. In New Zealand, carrots are usually harvested by mechanical harvester, a 'Toplifter', which grasps the leaves and pulls the roots from the ground. The carrots are topped (the leaves are cut off) and the roots are collected in large wooden bins (1 m^3 capacity). This method may cause less damage than the harvesting method used later in the season when carrots must be undercut, lifted and separated from the soil (using a machine like a potato digger). In both harvesting methods the carrots may have to drop more than 1 m before reaching the bottom of the bin. Carrots can be bruised and cracked from these drops. We recommend that a soft slide should be connected from the conveyer belt to the bin and drop heights reduced to minimize the impact and thus reduce the level of damage to carrots. We also recommend that chains, belts and chutes on harvest and grading equipment should be well padded to minimize bruising. Carrots should be harvested into clean bins to minimize the spread of disease, particularly if bins are stored prior to carrots being washed and graded.

3.2. Postharvest Handling and Treatment

3.2.1. Pre-washing Handling

Harvested carrots should be kept in a shady place or covered to prevent dehydration and loss of quality (Sherf and MacNab, 1986). During transportation, carrots should be covered with a tarpaulin. Overhead water sprinklers may be used to keep the carrots cool and wet so that the soil on the carrots is easy to remove during washing. A brief curing period at ambient temperature and high humidity to promote wound healing may help to control some storage diseases like licorice rot (caused by *Mycocentrospora acerina*) (Lewis et al. 1981).

3.2.2. Washing

Two methods of washing are used: the tumble (Colour Plate 15.3) and spray-brush methods. We found that the tumble method results in more damage to the carrots. There is a trend away from this method towards the use of spray-brush method (Flaherty, 1995). Gentle washing is likely to minimize establishment of storage rots. Handling damage from harvest, transport and washers may allow storage rots to establish during washing when damaged carrots, soil and water come into contact. Washing water should only be used once or changed often.

3.2.3. Sanitation

Chlorine is one of the most commonly used sanitizing agents for general disinfection of microorganisms in carrots (Haynes et al. 1990) although other chemicals, for example a mixture of chlorine/bromine (e.g. Nylate) or peroxide (e.g. Oxonia), are effective for control of storage rots (Cheah, unpublished data). The advantages of chlorine are that it leaves no chemical residue and is cost-effective. It is recommended that chlorine at 100 to 200 ppm used in the hydrocoolers to prevent the build-up of microbial numbers in the recirculating water (Haynes et al. 1990). When sodium or calcium hypochlorite, commonly used compounds for chlorination, are dissolved in water, several compounds are formed. As hypochlorous acid is the only effective form of chlorine for disinfection it is essential to maximize its level in relation to other forms in solution. This is done by controlling the pH of the chlorine solutions to between 6.5 and 7.0 (Ogawa et al. 1980; Haynes et al. 1990).

Recontamination of carrots after washing must be avoided. Wet hydrocooled carrots travelling over grading belts in packhouses may become recontaminated. Maintaining hygiene in hydrocoolers is important, as is regular cleaning of the grading and packing plants.

Punja and Gaye (1993) found increased inoculum levels of *T. basicola* along the handling chain, with the highest levels of fungal contamination occurring on the grading belts and the belts leading to packagers. A film of organic debris containing inoculum accumulated on the conveyor belts. Scrubbing and cleaning of the belts is recommended. These authors also recommended a sanitation or chemical dip treatment after grading.

3.2.4. Cooling

Hydrocooling is the preferred method of cooling (Colour Plate 15.4). Carrots are suited to hydrocooling, and the ability of this method to remove field heat in a short time suits its use in modern carrot packhouses. Flaherty (1995) reports a reduction from 25° C to 4° C in 25 minutes. Water is held at 1° C and carrots are cooled by a shower or dip in the refrigerated water. As the water is usually recycled, attention must be paid to maintaining water quality by monitoring chlorine levels and changing the water daily. In New Zealand initial field heat removal is sometimes achieved using cool water (10-12° C) from streams or artesian sources.

Forced air cooling is another cooling option. Although forced air cooling may take longer than hydrocooling (4-6 h compared to less than 30 min), this method dries carrots after washing and may help prevent establishment of storage rots. Hydrocooling systems can be installed on packing lines whereas forced air cooling takes washed carrots off the packing line into the cooler in an additional step.

In any cooling system the carrots must be graded and packed after cooling. These operations must be carried out quickly to prevent carrots warming to ambient temperatures. After packing, carrots must be again cooled prior to storage.

3.2.5. Handling Damage

Modern carrot handling systems involve mechanical harvesting and bulk handling. Some damage can be expected. This damage inevitably leads to splits, cuts and abrasions of the carrot surface. The damage allows access for storage rots. Punja and Gaye (1993) have shown that a wounding treatment applied either at harvest or 24 h later enhanced black root rot (*T. basicola*) development in storage.

In New Zealand, Nantes type carrots are less prone to handling damage than the Kuroda types. Galati and McKay (1995) reported increased carrot breakage and splitting at lower temperatures, particularly in cold, wet soil conditions. As the main seasons for exporting New Zealand carrots are autumn and winter, there is often no chance of avoiding harvesting in cold wet conditions.

3.2.6. Grading

Final quality control of carrots is carried out during grading. Carrots washed and graded before being stored had significantly less decay than carrots stored immediately (Lockhart and Delbridge, 1972). The washed carrots are conveyed to a size grader (graded by diameter and length) and then along a wide belt for visual grading. Any diseased and damaged carrots should be removed. Grader operators should be well trained so that they can carry out the grading properly.

3.2.7. Chemical Dips

The ability of chemical dips to prevent storage rots of carrots has been evaluated (Crisp, 1974; Wells and Merwarth, 1973; Ricker and Punja, 1991). A fungicide dip of benomyl or iprodione at harvest was effective for control of storage rots caused by B. cinerea, M. acerina and S. sclerotiorum (Crisp, 1974; Cheah et al. 1996a; Geeson et al. 1988), but this treatment did not reduce the incidence of crater rot caused by *Rhizoctonia carotae* (Geeson et al. 1988). However, Ricker and Punja (1991) found that dipping carrots in sodium orthophenylphenate reduced crater rot. Copper compounds (e.g. Bordeux mixture) were effective for control of bacteria soft rot when washed carrots were dipped in these solutions (Dye, 1953). He also found that carrots dried in bright sunshine for 20 min had reduced incidences of rot.

Other alternatives to chemicals have also been tested. Cheah et al. (1997) reported that chitosan, a naturally derived polysaccharide from deacetylation of crab and shrimp shell, applied as 2 and 4% solution coatings inhibited Sclerotinia rot on carrots. Ozone at 60 $\mu L/L$ gave a 50% reduction in growth of B. cinerea and S. sclerotiorum (Liew and Prange, 1994). A low dose of UV-C light induced production of 6-methoxymellein, which increased resistance to postharvest diseases in carrots (Mercier et al. 1993b).

3.3. Storage and Packaging

3.3.1. Storage Conditions

Carrots store best at 0° C and 98-100% relative humidity (Hardenburg et al. 1986). Robinson et al. (1975) showed that respiration was minimized at this temperature. Van den Berg (1987) reports that dessication and decay were reduced during long-term storage at 0° C and 98-100% relative humidity.

Carrots have a very long potential storage life. Mature carrots may be stored for 7-9 months (Hardenburg et al. 1986). Experience in New Zealand with Nantestype carrots suggests that they have a much shorter storage potential, particularly at the end of the winter after some months of ground storage (Brash, 1996).

Ehylene exposure induces the development of a bitter flavour due to isocoumarin formation (Hardenburg et al. 1986). Carrots should not be mixed with ethylene-producing commodities (e.g. apple) during storage.

Controlled atmosphere storage is of limited use for carrots and does not extend storage life beyond that in air (Kader, 1992).

3.3.2. Bulk Storage

Carrots are not generally stored in bulk in New Zealand. The temperate climate and use of ground storage ensure carrots are available for the domestic market all the year round.

Bulk stores must be set at temperatures close to 0° C and relative humidity as close as possible to 100%. Geeson et al. (1988) showed a cool store cooled by an ice bank had a humidity of 97-98% and was more suited to long term carrot storage than a conventionally cooled store. Extra effort must be paid to raising humidity in carrot stores through appropriate design of the refrigeration plant (large evaporator area, small temperature differential) and use of aerosol humidifying systems.

3.3.3. Storage for Export

New Zealand's export carrot trade uses single cardboard cartons lined with plastic bags. Each carton holds 10 or 20 kg of carrots. The liners are folded over after the carrots have been packed and the flaps on the carton are taped shut. The cartons are palletised and held in a cool store prior to loading in a refrigerated container for despatch by sea to Asia.

Carrots are very slow to cool in plastic lined cartons and must be cold when packed. It may take many days to cool carrots in cartons, particularly when in a pallet stack. Some of the complaints from export markets about storage rots may be related to warm carrots in cartons being loaded into refrigerated containers (the cartons form a solid stack inside the container).

Plastic liner bags prevent dessication and maintain a relative humidity very close to 100%. At these high humidities inside the package there is a high likelihood of condensation forming with small variations of temperature across the package (Patterson et al. 1993). If condensation comes into contact with the stored produce, postharvest life is shortened and disease may be encouraged, for example soft rot infection of potatoes (Pringle and Robinson, 1996). Suslow et al. (1996) and Story and Simons (1989) suggest that free moisture from condensation promotes carrot decay. Carrot storage life in cartons could be improved if condensation inside plastic liner bags was eliminated. Patterson et al. (1993) suggest that 'condensation control packaging' be introduced for this purpose. Condensation control packaging combines fibreboard, a moisture barrier, capillary moisture storage and a non-wettable fabric into a single package, and should be tested on carrots.

'White blush' symptoms are often noted at the export market destination. Cisneros-Zevallos et al. (1997) have shown that these symptoms develop in plastic bags containing baby, peeled carrots when moisture condenses on the bags, slowly causing the carrot surface to desiccate. A similar process occurs in cartons of whole carrots with plastic liner bags.

Some packhouse operators use brush washing to remove root epidermal layers and reduce 'white blush' symptoms, in a process known as 'polishing'. This harsh brushing of the epidermal layer may enhance the opportunity for storage rots to become established (see earlier section on silvering).

3.3.4. Domestic Market

Carrots for the local New Zealand market are stored and transported loose in 20 kg perforated plastic bags and plastic crates with plastic liner bags, or prepacked in 1-2 kg perforated plastic bags. Refrigerated transport is commonly used.

4. Conclusion

Most of the storage rots of carrots are diseases that have originated in the field and have carried over onto carrots after harvest. Physiological disorders also arise from poor handling of carrots between harvest and storage. Every step in the handling chain can influence the extent of disease and quality of the stored carrots. Control methods, therefore, should involve improved practices in the field, in packhouses and in cool storage. The best approach to minimize storage rots and reduce quality loss is to integrate all of the appropriate practices as outlined above.

The challenge facing the industry is to produce carrots with fewer or nil chemical inputs as public concern increases over food safety, environmental issues and chemical resistance. New technologies that utilize pesticide-free methods, e.g. essential oils (Dikshit et al. 1983), chitosan (Cheah et al. 1997), biological control agents (Cheah and Tran, 1995; Wilson and Wisniewski, 1989), host resistance (Mercier et al. 1993a) and inducing natural plant defences (Lamb et al. 1989), have potential to be incorporated into integrated management strategies to effectively minimize or eliminate carrot diseases and disorders.

Acknowledgements

We thank Ms Tracy Williams and Dr R. E. Falloon of Crop & Food Research for reviewing the manuscript.

References

Anoymous. 1986. Diseases and pests of carrot. Farmnote No. 89/86. Information and Media Services. Agriculture Western Australia. 2 pp.

Apeland, J. 1974. Storage quality of carrots after different methods of harvesting. *Acta Horticulturae* 38(2): 353-357.

Brash, D.W. 1996. Keeping carrots fresh for Asia. *New Zealand Commercial Grower* 51(4): 18-20.

Cheah, L-H. 1999. Influence of field diseases and cultural practices on storage rot of carrots. Proc. 12th Australasian Plant Pathology Conference: 199 pp.

Cheah, L-H. and T.B Tran. 1995. Postharvest biocontrol of Penicillium rot of lemons with industrial yeasts. *In:* Proc. 48th NZ Plant Protection Conference: 155-157.

Cheah, L-H and B.B.C Page. 1999a. Epidemiology and control of violet root rot of carrots. *In:* Proc. 52nd NZ Plant Protection Conference: 157-161.
Cheah, L-H and B.B.C Page. 1999b. Violet root rot: disease management strategies. *New Zealand Commercial Grower* **54**: 28-29.
Cheah, L-H., Brash, D.W. and A.P. Marshall. 1996a. Storage rots of carrots and their control. *In:* Proc. 49th NZ Plant Protection Conference: 314 pp.
Cheah, L-H., Marshall, A.P. and D.W. Brash. 1996b. First report of *Thielaviopsis basicola* on cool-stored carrots in New Zealand. Plant Disease 80: 821.
Cheah, L-H., Page, B.B.C. and R. Shepard. 1997. Chitosan coating for inhibition of Sclerotinia rot of carrots. *NZ J. of Crop & Horticultural Science* **25**: 89-92.
Cisneros-Zevallos, L., Saltveit, M.E. and J.M. Krochta. 1997. Hygroscopic coatings control white discolouration of peeled (minimally processed) carrots during storage. *Journal of Food Science* **62**(2): 363-366, 398.
Cook, R.J. 1988. Biological control and holistic plant care in agriculture. *Am. J. Altern. Agric.* **3**: 51-62.
Crisp, A.F. 1974. Control of wastage in cooled storage carrots. *Acta Horticulturae* **38**: 389-396.
Den Outer, R.W. 1990. Discolourations of carrot (*Daucus carota* L.) during wet chilling storage. *Scientia Horticulturae* **41**: 201-207.
Dikshit, A., Dubey, N.K., Tripathi, N.N. and S.N. Dixit. 1983. Cedrus oil—a promising storage fungitoxicant. *Stored Product Research* **19**: 159.
Dye, D. 1953. Control of soft rot (*Erwinia carotovora* (Jones) Holland) in carrots during transit and in storage. *NZ Science and Technology* **30**: 465-467.
El-Tarabily, K.A., Hardy, G.E., Sivasithamparam, K. and A.G. McKay. 1997. Amendment of soil with lime or gypsum and its effect on cavity spot disease of carrots (*Daucus carota* L.) caused by *Pythium coloratum*. *Australian Journal of Experimental Agriculture* **37**: 265-270.
Ewaldz, T. 1997. Prognosis of longtime storage ability of carrots based on test storage at high temperatures. Experiences of Swedish trials 1991-1993. *Vaxtskyyddsotiser* **61**: 4-7.
Flaherty, A. 1995. Super cooling sets the standard. *Grower* **124**(22): 10-13.
Galati, A. and McKay, A. 1995. Cavity spot disease of carrots. Farmnote No. 74/95. Information and Medical Services, Agriculture Western Australia 2 pp.
Galati, A., McKay, A. and S-C. Tan. 1995. Minimizing post harvest losses of carrots. Farm note No. 75/95. Information and Media Services, Agriculture Western Australia. 3 pp.
Geeson, J.D., Browne, K.M. and H.P. Everson. 1988. Storage diseases of carrots in East Anglia 1978-82, and the effects of some pre- and post-harvest factors. *Annals of Applied Biology* **112**: 503-514.
Hardenburg, R.E., Watada, A.E. and C.Y. Wang. 1986. The commercial storage of fruits, vegetables, and florist and nursery stocks. USDA Agricultural Research Service. Agriculture Handbook No. 66.
Haynes, Y., Joyce, D. and C. Yuen. 1990. Chlorination in postharvest horticulture. Farmnote No. 9/90. Information and Media Services, Agriculture Western Australia 3 pp.
Howard, R.J., Garland, J.A. and W.L. Seaman. 1994. *Diseases and Pest and Vegetable Crops in Canada*. MOU Printing Ltd, Ottawa 553 pp.
Kader, A.A. 1992. Postharvest technology of horticultural crops. University of California. Division of Agricultural and Natural Resources. Publication 3311.
Lamb, C.J., Lawton, M.A., Dron, M. and R.A. Dixon, 1989. Signals and transduction mechanisms for activation of plant defence against microbial attack. *Cell* **56**: 215.
Lewis, B.G., Davies, W.P. and B. Garrod. 1981. Wound-healing in carrot roots in relation to infection by *Mycocentrospora acerina*. *Annals Applied Biology* **99**: 35-42.
Liew, C-L. and R.K. Prange. 1994. Effect of ozone and storage temperature on postharvest diseases and physiology of carrots. *J. America Soc. Horticultural Science* **119**: 503-567.
Lockhart, C.L. and R.W. Delbridge. 1972. Control of storage diseases of carrots by washing, grading and postharvest fungicide treatments. *Can. Plant Dis. Survey* **52**: 140-142.

Mercier, J., Ponnampam, R., Berard, L.S. and J. Arul. 1993a. Polyacetylene content and UV-induced 6-methoxymellein accumulation in carrot cultivars. *J. Science of Food and Agriculture* **63**: 313-317.

Mercier, J., Ponnampalam, R. and M. Boulet. 1993b. Induction of 6-methoxymellein and resistance to storage pathogens in carrot slice by UV-C. J. *Phytopathology* **137**: 44.

Ogawa, J.M., Hoy, M.W., Manji, B.T. and D.H. Hall, 1980. Proper use of chlorine for postharvest decay control of fresh market tomatoes. *California Tomatorama. Informational Bulletin* **No. 27**: 1-2.

Patterson, B.D., Jobling, J.J, and S. Moradi. 1993. Water relations after harvest—a new technology. *In:* Proceedings of the Australasian Postharvest Conference, Gatton College, Queensland, 20-24 September. **1993**: 99-102.

Phan, C.T. and H. Hsu. 1973. Physical and chemical changes occurring in the carrot root during storage. *Canadian Journal of Plant Science* **53**: 629-634.

Pringle, R.T. and K. Robinson. 1996. Storage of seed potatoes in pallet boxes. 1. The role of tuber surface moisture on the population of Erwinia bacteria. *Potato Research* **19**: 205-222.

Punja, Z.K. and M.M. Gaye. 1993. Influence of postharvest handling practices and dip treatments on development of black root rot on fresh market carrots. *Plant Disease* **77**: 989-995.

Punja, Z.K., Chittaranjan, S. and M.M. Gaye. 1992. Development of black root rot caused by *Chalara elegans* on fresh market carrots. *Canadian J. of Plant Pathology* **14**: 299-309.

Ricker, M.D. and Z.K. Punja. 1991. Influence of fungicide and chemical salt dip treatments on crater rot caused by *Rhizoctonia carotae* in long-term storage. *Plant Disease* **75**: 476-474.

Robinson, J.E., Browne, K.M. and W.G. Burton. 1975. Storage characteristics of some vegetables and soft fruits. *Ann. Appl. Biol.* **81**: 399-408.

Rubatzky, V.E., Quiros, C.F. and P.W. Simon. 1999. Carrots and related vegetable *Umbelliferae*. CABI Publishing, UK, 294 pp.

Schwarz, A. 1977. Storage of carrots: a possible relationship between storage ability and soil conditions. *Acta Horticulturae* **62**: 345-350.

Segall, R.H. and A.T. Dow, 1973. Effects of bacterial contamination and refrigerated storage on bacterial soft rot of carrots. *Plant Disease Reporter* **57**: 896-899.

Sherf, A.F. and A.J. MacNab. 1986. *Vegetable Diseases and their Control.* John Wiley & Sons Inc., UK, 119-156.

Shibario, S.I., Upadhyaya, M.K. and P.M.A. Toivonen, 1998. Potassium nutrition and postharvest moisture loss in carrots (*Daucus carota* L.). *Journal of Horticultural Science and Biotechnology* **73**: 862-866.

Snowdon, A.L. 1991. A colour atlas of postharvest diseases and disorders of fruits and vegetables Vol 2: Vegetables. Wilfe Scientific Ltd, London, 269-293.

Sorenson, J.N., Jorgensen, U. and B.F. Kuhn. 1997. Drought effects on the marketable and nutritional quality of carrots. *J Sci. Food Agric.* **17**: 379-391.

Story, A. and D.H. Simons. 1989. Fresh produce manual: handling and storage practices for fresh produce. Second edition. Australian United Fresh Fruit and Vegetable Association 162 pp.

Subbarao, K.V. and J.C. Hubbard. 1996. Interactive effects of broccoli residue and temperature on *Verticillium dahliae* microsclerotia in soil and on wilt in cauliflower. *Phytopathology* **86**: 1303-1310.

Suslow, T.V., Mitchell, J. and M. Cantwell. 1996. Produce facts: carrots. *Perishables handling* **88**: 13-14. UC Davis, USA.

Tahvonen, R. 1985. The prevention of *Botrytis cinerea* and *Sclerotinia sclerotiorum* on carrots during storage by spraying the tops with fungicide before harvesting. *Annales Agriculturae Fenniae* **24**: 89-95.

Tamietti, G. and A. Matta. 1981. Atterazione die fittoni di carota durante il periodo invernate di conservazione in campo. *Riuista di Patologia Vegetale* **17**: 45-54.

Towner, D.B. and L. Berah. 1976. Core-rot: a bacteria disease of carrots. *Plant Disease Reporter* **60**: 357-359.

Tsao, P.H. and J.L. Brecker. 1966. Chlamydospores of *Thielaviopsis basicola* as surviving propagules in natural soils. *Phytopathology* **56**: 1012-1014.

Van den Berg, L. 1987. Influences of postharvest factors on postharvest reactions. *In*: Postharvest physiology of vegetables, (J. Weichmann, (Ed.), Marcel Dekker, New York, 597 pp.

Wells, J.M. and F.L. Merwarth. 1973. Fungicide dips for controlling decay of carrots in storage for processing. *Plant Disease Reporter* **57**: 697-700.

Wilson, C.L. and M.E. Wisniewski. 1989. Biological control of postharvest diseases of fruits and vegetables: an emerging technology. *Ann. Rev. Phytopathology* **27**: 425.

Wood, R.J. 1995. Carrots. *New Zealand Commercial Grower* **50**(1): 33.

Colour Plate 15.1. Sclerotinia rot (*S. sclerotiorum*) on cool-stored carrots.

Colour Plate 15.2. Black root rot (*Thielaviopsis basicola*) on cool-stored carrots.

Colour Plate 15.3. Tumble washer (top) and screen to remove soil and organic debris prior to hydrocooling.

Colour Plate 15.4. Hydrocooling of carrots after washing for rapid removal of field heat.

16

Minimal Processing of Fruits and Vegetables

Catherine Barry-Ryan and David O'Beirne
*Food Science Research Centre, Department of Life Sciences,
University of Limerick, Ireland*

Abbreviations

Anon.	anonymous
CO_2	carbon dioxide
cm^3	centimeters cubed
cfu/g	colony forming units per gram
CA	controlled atmosphere
°C	degrees Celsius
C_2H_4	ethylene
$C_6H_{12}O_6$	glucose
h	hour
LDPE	low density polyethylene
mg/l	micrograms per litre
µl/kg/hr	microlitres per kilogram per hour
mm	millimeters
min	minute
N_2	nitrogen
O_2	oxygen
ppm	parts per million
%	per cent
pp	poly-propylene
PVC	polyvinyl chloride

1. Introduction

A major change in the consumption patterns of fruit and vegetables has occurred in the last decade. Fresh fruit and vegetables are increasing in popularity at the expense of canned products and the demand for high-quality, easy to prepare, convenient food is steadily rising (Powrie and Skura, 1991). There is now a wide choice of prepacked fresh produce available, meeting the twin consumer demands for unadulterated freshness and maximum convenience (Day and Gorris, 1993).

Fruits and vegetables are living, respiring, edible tissue which continue to be metabolically active when harvested (Lee et al. 1995). Losses due to respiration, breakdown of organic substrates, and transpiration water, are not replaced once the fruit or vegetable is detached from the parent plant and so deterioration occurs. This physiological deterioration can be affected by intrinsic (i.e. climacteric vs non-climacteric commodities) and extrinsic (i.e. temperature, ethylene (C_2H_4), oxygen (O_2) and carbon dioxide (CO_2) concentrations) factors. Other types of deterioration which occur include chemical and enzymatic changes, microbial deterioration and attack by macroorganisms. Physical damage, such as improper harvesting, handling, processing and packaging (Powrie and Skura, 1991; Lee et al. 1995) can also effect the shelf-life of fruits and vegetables.

Minimal processing of fruits and vegetables may involve cooling, washing, trimming, peeling, shredding or slicing before packaging (Strungell, 1988). Such products include prepared salads, pre-cut vegetables, peeled fruit and whole vegetables, sold within 7-8 days of preparation, after storage at a low temperature. A minimally processed product must have a consistent, fresh appearance, be of acceptable colour and be reasonably free of defects.

Modified atmosphere packaging (MAP) is an increasingly popular food preservation technique which compliments refrigeration to increase shelf-life. The gaseous atmosphere surrounding the food is different from air; its use with minimally processed products enables the production of extended product ranges with reduced levels of preservatives and other additives as the MA controls respiration rate and indirectly microbial growth rate (Day and Gorris, 1993). The gas levels aimed for within these packages are typically in the range 2-4% O_2 and 3-10% CO_2. In setting up this delicate balance, the product's respiration rate and weight, the permeability of the packaging materials to O_2 and CO_2, and the storage temperature must be taken into account (O'Beirne, 1990).

This chapter reviews the impact of minimal processing steps, packaging and storage on the storage life and quality of ready-to-use products.

2. Minimal Processing

By definition, minimal processing would encompass any procedure, short of traditional complete preservation procedures (heat sterilisation, etc) that adds value to the vegetable (Floros, 1993). Minimally processed vegetables are those prepared for convenient consumption and distribution to the consumer in an almost fresh state (King and Bolin, 1989). Minimal processing refers especially to processes such as peeling, washing, slicing, shredding, etc which makes the product ready-to-use for the consumer (Strugnell, 1988).

2.1. Preliminary Processing

Raw material selection requires consideration of the intrinsic quality and susceptibility to damage during transport and storage. Barry-Ryan (1996) reported that the physiological age and cultivar type of raw material used for producing minimal processed products can affect their quality and storage-life. Lee et al. (1996) also reported that cultivars had distinct characteristics which affected its post-harvest behaviour, i.e. respiration rate and storage.

The first essential step in reducing the overall contamination of the raw material is by removing the outer leaf layers or surface dirt, depending on the vegetable type, before use. Marxcy (1978) reported that there was a higher microbial load on the outer leaves when compared to the inner leaf of lettuce (i.e. 6.3×10^4 and 3.2×10^1 cfu/g respectively, on plate count agar), due to the inner leaves being more protected from environmental contamination.

2.2. Peeling

In the preparation of ready-to-eat products, a minimum processing procedure such as coring and peeling is applied to most fruits and vegetables to remove the inedible parts. The peeling procedure is dependent on the raw material type and influences the quality of the finished product (Barry-Ryan, 1996). The outer epidermal layer on plants provide protection from external surroundings. The removal of this protective layer from the plant and the damage of underlying cells initiates certain biochemical reactions, such reactions involve the release of intercellular enzymes such as proteases, cellulases, peroxidases and lipoxygenases which affect the flavour, colour and texture of the vegetable once released (Buckenhuskes and Gierschner, 1990). Nutrients are also released which support microbial growth, which in turn affect the plant tissue through the action of their enzyme systems. The respiration rate increases when this protective layer is removed, this is due to the wound repair response and the increased availability of oxygen (Watada et al. 1986). The response of the tissue to the removal of this outer layer is the production of a new protective layer. This is especially true in abrasion peeled carrots, which quickly lose their bright orange colour due to the development of a white material on the surface. The degree of formation of this lignin type material, on the surface during storage, is directly related to the severity of abrasion during peeling (Bolin and Huxsoll, 1991a).

One of the earliest methods of peeling involved hand-held peelers or knives. This method yields high quality products, as there is little stimulation of a damage response, but it is time consuming and not suitable for bulk production. Currently used peeling methods employ either physical or chemical means to separate the outer skin from the inner parenchyma tissue. Abrasion, steam and lye peeling are common commercial methods.

Abrasion peeling using a carborundum disc and running water is the method of peeling commonly used for potatoes and root vegetables. It is an easy and efficient method of peeling, but consideration should be given to the correct load size, as low loading results in incomplete peeling and overloading causes excessive peeling (Setty et al. 1993).

Machines are available with various grit types, for example, coarse grits peel faster but leave a rough surface, while fine grits peel more slowly and smoothly. Other methods of commercial peeling include freeze, flame and vacuum peeling also acid, calcium chloride and ammonium salts have been used to peel various fruits and vegetables (Setty et al. 1993).

Barry-Ryan and O'Beirne (2000) reported that abrasion peeled carrots had higher respiration rates, greater microbial contamination and growth rates, higher pH values, higher rates of weight loss and shorter microbiological shelf-lives than hand peeled carrots. These results reflected the higher quality of the manually peeled carrots. Micrographs of the peeled surfaces confirmed that abrasion peeling inflicted greater damage.

2.3. Slicing

Many of the constituent fruits or vegetables in ready-to-eat products require further preparation such as slicing, shredding or dicing, before packaging. Prepared vegetables are more susceptible to spoilage than whole vegetables since processing disrupts the plant tissues, breaching protective epidermal layers and releasing nutrient rich vascular and cellular fluids (Adams et al. 1989). This causes an acceleration in the physiological breakdown of the plant tissue and a shortened storage life as these cellular fluids contain a wide range of enzymes and brings them into contact with their substrates.

Slicing plant tissue can result in loss in firmness. Kiwifruit slices lose 50% of their initial firmness in less than two days at 2° C. It was suggested that this textural breakdown of kiwifruit tissues during storage was due to the enzymatic hydrolysis of cell wall components (Varoquaux, 1991).

Discolouration occurs through the action of polyphenol oxidases present in the cellular fluids and texture breakdown can occur due to pectinolytic and cellulolytic activity. Dehydration of the large mass of exposed/damaged cells probably was responsible for the appearance of the whitish translucent tissue. During processing, carrots are peeled and sliced and this results in a surface discolouration due to the development of a white lignin type material on the cut surfaces during storage (Cisneros-Zevallos et al. 1995). Discolouration may result in minimally processed products due to the release of phenolic compounds and their oxidation by polyphenol oxidase (Barry-Ryan and O'Beirne, 1998b). Becker and Gray (1992) reported that when catechol was poured on the cut-lettuce surface a rapid darkening occurred, indicating the presence of polyphenol oxidase.

Oxidative rancidity in processed vegetable based products occurs through the oxidation of lipids and lipid soluble substances such as vitamins, phospholipids, flavours, and carotenoids. Rancidity of these various fatty constituents is caused by the action of lipases and lipoxygenases released (Buckenhuskes and Gierschner, 1990). Characteristic off odours and flavours develop after lipid oxidation. Colour change, loss of flavour, aroma and nutritional quality can also occur.

The extent of injury to the product, influenced by slicing, depends on a number of factors such as final piece size, sharpness of cutting surface, and the mechanical aspects of cutting action, and the mechanical properties of product being cut (Abe et al. 1993). These in turn have an impact on the physiological response of the product, its susceptibility to microbial spoilage and to some extent, to the incidence of enzymatic browning on the cut surfaces (Zhou et al. 1992).

Working with cut-lettuce, Bolin and Huxsoll (1991b) showed that slicing with a sharp blade was superior to either chopping with a sharp blade or slicing or chopping with a dull blade. Product storage life of cut lettuce was significantly reduced, in some instances by as much as 50%. Microscopic examination of lettuce leaves showed that noticeable exudation occurred only on the sliced surface which had more cell debris than leaves which were torn. Therefore, tearing lettuce instead of cutting should give better storage life. Smaller shred size reduced storage life, thin 3 mm slices of shredded lettuce respired more rapidly and had shorter shelf-life than salad-cut (9-20 cm^3) lettuce (Bolin and Huxsoll, 1991b).

Abe et al. (1993) showed that physiological changes associated with deterioration of partially processed carrots was dependent on the cut direction, the cut surface area and also that physiological changes in the xylem had a greater effect on respiration rate than changes in the phloem-cortex. The effect of shredding modes on the storage stability of partially processed pepper fruits also showed that the ratio of the wounded area and the wounded direction had an effect on deterioration. Increasing wound area increased the rate of deterioration, vertical as opposed to a horizontal cut also increased the rate (Zhou et al. 1992).

The respiration rate of whole carrots increased two to three-fold on slicing (Kahl and Laties, 1989; Priepke et al. 1976). Ethylene production by the kiwifruit was very low, about 5 μl/kg/h, and slicing caused the rate to increase to about 40 μl after 2 h and about 80 μl after 4 h at 20° C (Watada et al. 1986). The effects of slicing on the rate of respiration and ethylene production differs between climacteric and non-climacteric fruit and with physiological age of climateric fruit (Lee et al. 1995).

Carrot sticks that were prepared with a sharp culinary knife exhibited a whitish translucent appearance on the surface compared to carrot sticks sliced with a razor sharp blade. Scanning electron microscopy examination of the translucent tissue revealed that the knife tended to shear, separate and compress the cells and tissues of the root (Tatsumi et al. 1991).

The effects of slicing method on the quality and storage-life of modified atmosphere packaged carrot slices were determined using microscopy, sensory evaluation, microbial counts and a range of physical and chemical tests (Barry-Ryan and O'Beirne, 1998a). Slicing caused physical damage, physiological stress and enhanced microbial growth. The severity of these effects were in the order of blunt machine blade > sharp machine blade > razor blade. These findings provide insights into the magnitude and basis of slicing effects and also confirm the importance of gentle processing and the use of a sharp blade.

2.4. Final Treatment

Electron microscopy of both surfaces of unwashed lettuce showed them to be liberally coated with bacteria and other debris. Washing in tap water reduced total aerobic counts on exposed surfaces, although substantial numbers remained in hollows at the junction of epidermal cells and in folds in the epidermis (Adams et al. 1989).

Solutions of a small number of antimicrobial materials, e.g. chlorine, citric acid and ascorbic acid, have been used as dips to reduce the initial microbial load. Huxsoll and Bolin (1989) showed that thorough washing to remove the free cellular contents that are released by cutting was also found to be important in prolonging the shelf-life of cut-lettuce. Chemical treatments are commonly used, but must be effective for maintaining one or more quality factors, but not degrade any other quality factor such as product wholesomeness (Huxsoll and Bolin, 1989).

Chlorine is universally recognized as a convenient, relatively inexpensive, efficient destroyer of microorganisms. It is often added to industrial process waters to destroy bacteria and inhibit algal growth. Depending on the operation, the chlorine can be added to the water as a solid (calcium hypochlorite or trichlorocyanuric acid), as a liquid (sodium hypochlorite) or as a gas in which case it can be injected directly into the liquid being treated (Anonymous, 1973). The use of chlorine dips or sprays have been effective in controlling bacterial contamination of vegetables. The antimicrobial activity of hypochlorite solution is related to the concentrations of undissociated hypochlorite.

Work done with shredded lettuce and carrot discs showed that a chlorine solution (100 ppm) reduced the initial microbial loads. For the shredded lettuce dipped in chlorine, the microbial load remained one Log cycle lower than that of lettuce dipped in unchlorinated water. The same effect was observed initially with the carrot discs, but the beneficial effect of the chlorine was lost after six days of storage (Barry-Ryan, 1996).

Brackett (1987) found that chlorine levels in the range of 10-40 mg/l were sufficient to destroy *Listeria monocytogenes* cells. Commercial production of ready-to-use shredded lettuce utilises 10-150 ppm chlorine in the wash water to produce a product with high quality and long storage life (Dowling, 1991).

There is now a range of methods available to monitor the residual chlorine in the wash water, including indicators, titrimetric and colourmetric techniques and specially designed equipment (Morrow and Martin, 1977; Anonymous, 1977). For example Water-Chex, a commercially available product, consists of a plastic sachet which is immersed in the water to be tested and a fixed volume drawn in. A colour change occurs which may be read to determine the residual chlorine in solution (Anonymous, 1973).

Residual acid hypochlorite remaining after washing showed no effective antimicrobial activity in lettuce stored at abuse temperature. Reducing the pH of the chlorinated wash water in the range 4-7 produced a slightly greater lethal effect, though at pH 4 the hypochlorite was unstable. Increasing the washing period (5 min), with hypochlorite showed no improvement, even though 60% of the original free available chlorine remained after 30 min washing (Adams et al. 1989). It is noteworthy however, that chlorine only delays microbial spoilage and does not exhibit any beneficial effects in biochemical and physiological disorders (Varoquaux, 1991).

Antioxidants are a range of substances which can inhibit or interfere with free radical formation. The antioxidants most commonly used for vegetable products include citric acid, ascorbic acid and ascorbyl palmitate, which can be applied by dipping (Dorko, 1994).

In work using citric acid/ascorbic acid dips as antioxidants it was found that these solutions substantially reduced aerobic plate counts, anaerobic plate counts, and bile tolerant-lactose fermenting organisms (coliforms) a day after dipping and after storage for 7 days (MAP or vacuum packaging) and 14 days (O'Beirne and Ballentyne, 1987). Similarly, immediately after treatment with 'Tater White' which contains sodium meta-bisulfite, sodium citrate, and sodium erythrobate, the counts were lower than on the untreated lettuce and the difference was even greater after 48 h (Marxcy, 1978). Tatsumi et al. 1993, showed that treatment of carrots with a sodium chloride solution reduced the amount of white tissue development on the carrot surface.

Depending on the product, rinse water may have to be removed before packaging to prevent extensive microbial growth. It was also determined that if the lettuce surface was left wet, its storage life was drastically reduced. A common method for removing water is centrifugation. Centrifuging to a point of slight desiccation can be useful in prolonging shelf-life, but at higher levels, changes in texture are apparent. Other methods of removing water include fixed drying air and blotting with a porous material (Floros, 1993). Centrifugation may remove soluble components from lettuce other than added rinse water. When unrinsed lettuce was centrifuged there was an immediate weight loss, evidently caused by the loss of cellular fluids released by rupture of the cells during cutting (Bolin and Huxsoll, 1991b).

3. Storage

Modern retail packaging of fresh horticultural produce is expected to meet a wide range of requirements including prevention of mechanical damage, minimisation of weight loss and shrinkage.

Unlike other chilled perishable foods that are packaged, fresh fruit and vegetables continue to respire and transpire after harvesting, and consequently any subsequent packaging must take into account this respiratory activity. Respiration is a metabolic process which is defined as the oxidative breakdown of complex materials such as starch and sugars, to simpler molecules such as carbon dioxide and water with the concurrent production of energy (Mannapperuma et al. 1991):

$$C_6H_{12}O_6 + 6O_2 \rightarrow 6CO_2 + 6H_2O + energy$$

Transpiration is the loss of water by the product and induces wilting, shrinkage, loss of firmness, flavour and succulence (Moleyar and Narasimham, 1994; Ben-Yehoshua, 1987). Mechanical damage and physical injury, bruising, scratches and surface cuts, greatly accelerate the rate of water loss from fruit and vegetable tissue (Floros, 1993).

The rates of respiration and transpiration are a good indication of shelf-life, the higher the rates the shorter the storage-life, and thus can be used to monitor deterioration during processing and storage. Since the objective of short-term preservation technologies is to maintain the fresh appearance, taste, flavour and other qualities of the horticultural products, respiration and transpiration rates should be kept as low as possible. To prolong the shelf-life of these products it is necessary to minimize the synthesis of ethylene, which may also be increased by processing and damage.

Temperature is the most important environmental factor in the postharvest life of fruits and vegetables (Lee et al. 1995). Its effect is dramatic not only on respiration and transpiration rates, but also on other biological and biochemical reactions as well and so should be reduced as soon as possible after harvesting (Raison, 1980; Kader, 1987). Rizvi (1981) showed that a decrease of 10° C usually decreases the respiration rate by a factor of 2 to 3. The Guidelines for Handling Chilled Foods (IFST, 1987) recommends a storage temperature range of 0° to 8° C for salad vegetables depending on their susceptibility to chilling injury. Bolin et al. (1977) showed that numerous factors affected the storage stability of shredded Iceberg lettuce, with temperature being most important. Pouches of shredded lettuce stored at 2° C retained a marketable quality 2.5 times longer than those held at 10° C. A significant increase in the ratio of gram positive to gram negative microorganisms occurred with increased time at the abuse temperature of 30° C. A ratio of gram positive to gram negative of 0.5 or greater is considered as indicative of temperature abuse (Manvell and Ackland, 1986).

3.1. Modified Atmosphere Packaging

The terms 'controlled atmosphere' (CA) and 'modified atmosphere' (MA) mean that the atmosphere composition surrounding a perishable product is different from that of normal air. Both commonly involve manipulation of CO_2, O_2 and nitrogen levels; however, other gases such as carbon monoxide, ethylene, propylene and acetylene are sometimes included (Smock, 1979). The gas levels aimed for within the package are typically in the range 3-5% O_2 and 3-10% CO_2 (Day, 1991). Respiration and weight of the product, permeability of the packaging material to the different gases and storage temperature have to be taken into account when setting up this delicate balance (O'Connor et al. 1992).

The following model permits good predictive estimations of a generated atmosphere at equilibrium, when the diffusions of O_2 and CO_2 through the film balance the consumption of O_2 and production of CO_2 by the respiring plant (Varoquaux, 1991; Exama et al. 1993).

$$X(O_2\%) = KSxo/(KS \div \alpha) + \alpha mxo/(KS \div \alpha m) - e^{-(KS + \alpha m/v)t}$$

X = concentration O_2 (%) at time t
xo = initial concentration of O_2
K = O_2 diffusivity through the film
S = surface area of the film
V = headspace
m = weight of plant tissue
t = storage time
α = proportionality between respiration intensity and O_2 concentration (Varoquaux, 1991).

However, if the respiration rate is too slow to prevent discolouration (enzymic browning) or retard physiological changes, flushing the package with a pre-selected gas mixture containing 100% nitrogen, or low O_2 and elevated CO_2 concentrations using N_2 as an inert filler gas to make up the remaining gas volume, may have important benefits on shelf-life (Zagory and Kader, 1988).

MAP reduces the respiration rate of produce thus slowing down the process of physiological ageing. MAP also inhibits the action of the ripening hormone ethylene, slows down the various compositional changes associated with ripening such as softening and protects the colour of green vegetables. The use of low levels of O_2 and elevated levels of CO_2 have also been shown to reduce the rate of both ethylene synthesis and action (Brecht, 1980; Lee et al. 1995). In addition, it also reduces microbial spoilage by maintaining product condition (Carlin et al. 1990). Loss of antioxidant vitamins, such as Vitamin C, is reduced under MA conditions (Barry-Ryan and O'Beirne, 1998b). While chilling alone will extend shelf-life, chilling combined with MAP is substantially more effective. The required combinations of temperature, O_2 and CO_2 levels vary not only with the vegetable type but also with the variety, origin and season. Care must be taken that too high a CO_2 level is not used as it can cause physiological damage (Herner, 1987). For example, Brecht (1980) reported that CO_2 levels above 1% induced brown stain in Crisphead lettuce and >15% CO_2 has been shown to produce off-flavours in several fruits. Too low a concentration of O_2 can result in anaerobic respiration and thus the development of off-odours (Weichman, 1987) and in some products it could facilitate the growth and toxin production of *Clostridium botulinum*, even at temperatures less than 4° C.

Preliminary work done by Day (1996) and Barry-Ryan and O'Beirne (1998b) using alternative novel MAs, such as high oxygen (80% and 20% nitrogen) or argon (60% and 40% nitrogen), in a barrier film for selected prepared produce items suggests that these atmospheres are promising and may have beneficial effects on quality. This work established some beneficial effects on the quality, enzymatic activity and other oxidative processes of these minimally processed vegetables using novel atmospheres.

3.2. Packaging Types

A modified atmosphere is created naturally in a sealed package, as a direct result of counterbalancing the O_2 uptake and CO_2 production of the produce with diffusion of gases across the packaging film. Eventually, at a given temperature, the two gases approach steady-state when the rate of gas permeation through the package film equals the rate of respiration (Moleyar and Narasimham, 1994). Packaging produce reduces damage and contamination as well as the production and maintenance of a MA (Hardenburg, 1971). Everson et al. (1992) showed that although packaging can retard undesirable textural changes such as increase in toughness and fibrousness of asparagus during ambient shelf-life, the use of an insufficiently permeable films can lead to over modification of the pack and result in anaerobic respiration and off-odour development.

The challenge facing packaging suppliers has been to develop not just a single packaging material of the correct permeability, but a series of materials whose permeability can be adjusted to the specific requirements of the product to be packed. Other features to be considered may include transparency, seal integrity, durability and microwaveability. Packages are also barriers to movement of water vapour and can aid in the maintenance of high relative humidity and turgor of fruit and vegetables (Myers, 1989). Gas permeability is a function of molecular structure, including plasticiser used, film thickness, coatings and other treatments (O'Beirne, 1987). The selection of a film must take into account the respiration rate of the particular produce. Table 1 gives the classification of fruits and vegetables based on their respiration rates (Robertson, 1993; Moleyar and Narasimham, 1994).

MAP utilizes polymeric films of differential permeabilities to O_2, CO_2, C_2H_4 and water vapour to extend shelf-life. Plastic polymeric films are the major materials used for retail packaging of horticultural produce. The most commonly used plastic films include monolayer polyvinyl

Table 1. Classification of fruits and vegetables according to their respiration rate

Respiration Rate $[CO_2$ production $(ml\ kg^{-1} h^{-1})]$		Commodity
Low	(<10)	dry onions, pineapples, lettuce (Kordaat), cabbage, celery, cucumber, beetroot and tomato.
Medium	(10-20)	carrots, potatoes, peppers, parsnip, cabbage (Primo), mango and lettuce (Kloek).
High	(20-40)	cauliflower, strawberries, romaine, Brussels sprouts, blackberries and asparagus.
Very High	(40-60)	spinach, watercress, broad beans, sweetcorn and raspberries.
Extremely high	(>60)	broccoli, mushrooms, carrots (Julienne-cut) and peas in pod.

Adapted from Robertson, 1993; Moleyar and Narasimham, 1994.

chloride (PVC) for tray overwrapped produce and perforated thin low density polyethylene (LDPE) for bagged produce (Greengrass, 1993). Transparent packaging films are desirable in most cases, but due to storage at low temperatures, condensation can build up on the film. However, this may be overcome by the use of additives or coatings on the film which impart antifog properties but do not affect the permeability (Robertson, 1993). The recent developments in co-extrusion technology have made it possible to manufacture films with designed transmission rates of oxygen (Brody, 1992). Micro-porous films with very high permeabilities are also available commercially (Pratt, 1990).

Tray packs comprise a lower, rigid tray with a transparent plastic film overwrap to contain the produce. The trays are generally manufactured from white or un-pigmented PVC or poly-propylene (PP), the overwrap film must be either heat shrinkable, PVC or a stretch film, LDPE. Temperature responsive gas permeable films have been developed which are based on side chain crystallizable polymers that can match or exceed the increasing respiration rates of the fresh produce which has been exposed to temperature fluctuations (Robertson, 1993). The possibility of using MAP trays directly in microwave ovens for subsequent cooking of fresh vegetables has recently been of commercial interest. Fresh whole or prepared vegetables could be packaged in an appropriate tray and lidding film system to develop a MA and extend shelf-life. The tray could be placed directly into the microwave oven and the lidding film perforated to allow the release of steam. Since fresh vegetables have a high moisture content, cooking time would be short and the temperature would not exceed 100° C, therefore inexpensive plastic trays could be used.

The availability of absorbers and adsorbers of O_2, CO_2 and C_2H_4 and water provides additional tools for maintaining a desired atmosphere within a package (McGlasson, 1992), though these have found little commercial use. These gas scavengers are placed in the products package. They are made by enclosing small amounts of the reactive chemical inside highly permeable, opaque pouches (Kader et al. 1989). For example calcium chloride can be used to remove excess water or powdered iron as an oxygen scavenger (Day, 1990).

Vacuum packaging involves packaging of the product in a film impermeable to O_2, air is evacuated and the package is sealed. O_2 levels in the pack are reduced to less than 1% (Day, 1992).

4. Perspectives

MAP of prepared fruits and vegetables has become an area of very active research, development, and commercial application worldwide. The reasons for this include the emergence of suitable packaging films, expansion of the area devoted to chilling at supermarket level, interest in the consumers in fresh foods, and the parallel development of MAP in fresh and cured meats, in fish, and bakery products.

Continued research is required to fill the gaps in our understanding of the nutritional, physiological and microbiological consequences of processing fresh produce. There is also a need for further work into the modification of the permeability properties of the common polymeric films to make them more suitable for MAP. Continued research into the possibility of the use of alternative atmospheres for ready-to-use horticultural products is also required.

References

Abe, K., Yoshimura, K. and I. Takashi. 1993. Effects of cutting direction on storability and physiological changes in processed carrots. *Journal of the Japanese Society of Food Science and Technology* **40**: 101-106.

Adams, M.R., Hartley, A.D. and L.J. Cox. 1989. Factors affecting the efficacy of washing procedures used in the production of prepared salads. *Food Microbiology* **6**: 69-77.

Anonymous. 1973. Determination of chlorine in water. *Chemical Processing* **19**: 13-15.

Anonymous. 1977. Quick method for measuring pH and chlorine gives accurate readings in less than an hour. *Food Engineering* **49**: 193.

Barry-Ryan, C. 1996. Factors in deterioration of vegetables processed using novel mild techniques. Ph.D. Thesis, University of Limerick, Ireland, p. 131-138.

Barry-Ryan, C. and D. O'Beirne. 1998a. Quality and shelf-life of fresh cut carrot slices as affected by slicing method. *Journal of Food Science* **63**: 851-856.

Barry-Ryan, C. and D. O'Beirne. 1998b. Novel high oxygen and noble gas modified atmosphere packaging for extending the quality and shelf-life of fresh prepared produce. *In*: Proceedings of the 19th International Conference on 'Advances in the refrigeration systems, food technologies and cold chain.' 23-26 September 1998, p. 39.

Barry-Ryan, C. and D. O'Beirne. 1999. Ascorbic acid retention in shredded Iceberg lettuce as affected by minimal processing. *Journal of Food Science* **64**: 206-208.

Barry-Ryan, C. and D. O'Beirne. 2000. Effects of peeling method on the quality of ready-to-use carrots. *International Journal of Food Science and Technology* **35**: 0-11 *(in press)*.

Becker, R. and G.M. Gray. 1992. Evaluation of a water-jet cutting system for slicing potatoes. *Journal of Food Science* **57**: 132-137.

Ben-Yehoshua, A. 1987. Transpiration, water stress and gas exchange. In: *Postharvest Physiology of Vegetables*, J. Weichman. (Ed.), Marcel Dekker, New York, NY, p. 113.

Bolin, H.R. and C.C. Huxsoll. 1991a. Effect of preparation procedures and storage parameters on quality retention of salad-cut lettuce. *Journal of Food Science*, **56**: 60-64.

Bolin, H.R. and C.C. Huxsoll. 1991 b. Control of minimally processed carrot (*Daucus carota*) surface discolouration caused by abrasion peeling. *Journal of Food Science* **56**: 416-418.

Bolin, H.R., Stafford, A.E. King (Jr), A.D. and C.C. Huxsoll. 1977. Factors affecting the storage stability of shredded lettuce. *Journal of Food Science* **42**: 1319-1321.

Brackett, R.E. 1987. Antimicrobial effect of chlorine on *Listeria monocytogenes*. *Journal of Food Protection* **50**: 999-1003.

Brecht, P.E. 1980. Use of controlled atmospheres to retard deterioration of produce. *Food Technology* **34**: 45-50.

Brody, A.L. 1992. New developments in reduced oxygen packaging of fruits and vegetables. *Asia Pacific Food Industry* **4**: 76-84.

Buckenhuskes, H. and K. Gierschner. 1990. Manufacture of delicatessen salads: special requirements for fruits and vegetables. *Fleischwirtsch* **70**: 1044-1047.

Carlin, F., Nguyen-the, C. Hibert, G. and Y. Chambroy. 1990. Modified atmosphere packaging of fresh 'ready-to-use' grated carrots in polymeric films. *Journal of Food Science* **55**: 1033-1038.

Cisneros-Zevallos, L., Saltveit, M.E. and J.M. Krochta. 1995. Mechanism of surface white discoloration of peeled (minimally processed) carrots during storage. *Journal of Food Science* **60**: 320-323, 333.

Day, B.P.F. 1990. Extension of shelf-life of chilled foods. *In*: Proceedings of the Internatuional Conference on Modified Atmosphere Packaging. Campden food and drink research association. October 1990.
Day, B.P.F. 1991. New atmospheres for international packaging. *Food Manufacture International* **7**: 28-31.
Day, B.P.F. 1992. Guidelines for the manufacture and handling of modified atmosphere packed food products. Technical manual No. 34, Campden Food and Drink Research Association, Chipping, Campden, UK.
Day, B.P.F. 1996. Novel modified atmosphere packaging for fresh prepared produce. The European Food and Drink Review Spring 1996, 73-80.
Day, B.P.F. and L.G.M. Gorris. 1993. Modified atmosphere packaging of fresh produce on the West-European market. *Zeitschrift fur Lebensmittel* **44**: 32-37.
Dorko, C. 1994. Antioxidants used in foods. *Food Technology* **48**: 33.
Dowling, W. 1991. Preservation of fruit and vegetables. UK Patent Application, GB2258993A.
Everson, H.P., Waldron, K.W., Gleeson, J.D. and K.M. Browne. 1992. Effects of modified atmospheres on textural and cell wall changes of asparagus during shelf-life. *International Journal of Food Science Technology* **27**: 187-199.
Exama, A., Arul, J., Lencki, R.W., Lee, L.Z., and C. Toupin. 1993. Suitability of plastic films for modified atmosphere packaging of fruits and vegetables. *Journal of Food Science* **58**: 1365-1370.
Floros, J.D. 1993. The shelf-life of fruits and vegetables. *In*: Shelf-life Studies of Food and Beverages, G. Charalambous (Ed.), Elsevier Science Publishers, London, 195-216.
Greengrass, J. 1993. Films for MAP of foods. *In*: Principles and Applications of Modified Atmosphere Packaging of Foods. R.T. Parry (Ed.), Blackie Academic and Professional, London, 63-100.
Hardenburg, R.E. 1971. Effect of in-package environment on keeping quality of fruits and vegetables. *HortScience* **6**: 178.
Herner, R.C. 1987. High carbon dioxide effects on plant organs. *In*: Postharvest Physiology of Vegetables, J. Weichman (Ed.), Marcel Dekker, New York, NY, 239-253.
Huxsoll, C.C. and H.R. Bolin. 1989. Processing and distribution alternatives for minimally processed fruits and vegetables. *Food Technology* **43**: 124-128.
IFST, 1987. *Guidelines for the handling of chilled foods*. Institute of Food Science and Technology, London, UK.
Kader, A.A. 1987. Respiration and gas exchange of vegetables. *In*: Postharvest Physiology of Vegetables. J. Weichman (Ed.), Marcel Dekker. New York, NY, 25-43.
Kader, A.A., Zagory, D. and E.L. Kerbel. 1989. Modified atmosphere packaging of fruits and vegetables. *Critical Review of Food Science and Nutrition* **28**: 1-30.
Kahl, G. and G.G. Laties. 1989. Ethylene induced respiration in thin slices of carrot root. *Journal of Plant Physiology* **134**: 496-503.
King, A.D. and H.R. Bolin. 1989. Physiological and microbiological storage stability of minimal processed fruit and vegetables. *Food Technology* **43**: 132-135.
Lee, L., Arul, J., Lencki, R. and F. Castaigne. 1995. A review on modified atmosphere packaging and preservation of fresh fruits and vegetables: physiological basis and practical aspects—Part 1. *Packaging Technology and Science* **8**: 315-331.
Lee, L., Arul, J., Lencki, R., and F. Castaigne. 1996. A review on modified atmosphere packaging and preservation of fresh fruits and vegetables: physiological basis and practical aspects—Part 2. *Packaging Technology and Science* **9**: 1-17.
Mannapperuma, J.D., Singh, R.P. and M.E. Montero. 1991. Simultaneous gas diffusion and chemical reaction in food stored in a modified atmosphere. *Journal of Food Engineering* **14**: 167-183.
Manvell, P.M. and M.R. Ackland. 1986. Rapid detection of microbial growth in vegetable salads at chill and abuse temperatures. *Food Microbiology* **3**: 59-65.
Marxcy, R.B. 1978. Lettuce salad as a carrier of micro-organisms of Public Health significance. *Journal of Food Protection* **41**: 435-438.

McGlasson, W.B. 1992. Modified atmosphere packing matching physical requirement with physiology of produce. *Food Australia* **44**: 168-170.

Moleyar, V. and P. Narasimham. 1994. Modified atmosphere packaging of vegetables: an appraisal. *Journal of Food Science Technology* **31**: 267-278.

Morrow, J.J. and J.B. Martin. 1977. Effective measurement of chlorine residual. *Effluent and water treatment Journal* **15**: 238-242.

Myers, R.A. 1989. Packaging considerations for minimally processed fruits and vegetables. *Food Technology* **43**: 129-131.

O'Beirne, D. 1987. Modified atmosphere packaging—new technology for food producers. Farm Food (Teagasc): 7-8.

O'Beirne, D. 1990. MAP of fruits and vegetables, *In*: Chilled foods: The state of the Art, T.R. Gromley (Ed.), Elsevier Science Publishers, London, 183-191.

O'Beirne, D. and A. Ballentyne. 1987. Some effects of modified atmosphere packaging and vacuum packaging in combination with antioxidants on quality and storage life of chilled potato strips. International Journal of Food Science Technology 22: 515-523.

O'Connor, R.E., Skarshewski, P. and S.J. Thrower. 1992. Modified atmosphere packaging of fruits, vegetables, seafood and meat: state of the art. *ASEAN Food Journal* **7**: 127-134.

Powrie, W.D. and B.J. Skura. 1991. Modified atmosphere packaging of fruits and vegetables. *In*: Modified Atmosphere Packaging of Foods. B. Ooraikal, M.E. Stiles, and W. Ellis (Eds.), Horwood Sussex, UK, 169-243.

Pratt, A. 1990. Fresh applications. *Packaging Today* **4**: 44-48.

Priepke, P.E., Wei, L.S. and A.I. Nelson. 1976. Refrigerated storage of pre-packaged salad vegetables. *Journal of Food Science* **46**: 379-382.

Raison, J.K. 1980. Effect of low temperature on respiration. *In*: Biochemistry of Plants. D.D Davis (Ed.), Academic Press, New York, NY, p. 613.

Rizvi, S.S.H. 1981. Requirements for foods packaged in polymeric films. *Critical Reviews of Food Science and Nutrition* **14**: 111-134.

Robertson, G.L. 1993. Packaging of horticultural products. *In*: Food Packaging: Principles and Practice. Marcel Dekker, New York, NY, 470-506.

Setty, G.R., Vijayalakshmi, M.R. and A.U. Devi. 1993. Methods for peeling fruits and vegetables: A critical evaluation. *Journal of Food Science Technology* **30**: 155-162.

Smock, R.M. 1979. Controlled atmosphere storage of fruits. *Horticultural Review* **1**: 301-336.

Strugnell, C. 1988. Increasing the shelf-life of prepacked vegetables. *Irish Journal of Food Science Technology* **12**: 81-84.

Tatsumi, Y., Watada, A.E. and W.P. Wergin. 1991. Scanning electron microscopy of carrot stick surface to determine cause of white translucent appearance. *Journal of Food Science* **56**: 1357-1359.

Tatsumi, Y., Watada, A.E. and P.P. Ling. 1993. Sodium chloride treatment or waterjet slicing effects on white tissue development of carrot sticks. *Journal of Food Science* **58**: 1390-1392.

Varoquaux, P. 1991. Ready to use fresh fruit and vegetables. *La Revue Generale du Froid* **33**: 1-11.

Watada, A.E., Abe, K. and N. Yamuchi. 1986. Physiological activity of partially processed fruits and vegetables. *Food Technology* **40**: 82-85.

Weichman, J. 1987. Low O_2 effect on plant organ. *In*: Postharvest Physiology of Vegetables. J. Weichman (Ed.), Marcel Dekker, New York, NY, p. 13.

Zagory, D. and A.A. Kader. 1988. Modified atmosphere packaging of fresh produce. *Food Technology* **42**: 70-77.

Zhou, Y.F., Abe, K. and T. Iwata. 1992. Effect of shredding mode on the deterioration of the quality of partially processed pepper fruits. *Society of Food Science and Technology* **39**: 161-166.

Index

A

abiotic stresses 291
absorption concentrations 70
ACC 263, 266, 267, 268, 269, 270, 273, 274, 275, 276, 277, 283, 307, 327, 332, 333
ACC oxidase 267, 268, 269, 273, 274, 275, 276, 283
ACC synthase 269, 273, 275, 276, 283, 321
acceptability 122, 125, 319
acceptance 8, 104, 125, 127, 129, 158, 309
achenes 211, 217, 220
active carbonate 24, 25
adjuncts 122, 128
albinism 218
alcohol 110, 111, 112, 113, 114, 115, 116, 117, 118, 119, 120, 121, 122, 123, 124, 125, 127, 128, 129, 130, 131, 214, 230, 231, 296, 302
alcohol acyltransferase 214, 231
algorithm 72
ammonium proportion 69, 70
ammonium thiosulphate (ATS) 92, 101, 103
ammonium toxicity 69
anaerobic fermentation 117, 120
anion concentration ratios 68
anthocyanins 212, 215, 225, 228, 233, 237
anthracnose 143, 219
apple 17, 21, 27, 28, 31, 34, 38, 40, 94, 95, 96, 97, 102, 103, 104, 105, 109, 110, 112, 114, 116, 178, 179, 180, 181, 182, 183, 184, 185, 186, 187, 189, 190, 191, 192, 230, 272, 291, 292, 326, 340
Apples, pears, plums, cherries, berries 111
apricot 98, 104, 107, 109
Asian pear 103
attitudes 3, 9, 124, 210
autocatalysis of C_2H_4 production 267, 273, 274, 279, 280, 282
automation 71, 77
auxin 178, 211, 230, 231, 238, 291

B

biennial bearing 89, 93, 94, 96, 100, 103, 104, 105, 106
bioregulators 106
biotic 238, 248, 291
blending 114, 120, 122
blossom burners 97
blossom desiccants 97, 102, 107
blossom-end rot 37, 53, 68, 95, 96, 101, 105, 156
blotchy ripening 52
boron 16, 28, 30, 31, 32, 33, 34, 35, 36, 65, 181, 218
Botrytis 217, 218, 223, 225, 228, 229, 231, 245, 247, 248, 249, 253, 265, 274, 276, 277, 278, 279, 280, 281, 283, 284, 286, 310, 323, 329, 341
Botrytis cinerea 218, 225, 228, 229, 231, 245, 247, 248, 249, 253, 277, 278, 283, 286, 310, 329, 341
Botrytis Infections 276, 277
Botrytis infections 265, 265-289, 274, 277, 279, 284
breeding 138, 142, 150, 152, 154, 158, 160, 162, 163, 166, 230, 245, 246, 291, 304
budbreaking chemicals 158, 162, 163

C

CA 9, 209, 223, 224, 225, 228, 232, 263, 265, 274, 276, 281, 282, 283, 284, 285, 287, 307, 316, 322, 345, 352
calcium 5, 9, 24, 25, 26, 32, 33, 34, 35, 40, 49, 53, 55, 59, 67, 71, 74, 76, 78, 79, 81, 87, 90, 107, 123, 136, 141, 144, 146, 152, 162, 175, 176, 177, 178, 179, 180, 181, 182, 183, 184, 185, 186, 212, 218, 290, 332, 334, 348, 350, 355
calibration 17, 18, 28, 33, 35
cation saturation 13, 26, 33
cellulase 213, 227, 238
Cheah 266, 285, 327, 328, 329, 330, 331, 332, 333, 334, 336, 338, 339
chemical growth regulators 307, 311
chemical thinning 89, 90, 96, 97, 98, 99, 100, 102, 103, 104, 105, 106, 107, 108, 109, 110
chilling 5, 155, 158, 159, 160, 161, 162, 163, 165, 182, 186, 250, 251, 252, 253, 254, 255, 274, 275, 276, 279, 280, 282, 283, 284, 285, 308, 309, 310, 311, 312, 314, 315, 316, 317, 321, 322, 323, 324, 325, 326, 339, 352, 353, 355
chilling injury 5, 182, 250, 251, 252, 253, 254, 255, 309, 310, 311, 312, 314, 315, 316, 317, 322, 323, 324, 325, 326, 352
choice 2, 3, 4, 8, 17, 98, 127, 129, 248, 345

citrus 21, 26, 27, 33, 94, 98, 103, 104, 106, 107, 108, 214, 235, 236, 237, 238, 239, 241, 242, 244, 245, 246, 247, 249, 250, 251, 252, 253, 254, 255, 276
climacteric of respiration 266, 267, 269, 271
closed systems 64, 77
CO_2 7, 16, 24, 56, 57, 58, 80, 111, 112, 116, 117, 119, 122, 129, 130, 211, 223, 224, 225, 226, 228, 230, 239, 241, 242, 249, 264, 267, 274, 276, 279, 281, 283, 284, 285, 307, 314, 315, 316, 317, 322, 345, 346, 352, 353, 354, 355
coffee 92, 105, 196
consumer acceptance 125, 129, 308
consumers 1, 3, 4, 7, 8, 42, 111, 124, 127, 128, 129, 134, 136, 151, 160, 164, 165, 196, 197, 200, 201, 210, 264, 281, 355
convenience 4, 112, 345
cooling 116, 209, 220, 221, 222, 223, 224, 226, 227, 228, 229, 231, 232, 280, 328, 330, 335, 339, 346
correlation 14, 17, 18, 22, 25, 29, 33, 35, 36, 213, 222, 239, 251, 252
cost of production 123
crop loading 94
currants 111
cut flower 46
cuticle cracking 53

D

dealcoholization 112, 114, 122, 129, 130
dealcoholized 111, 112, 114, 117, 121, 125, 128, 129, 131
dealcoholized, low, and reduced-alcohol wine 111, 112
defects 136, 212, 215, 216, 346
denitrification 69
developing countries 7, 189, 190, 191, 196, 205, 206, 236
diagnosis 13, 14, 32, 33, 84, 330
dialysis 112, 115, 116, 121, 122, 129
dilution 71, 74, 76, 112, 116, 120, 136, 149
dilution ratio 71, 74, 76
disinfection of the drain solution 77
distillation 112, 114, 115, 117, 119, 122
distillation columns 112
dosages of fertilizers 74
dosages of the nutrients 73
drain solution 55, 64, 77
duty 123, 124

E

economic considerations 123
economic disparity 7
edible coatings 224

efficiency 94, 98, 114, 120, 148, 157, 190, 201, 202, 316, 320, 326
electrical conductivity 37, 44, 45, 51, 79, 80, 83
elevated CO_2 223, 274, 353
ellagic acid 215, 230
endogenous growth substances 321
environment 3, 5, 14, 34, 41, 43, 44, 46, 47, 48, 49, 51, 54, 55, 57, 60, 61, 64, 65, 67, 68, 69, 70, 71, 78, 81, 84, 116, 124, 136, 137, 142, 176, 226, 239, 240, 245, 251, 265, 278, 279, 280, 284, 294, 308, 313, 314, 324, 325, 326, 330, 357
environmental 5, 7, 8, 27, 92, 102, 136, 151, 154, 158, 181, 205, 209, 211, 212, 230, 251, 289, 307, 308, 310, 311, 313, 320, 322, 324, 338, 347, 352
environmental issues 7, 8, 338
ethical issues 3, 8
ethylene 6, 9, 37, 145, 147, 152, 155, 176, 179, 182, 211, 227, 230, 231, 233, 238, 239, 241, 242, 243, 249, 255, 263, 265, 268, 269, 273, 274, 275, 277, 278, 279, 280, 281, 282, 283, 284, 285, 286, 287, 309, 313, 316, 320, 321, 322, 323, 324, 325, 326, 336, 345, 346, 349, 352, 353, 357
ethylene biosynthesis 263, 274, 282, 284, 286, 287, 326
exogenous enzyme 117

F

fashion 4, 211
Fe-chelates 55
fermentation 295, 296, 306
fermenting 111, 118, 295, 351
fertilizers 15, 26, 37, 39, 40, 42, 43, 71, 72, 73, 74, 75, 76, 154, 162, 167, 175, 180, 290, 331
flavour 3, 4, 120, 121, 122, 128, 130, 134, 135, 210, 212, 213, 215, 218, 222, 225, 228, 232, 233, 238, 336, 347, 348, 352
flavour adjustment 122
flavour components 121
flower quality 51, 82
food safety 1, 3, 5, 143, 151, 338
forced air cooling 209, 221, 222, 227, 335
formulae 39, 71, 73
fractional distillation column 119
Fragaria x ananassa Duch 209, 211
freeze concentration 112, 115
fresh produce 1, 3, 4, 5, 6, 8, 228, 233, 326, 340, 345, 355, 356, 357, 358
fruit cell division 95
fruit development period 159, 162, 163
fruit juice 52, 119, 120, 129
fruit quality 51, 52, 59, 82, 89, 99, 100, 106,

108, 120, 121, 157, 158, 159, 162, 164, 165, 175, 176, 179, 180, 182, 184, 185, 186, 210, 217, 223, 273, 274, 282, 285
fruit set 69, 89, 90, 91, 92, 93, 94, 102, 103, 104, 105, 106, 107, 108, 109, 138, 140, 141, 155, 159, 179
fruit size 53, 89, 90, 91, 94, 95, 96, 99, 100, 101, 103, 104, 105, 106, 108, 109, 138, 161, 164, 180, 181, 212, 228
fruit softening 186, 219, 264, 265, 271, 272, 281
fruit-wine 111
furaneol 214, 225

G

gas composition 311, 316
genetic 239, 244, 284, 304, 305, 309
genetically modified organisms 3, 4, 8
gibberellin 91, 107, 109, 211, 238, 311, 325
glassiness 54, 81
glucose oxidase 111, 117, 118, 119, 129, 130
gold specks 52, 68, 82
grades 215, 216, 227, 232
grape wine 112, 127
green spot 53
Grey mould rot 218
growth regulators 5, 89, 91, 94, 96, 97, 101, 103, 107, 108, 109, 152, 165, 307, 311, 321, 323

H

HACCP (hazard analysis and critical control points 7
hand thinning 89, 90, 94, 95, 96, 97, 102, 103, 104, 105, 106, 108, 109
Hard Rot 219
harvest-picking 217
harvester 220, 227, 231, 232, 333
health 89, 92, 96, 124, 127, 136, 137, 151, 163, 205, 247, 306, 358
high pressure fast cooling 209, 222
high temperature 148, 221, 273, 284
horticultural produce 1, 3, 4, 6, 7, 8, 308, 351, 354
humic substances 54, 56, 61, 79, 83
hybridization 209, 269
hydrocooling 166, 209, 221, 228, 330, 335, 344
hydroperoxide lyase 214, 231
hydroponics 37, 38, 39, 40, 41, 42, 43, 44, 46, 47, 48, 50, 51, 52, 54, 55, 56, 57, 59, 63, 64, 65, 67, 68, 71, 77, 78, 80, 81, 82, 83, 84, 85

I

internal C_2H_4 concentration 266, 280
internal fruit rot 53
interpretation 14, 17, 18, 22, 33, 34, 35, 36, 37, 77, 78, 83
introduced varieties 157, 158, 160, 171

K

Kenya 189, 190, 191, 192, 194, 195, 196, 197, 199, 205, 206, 207
kiwifruit 16, 21, 29, 31, 32, 33, 36, 105, 108, 120, 263, 264, 265, 266, 267, 268, 269, 270, 271, 272, 273, 274, 275, 276, 277, 278, 279, 280, 281, 282, 283, 284, 285, 286, 287, 348, 349

L

labelling 124, 125
Leather rot 219, 233
light 49, 60, 68, 80, 89, 93, 95, 97, 103, 104, 125, 127, 131, 135, 138, 144, 145, 150, 153, 158, 165, 212, 215, 219, 250, 310, 311, 314, 315, 316, 318, 319, 320, 322, 324, 325, 326, 331, 336
lipoxygenase 214, 231
liquid-liquid extraction 117
long-term storage 180, 184, 223, 236, 264, 265, 281, 294, 302, 308, 321, 322, 336, 340
Loss of stored 302
low chilling requirement 158, 160, 161, 162
low oxygen 254, 265, 274, 284, 286, 323
low temperature 6, 147, 180, 181, 182, 268, 269, 271, 274, 275, 279, 281, 284, 285, 286, 308, 309, 310, 311, 314, 315, 316, 317, 319, 320, 321, 322, 324, 325, 326, 346, 358

M

magnesium 25, 26, 27, 32, 33, 34, 35, 40, 76, 141, 175, 176, 178, 180, 181, 218
malformation 217
malic and citric acids 214
market driven 9
market performance 125, 127
marketing issues 111, 124
marketing system 198, 203
marketplace 2, 4, 9
markets 1, 3, 8, 124, 127, 128, 155, 159, 164, 166, 197, 199, 200, 201, 202, 203, 204, 205, 206, 227, 229, 244, 294, 301, 302, 327, 330, 337
maturity indices 215, 264
mechanical thinning 96

membrane   112, 115, 121, 122, 123, 128, 179, 181, 238, 248, 249, 250, 251, 309, 310, 311, 324, 326
methods assessing available nutrient   18
micronutrient concentrations   43, 62, 64, 65, 72, 76
minimal   4, 63, 65, 68, 114, 115, 123, 125, 235, 249, 345, 346, 347, 356, 357
minimal processing   4, 345, 346, 356
modifications incorporate   112
modified atmosphere   209, 223, 224, 227, 228, 231, 232, 233, 324, 326, 346, 349, 352, 354, 356, 357, 358
modified atmosphere packaging   209, 223, 228, 232, 233, 324, 326, 346, 352, 356, 357, 358
molecular biology   4
mouthfeel   119, 122, 130
Mucor Rot   220

N

naphthalene acetamide (NAAm)   90
naphthalene acetic acid (NAA)   90
natural soil solutions   62
net uptake   21, 69, 176
nitrification   18, 57, 61, 69
nitrogen   5, 18, 19, 33, 34, 35, 36, 37, 42, 61, 62, 67, 69, 70, 72, 73, 79, 80, 81, 140, 150, 175, 177, 179, 180, 182, 183, 184, 185, 186, 215, 289, 293, 310, 323, 324, 345, 352, 353
nutrient availability   54
nutrient cation ratios   67
nutrient deficiencies   46, 218
nutrient demand   20, 21
nutrient film technique   39, 41, 80, 82
nutrient solution   37, 38, 39, 40, 41, 42, 43, 44, 45, 46, 47, 48, 49, 50, 51, 52, 53, 54, 55, 56, 57, 58, 59, 60, 61, 62, 63, 64, 65, 66, 67, 68, 69, 70, 71, 72, 73, 74, 75, 76, 77, 78, 79, 80, 82, 83, 84, 85
nutritional value   1, 3, 49, 212, 214, 215, 227, 237, 240

O

off-flavours   224, 225, 353
offseason production   157
olives   29, 90, 104, 108
on-farm processing   199, 203, 204
organic produce   3
overripeness   217
ozone, carbon monoxide   225

P

packaging   4, 6, 8, 125, 149, 197, 200, 203, 204, 209, 210, 217, 222, 223, 224, 226, 228, 231, 232, 233, 240, 244, 324, 326, 328, 336, 337, 346, 348, 351, 352, 354, 355, 356, 357, 358
peach   16, 21, 25, 27, 31, 34, 98, 104, 105, 106, 109, 156, 157, 158, 159, 160, 161, 162, 163, 164, 165, 166, 167, 169, 170, 215, 317
pear   5, 17, 21, 25, 28, 31, 33, 93, 98, 102, 103, 107, 108, 109, 323
pectate lyase   213, 230
pectinesterase   213
peeling   276, 295, 346, 347, 348, 356, 358
pelargonic acid   92
persimmon   106
pH buffers   61
pH range   54, 55, 56
phenylalanine-ammonio-lyase (PAL)   212
phosphate forms   56
phosphorus   19, 23, 32, 33, 34, 35, 36, 56, 76, 81
physiological disorders   52, 54, 175, 176, 178, 180, 181, 182, 184, 185, 186, 217, 238, 241, 250, 265, 289, 291, 292, 293, 328, 330, 331, 338, 351
plum   92, 98, 104, 105
pollution   8
polygalacturonase   4, 213, 231
polyphenols   154, 212, 214, 215, 229, 348
positioning   124, 125
postharvest factors   1, 5, 6, 341
postharvest quality control   278
potassium   23, 32, 33, 34, 35, 37, 40, 65, 71, 78, 80, 81, 115, 149, 175, 176, 178, 180, 181, 182, 184, 186, 215, 289, 293, 310, 332, 340
potato   22, 189, 194, 196, 197, 198, 199, 200, 201, 202, 203, 204, 205, 206, 207, 304, 333, 340, 358
pre-conditioning with C_2H_4 281
pre-postharvest continuum   6
precooling   221, 222, 223, 226, 228, 231, 232, 280, 286
preharvest   1, 2, 4, 5, 8, 9, 107, 108, 145, 156, 158, 167, 175, 176, 182, 183, 184, 185, 186, 200, 217, 219, 253, 254, 274, 286
preharvest factors   5, 9, 158, 167
product descriptors   3
production areas   8, 89, 148, 204, 205
prognosis   13, 14, 339
promotion   124, 125
promotion and positioning   125
propagation practices   291
propylene   267, 268, 271, 272, 273, 283, 284, 285, 286, 287, 345, 352, 355
pruning   5, 94, 98, 103, 105, 162, 165, 167, 244, 294, 302, 305

Q

quality assurance 3, 7, 151
quality standards 2, 3, 7, 9, 163, 164, 205, 216, 221

R

recycling 38, 77
reduced-alcohol 111, 112, 114, 118, 119, 120, 122, 123, 127, 128, 130
relative humidity 5, 133, 136, 141, 148, 149, 165, 182, 209, 221, 235, 240, 241, 242, 244, 251, 311, 313, 314, 327, 329, 332, 336, 337, 354
relative proportions 56, 66, 72
respiration 6, 7, 159, 176, 177, 179, 182, 186, 210, 211, 217, 221, 223, 224, 225, 231, 236, 238, 239, 241, 255, 264, 266, 267, 268, 269, 270, 271, 277, 282, 284, 286, 294, 309, 310, 311, 314, 316, 317, 318, 332, 336, 346, 347, 348, 349, 351, 352, 353, 354, 355, 357, 358
respiration rates 7, 211, 224, 309, 332, 348, 354, 355
retailers 3, 4, 6, 8, 9, 151, 166, 201, 244
reverse-osmosis 111, 115, 121
RH 133, 136, 148, 149, 209, 215, 221, 226, 235, 241, 243
Rhizopus Rot 219
Rhizopus rot 220
ripening 6, 9, 10, 14, 52, 107, 139, 140, 145, 146, 147, 152, 153, 155, 156, 162, 176, 179, 182, 186, 211, 212, 213, 214, 219, 221, 227, 229, 230, 231, 232, 233, 238, 239, 242, 243, 251, 263, 264, 265, 266, 267, 269, 270, 271, 272, 273, 274, 276, 278, 279, 280, 281, 282, 283, 284, 285, 286, 287, 353
rockwool 39, 40, 41, 47, 65, 67, 78, 79, 82, 83, 84, 85
room cooling 209, 221
russeting 52, 53, 90, 100

S

sales tax 123
salinity effects 48
salinity threshold value 37, 45
salinity yield decrease 37, 46
salt concentration 41, 43, 44, 45, 46, 47, 49, 53, 54, 59, 66, 71, 72
sample handling 17
sampling depth 15
sampling methods 15
sampling time 17
savings 123, 220
seed numbers 100
seedling orchards 157, 160
semipermeable membranes 115
sensory properties 113, 121, 122, 130
sensory quality 121, 122, 128, 223
shelf life 52, 53, 54, 68, 149, 150, 156, 164
short-term storage 223, 307, 321, 322
shrivelling 148, 217
silicon 65, 81
slicing 346, 348, 349, 350, 356, 358
social and technical factors 9
social drivers 4
sodium dinitro-ortho-cresylate (DNOC) 90
soil sampling 14, 15
soilless culture 38, 39, 41, 42, 44, 54, 55, 69, 78, 79, 80, 81, 82, 83, 84, 85
soluble solids 100, 162, 164, 165, 167, 179, 183, 209, 213, 215, 235, 237, 238, 252, 263, 264, 265, 266, 274
spinning cone column 111, 113, 129, 131
spray application technology 92
standards 2, 3, 7, 9, 133, 152, 160, 163, 164, 204, 205, 215, 216, 221, 232
steps of soil testing 14
stock solutions 40, 71, 72, 74, 76, 77
storage 6, 8, 9, 36, 135, 136, 147, 149, 150, 151, 152, 154, 155, 164, 175, 177, 178, 179, 180, 181, 182, 183, 184, 185, 186, 203, 204, 211, 212, 219, 222, 223, 224, 225, 228, 229, 230, 231, 232, 233, 236, 237, 238, 239, 241, 242, 244, 245, 250, 251, 252, 253, 254, 255, 256, 263, 264, 265, 271, 272, 274, 275, 276, 277, 278, 279, 280, 281, 282, 283, 284, 285, 286, 287, 289, 291, 294, 295, 296, 301, 302, 303, 305, 306, 307, 308, 309, 310, 312, 313, 314, 315, 316, 317, 318, 319, 320, 321, 322, 323, 324, 325, 326, 327, 328, 329, 330, 331, 332, 333, 334, 335, 336, 337, 338, 339, 340, 341, 346, 347, 348, 349, 350, 351, 352, 353, 355, 356, 357, 358
storage methods 285, 294, 308, 321, 322
strawberry 21, 31, 32, 34, 209, 210, 211, 212, 213, 214, 215, 216, 217, 218, 219, 220, 221, 222, 223, 224, 225, 226, 227, 228, 229, 230, 231, 232, 233, 317, 324
strawberry production 210, 227, 229
stress ethylene 274, 284
substrate analysis 77, 78
subtropical production systems 158
subtropics 157, 158, 159, 160, 161, 167
sucrose 212, 213, 315, 319
sugar reduction 119
supermarkets 3, 8, 166, 197
suppliers 3, 354
sweet potato 22, 194, 196, 197, 198, 199, 200, 201, 202, 203, 204, 205, 206, 207, 304
systems approach 9, 190, 192, 205

364 *Index*

T

Tan Brown Rot 220
target values 72, 73, 74, 75, 76
tariffs 123
temperate fruit 157, 158, 159, 164, 166
the tax 123
Thermoregulation of C_2H_4 Biosynthesis 268
thinning 5, 89, 90, 91, 92, 94, 95, 96, 97, 98, 99, 100, 101, 102, 103, 104, 105
titratable acidity 182, 186, 212, 214, 215
total carbonate 24
transaction costs 190, 191, 201, 202, 204
transpiration 6, 81, 82, 177, 178, 217, 236, 237, 239, 240, 241, 242, 309, 313, 346, 352, 356
transport 8, 49, 116, 166, 177, 178, 181, 182, 184, 199, 200, 201, 202, 204, 217, 219, 221, 223, 224, 225, 226, 227, 228, 235, 286, 293, 330, 334, 338, 346

U

ULO 263, 265, 274, 276, 281, 282, 283

unripe 112, 134, 211, 217, 264
uptake ratio 60, 63, 64, 68, 69

V

vitamin C 1, 52, 53, 54, 150, 212, 214, 215, 238, 353

W

washing 221, 241, 243, 244, 248, 294, 328, 330, 331, 334, 335, 338, 340, 344, 346, 350, 351, 356
waste 4, 8, 296, 303
water extract 78
wounding 6, 274, 276, 279, 280, 283, 284, 329, 335

Y

yeasts 120, 135, 253, 338

Z

Zinc, Copper, Manganese and Iron 27

UNIVERSITY OF LINCOLN